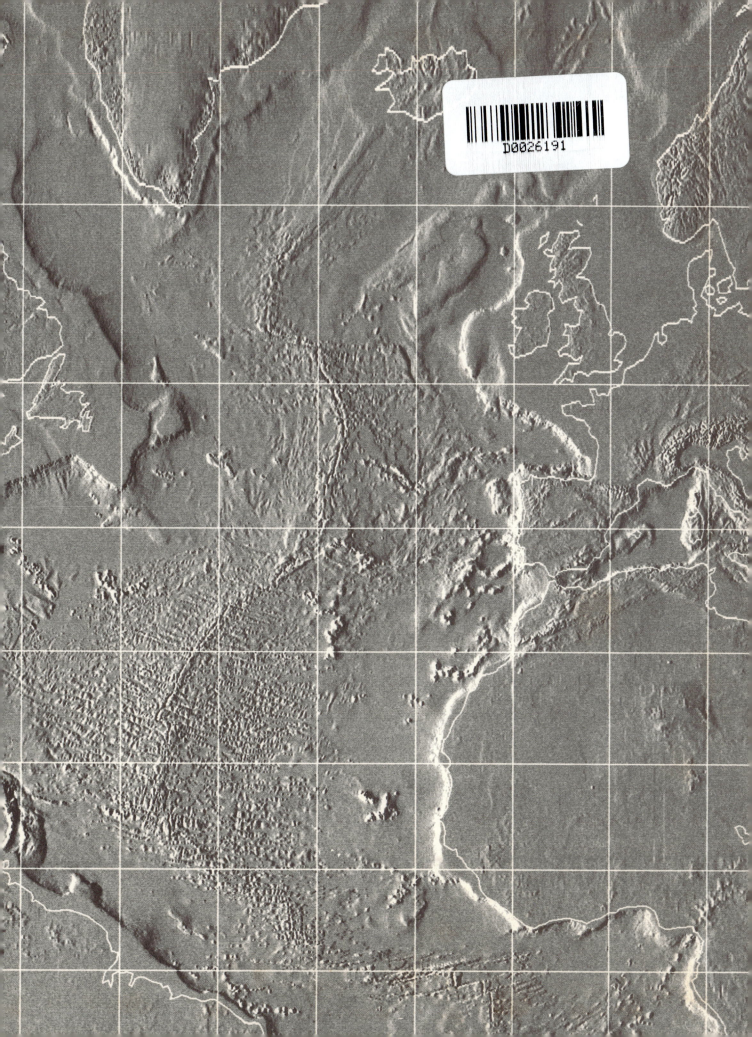

Oceanography

A View of the Earth

FIFTH EDITION

M. Grant Gross

Prentice Hall
Englewood Cliffs, New Jersey   07632

Library of Congress Cataloging-in-Publication Data

Gross, M. Grant (Meredith Grant), 1933–
    Oceanography, a view of the earth / M. Grant Gross.—5th ed.
        p.   cm.
    Includes bibliographical references.
    ISBN 0-13-629742-0
    1. Oceanography.   I. Title.
GC16.G7   1990
551.46—dc20   89-36413
                                                     CIP

Editorial/production supervision: Cristina Ferrari
Interior design: Andy Zutis
Art supervision: Christine Gehring-Wolf
Manufacturing buyer: Paula Massenaro
Photo research: Tobi Zausner
Cover photo: George Schwartz/FPG

© 1990, 1987, 1982, 1977, 1972 by Prentice-Hall, Inc.
A Division of Simon & Schuster
Englewood Cliffs, New Jersey 07632

Printed in the United States of America
10   9   8   7   6   5   4   3   2   1

ISBN 0-13-629742-0

Prentice-Hall International (UK) Limited, *London*
Prentice-Hall of Australia Pty. Limited, *Sydney*
Prentice-Hall Canada Inc., *Toronto*
Prentice-Hall Hispanoamericana, S.A., *Mexico*
Prentice-Hall of India Private Limited, *New Delhi*
Prentice-Hall of Japan, Inc., *Tokyo*
Simon & Schuster Asia Pte. Ltd., *Singapore*
Editora Prentice-Hall do Brasil, Ltda., *Rio de Janeiro*

Contents

6
Ocean and Climate   133

7
Currents   165

14

Benthos   343

15

Sediments   375

Appendices

Index

Preface

The Fifth Edition, like its predecessors, seeks to introduce students to the ocean, how it works, and how it influences our lives. The level of difficulty remains unchanged—a high school background in science and mathematics is assumed.

A second objective is to introduce students to science as a process of continual inquiry and open testing of ideas. In modern western society, science is also an important framework for structuring our understanding of the world around us. In short, to understand our world, one needs to know how science works. Oceanography provides an example of a rapidly expanding science.

The role of technology in supporting scientific study of the Earth is another major theme. For instance, the ability to observe Earth processes on a planetary scale has revolutionized how we observe the ocean. Now we can improve our predictions of hurricanes or shifts in weather patterns. Examples of such advances are presented and discussed throughout the book. Rapid expansion of computing power and its effect on science and its application is also highlighted throughout the book.

FEATURES OF THE FIFTH EDITION

UP-TO-DATE

The Fifth Edition has been extensively revised to cover recent advances in ocean science and technology. Particular attention is paid to advances in applications of ocean science. Sections dealing with climate change, aquaculture, and new technology have been rewritten. References have been extensively expanded to include recent publications.

EASY TO STUDY FROM

This edition has been extensively revised to make it easier for students to understand on their own. More examples are provided. More bridges are provided to put the material in context.

The chapters have been rewritten and reorganized to make it easier to assign the chapters in different orders. Each one essentially stands alone.

Graphics have been revised and expanded, using color, to make them easier for students to grasp their significance. Explanatory drawings and labels have been added to help students interpret photographs.

Review questions have been revised and expanded. The chapter outlines have also been extensively revised.

STUDENT INTEREST

Much of the material in the book has been revised to broaden the student's perspective. Special topics highlight the contributions of important people and present major advances in ocean science. Particular attention is given to applications of ocean science and their importance to society.

CLARITY

The text in this edition has been thoroughly revised, sentence by sentence. The intent has been to make the text easy to understand. Mathematical equations have been replaced by other formulations to assist students with limited mathematical skills. Technical words and phrases have been reduced. They are usually defined when they are first introduced. The Glossary has been updated and expanded to make it more useful to students new to science.

BALANCED COVERAGE

Like previous editions, the Fifth Edition provides a balanced coverage of the subfields of ocean sciences. The biology coverage has been expanded and reorganized. The chapter on sediments has been extensively revised. The history of the ocean continues to be a major theme, integrating several different aspects of ocean sciences.

ACKNOWLEDGEMENTS

This edition, like the ones before, carries an enormous debt of gratitude to all my colleagues and students who taught me as we worked and learned together. The list has grown long through the various editions, but my gratitude is still very real.

Particular thanks for assistance on this edition goes to Jean Snider who read the various drafts of new and revised chapters and commented on them with insight and patience. I am also indebted to Ken Tenore and the staff of the Chesapeake Biological Laboratory who hosted my sabbatical year. Lastly, I want to acknowledge the helpful comments from Prentice Hall reviewers: Richard Dame, University of South Carolina, Coastal Carolina College; Dirk Frankenburg, University of North Carolina at Chapel Hill; Robert Meade, California State University, Los Angeles; and Bernard Pipkin, University of Southern California.

Oceanography

A View of the Earth

A 12th century map of the Arab world by Al Idrisi shows the Mediterranean on the left and the Northern Indian Ocean on the right. Africa is on the lower left. (Courtesy The Granger Collection/Photo Researchers.)

CHAPTER 1

History of Oceanography

OBJECTIVES

1. To grasp the role of the ocean in human history;

2. To understand the development of ocean science;

3. To understand the importance of technological advances in ocean science;

4. To recognize and understand recent changes in the legal status of the ocean.

O ceanography—scientific study of the ocean—is a modern activity, but dealing with the ocean has been important in people's daily lives for many centuries. Sailors and traders explored the ocean and its shores, looking for new lands to settle, new trade routes to ply, or new products to buy and sell. Most of these discoveries were not recorded, in part because they were much too valuable to share.

Today, oceanographers study the ocean (Fig. 1-1) to develop better ways to predict the behavior of weather systems, such as the monsoons (heavy summer rains) which are necessary for growing rice in India and Southeast Asia, to devise methods to extract more food from the sea, or to find new sources for minerals or oil and gas on the ocean bottom. As we learn more about our nearby neighbors and the Earth's moon, we see what a unique feature the ocean is (Fig. 1-2, Fig. 1-3).

In short knowing how the ocean works, permits us to improve and protect our lives in many ways. In this chapter, we discuss:

FIGURE 1-1
Seen from space, Earth is a blue sphere (ocean) with clouds (white areas) covering more than half its surface. Land is scarce in this southern hemisphere view. Desert areas of Africa are rust colored. The ice-covered Antarctic continent (white) is at bottom center. (Courtesy NASA.)

FIGURE 1-2
The Moon's surface shows scars from numerous meteorite impacts over its 4.5 billion years of existence. There is no water to erode such features and its crust is not recycled by mountain building as it is on Earth. (Courtesy NASA.)

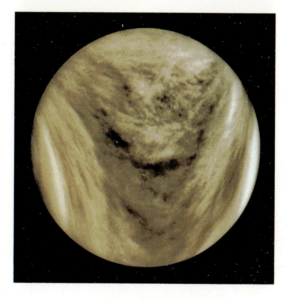

FIGURE 1-3
The surface of Venus is obscured by dense clouds of water vapor, sulfuric acid and carbon dioxide. While Venus resembles Earth in many ways, its history is markedly different from Earth's where the water has collected in ocean basins and most of the carbon dioxide is deposited as limestones or organic carbon. We do not know if there is mountain building on Venus. (Courtesy NASA.)

The ocean's role in human history;
History of ocean studies;
Uses of ocean sciences in exploiting resources;
Importance of new instruments in observing and studying the ocean; and
Changing political status of the ocean.

EARLY USE OF THE SEA

The sea played an important role in human affairs long before human history was written down. Shells in refuse piles of ancient coastal villages show that seafood was important in the villagers' diets. Some refuse piles also contain bones of deep-sea animals, which suggests that boats may have been used in fishing well offshore, even though we have no direct evidence of such seafaring capabilities.

Transportation by sea must also have been an important activity in early human history, but little evidence remains of these ancient maritime activities. Wood, skins, reed—materials commonly used in primitive boats—are rarely preserved. The earliest evidence comes

FIGURE 1-4
Large sailing canoes (center) were used by Polynesians for their trans-Pacific voyages. They were still being used when Europeans first arrived. (Courtesy Bishop Museum.)

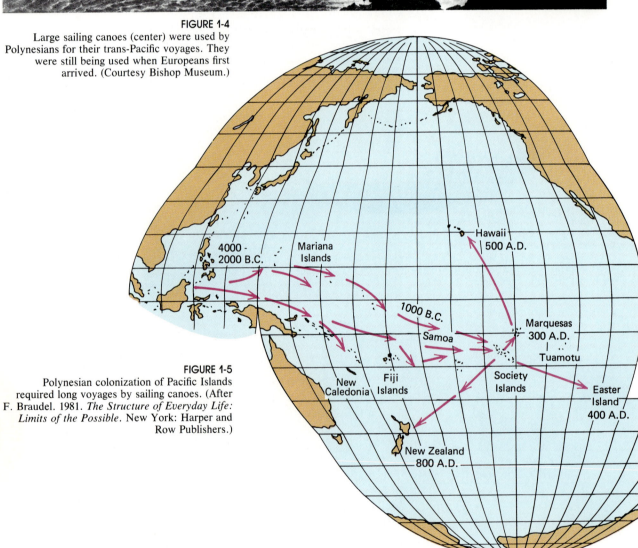

FIGURE 1-5
Polynesian colonization of Pacific Islands required long voyages by sailing canoes. (After F. Braudel. 1981. *The Structure of Everyday Life: Limits of the Possible*. New York: Harper and Row Publishers.)

from rock carvings in Norway that show boats similar to those later used by the Vikings. The earliest models and ships known come from Egyptian tombs and Viking graves. All suggest that these peoples were highly skilled sailors.

Four kinds of boats appear in rock carvings or are mentioned in ancient texts: dugouts made from logs that were hollowed and shaped by fire and simple tools, reed boats constructed of bundles of reeds lashed together, boats made of split sections of thin bark sewn together and stretched over a wooden frame (the birch bark canoe of American Indians), and skin boats made of sewn animal hides stretched over a wood frame. Boats similar to these were used extensively within recorded history. Some have been reconstructed from drawings and descriptions. Successful open-ocean voyages made in them show that they were quite capable of extensive voyages.

Early human migrations also argue for extensive seafaring. Around 10,000 years ago, western Europe was colonized by peoples from Africa who must have crossed the Strait of Gibraltar, which at that time was never less than 10 kilometers (6 miles) wide. Seaborne commerce was also developed early in human history. Around 1000 B.C. the *Phoenicians* had an extensive navy and dominated trade in the Mediterranean and adjacent waters in the North Atlantic. They sailed as far as England to get tin for making bronze.

Polynesians constructed elaborate double-hulled vessels (Fig. 1-4). The largest of these had living quarters for people and animals. They were used in transoceanic voyages to colonize the islands of the Pacific, including Hawaii (Fig. 1-5). The Micronesians colonized many of the larger islands of the western Pacific.

We know little about the seafaring traditions of these peoples, as there are no written records. In many cases our information comes from the accounts of the first European explorers to contact them. There is ample evidence, however, that they were skillful sailors and experienced navigators. One example of their navigational skills is the stick charts (Fig. 1-6) used by the *Micronesians*. Shells mark locations of islands, and the bamboo strips show wave patterns. Stars, cloud patterns, and winds were also used to navigate between islands.

FIGURE 1-6
Micronesian navigators used stick charts to sail the Pacific. Islands are represented by shells. Prevailing wave directions are shown by bamboo strips. Effects of islands on wave patterns are shown by the curved bamboo strips around the islands in the lower left. (Photograph courtesy Library of Congress.)

CHINESE OCEAN EXPLORATION

Between 1405 and 1433, early in the Ming dynasty (1368–1644), the Chinese undertook seven voyages in the Pacific and Indian oceans. These were the largest peacetime voyages ever undertaken, involving 37,000 men and 317 ships.

The Chinese ships were far bigger than any in western Europe. The largest had nine masts, was 135 meters (444 feet) long, and had a beam (width) of 55 meters (180 feet). They employed modern features, such as transverse bulkheads, which divided the ships into several watertight compartments. Thus if a ship's hull were slightly damaged, the water could be confined to one compartment. (Such construction is now standard in ships.) Magnetic compasses and detailed navigation charts were also used.

These expeditions were unlike any before or since. They did not seek to conquer, to collect treasure, to make religious converts, or to gather scientific information. Instead, they were to extend Chinese influence. The Chinese tradition can best be summarized as "live and let live." China was then far more advanced technologically than the rest of the world and was essentially self-sufficient for food and raw materials. In their eyes, the Chinese had nothing useful to learn or to gain from the outside world. Thus these expeditions were undertaken simply to display the splendor and power of the Ming dynasty.

Under the Chinese system, tributary states brought gifts to acknowledge China as the most civilized country in the world. To demonstrate their superior position, the Chinese felt obligated to give back more than they received. Tributary states were thus a financial drain. Expanding their influence simply increased the financial burden on the country. The expeditions brought back animals unknown to the Chinese, such as giraffes, to add to the imperial zoo. Otherwise these voyages contributed little to China.

Bowing to internal politics and economic pressures, the emperor ended these voyages, and China turned to internal developments. For example, the 2500-kilometer-long (1500 miles) *Great Wall* was strengthened to repel the Tartars and Mongols on China's northwest frontier. The 1600-kilometer-long (1000 miles) *Grand Canal* was also repaired and rebuilt so that boats could use it all year. This ended the need for the coastwise sailing to transport goods and grain around the country. By the early 1500s, when western Europe was beginning to expand its horizons through ocean exploration, the Chinese government was outlawing seafaring, withdrawing inward, and beginning a period of isolation that lasted until the mid-nineteenth century. The great Chinese expeditions made no lasting contribution to improving our understanding of the ocean.

ANCIENT OCEAN EXPLORATION

Many ocean features and processes were well known in Europe in antiquity. Some knowledge came from the stories of explorers and seafarers. But little was written down, perhaps because the knowledge was so valuable. Consequently, we know little of the extent of their knowledge.

Much of what we do know comes from the writings of philosophers and theologians, based on their analyses of the observations and reports of seafarers and explorers. *Aristotle* (384–322 B.C.), for instance, noted that the sea neither dries up nor overflows. He concluded, therefore, that the amount of rainfall must equal evaporation over the earth. *The Venerable Bede* (673–735), English historian and theologian, knew that the moon controls the tides. Tables of predictions of the tides at London Bridge were issued in the late twelfth or early thirteenth century.

European use of magnetic compasses was first recorded in the thirteenth century. The oldest surviving chart for ship navigation dates from A.D. 1275. It provides compass directions and indicates distances.

Various other devices were used for navigating at sea. For example, to determine distance a sandglass was used in conjunction with a log attached to a line to determine speed through the water. The log was thrown overboard and the line allowed to run out until all the sand had run through the glass. Another device used to determine water depth and position was the *lead line*. A weight with sticky wax on the bottom was attached to a line and thrown overboard. The amount of line fed out before the weight touched the bottom indicated water depths. The type of material recovered on the wax could be compared with notations on a chart to obtain some idea of location. Navigation charts still show areas of sand, mud, and shells on the bottom.

EUROPEAN OCEAN EXPLORATION

Publication in 1410 of a Latin edition of *Ptolemy's* (ca. A.D. 140) map greatly influenced thinking in western Europe. His maps were immensely influential. His influence was so great that it took centuries before all the mistakes in Ptolemy's maps were finally removed from newly made maps. Ptolemy's maps were simply copied uncritically until finally replaced by the results of new exploration.

The great age of European ocean exploration began early in the fifteenth century. It was probably stimulated by the knowledge gained from the rediscovery of Greek and Arab geography in the Islamic libraries when southern Spain was recaptured by Christian armies. Along with Ptolemy's maps came two ideas well known to the Greeks: (1) The earth was a sphere, and (2) the ocean was navigable. These ideas came to Europe from studies of the manuscripts in the Islamic libraries.

Portugal led in the early European exploration of the Atlantic. The Canary Islands off northwest Africa were explored in 1416. The Azores, in the middle of the Atlantic, were discovered by Europeans between 1427 and 1432. Perhaps the most influential individual was Portugal's *Prince Henry the Navigator* (1392–1460), who established a center for seafaring in southern Portugal. There he bought the most learned people to teach navigation to Portuguese sea captains. His efforts greatly stimulated Portuguese exploration.

In the voyages that followed, the west coast of Africa was explored. Portuguese navigators reached the tip of South Africa (near present-day Cape Town) in 1488. *Vasco da Gama* (1460–1524) reached India in May 1498. His discovery opened up profitable trade routes from Portugal and India.

Development of the three-masted ship (Fig. 1-7) was a major reason for the success of western European exploration of the sea. The ships were large enough to carry men and supplies for long trading and exploring voyages. New sail designs were sturdy enough to contend with the stormy North Atlantic. Previously most seafaring had been limited to the Mediterranean. There the longest voyages were only a

FIGURE 1-7
Three-masted ships were used extensively by Europeans to explore the New World. Here the ship *Half Moon,* commanded by Henry Hudson, is shown exploring the Hudson River, which was named after him. (Courtesy Library of Congress.)

few days out of sight of land. Most ships were beached at night, while the crew and passengers slept ashore.

Part of the impetus for ocean exploring was a need to replace the overland trade routes to China and India (called the *Silk Route*), the sources of luxuries and spices for cooking and for medicinal purposes. The decline of the Mongol Empire in central Asia and the fall of Constantinople to the Turks in 1453 had cut off these ancient trade routes. The eastern Mediterranean was dominated by hostile Islamic peoples. In addition, transport by sea was far cheaper than on land. Roads were so poor that moving goods only a few tens of miles doubled their cost.

Newly developed cannons carried aboard these ships made them formidable weapon systems. This led directly to the domination of the ocean by western European navies and to the conquest and colonization of coastal areas. Eventually the western European nations dominated the continents and oceans for more than 400 years.

The many explorers and exploring expeditions at this time are beyond the scope of this book, but a few are worthy of special mention. *Ferdinand Magellan* (1480?–1521) was the first to sail around the world, crossing the Pacific in 1520–1521. *Henry Hudson* (d. 1611), among many others, explored North America as the major western European nations established colonies and divided up the New World. Queen Elizabeth I (1533–1603) of England encouraged development of British maritime traditions. During her reign, *Sir Francis Drake* (1540–1596) sailed around the world, capturing Spanish ships and cargoes as he went.

During these centuries of exploration, most of the advances in knowledge about the ocean grew out of work on very practical problems. Improvements in navigation permitted accurate mapping of ocean shorelines, which was extremely important to sailors. National offices for ocean mapping were established by France in 1770 and by England in 1795.

Captain James Cook (1728–1799) made three voyages between 1768 and 1780, exploring and mapping the Pacific islands. His observations used the latest navigational instruments, including the newly developed chronometer to determine longitude. His maps were the most accurate made up to that time. He helped to fill the voids in knowledge about the Southern Ocean.

INDIVIDUAL SCIENTIFIC INQUIRIES

Knowledge about the ocean was especially important in the American colonies. One of the earliest published maps of the Gulf Stream—the strong current system along the U.S. Atlantic coast—was published by *Benjamin Franklin* (1706–1790) just before the American Revolution. Franklin had noted that the royal mail ships from England took much longer to make the voyage than did American ships.

His cousin, Timothy Folger—a Nantucket whaler—told him about the Gulf Stream, which was well known among American whalers and presumably among other American ship captains as well. Franklin's map of the Gulf Stream (Fig. 1-8) and his instructions for

FIGURE 1-8
Benjamin Franklin's map of the Gulf Stream, published in 1770, was the earliest map of this current system. (Courtesy Woods Hole Oceanographic Institution.)

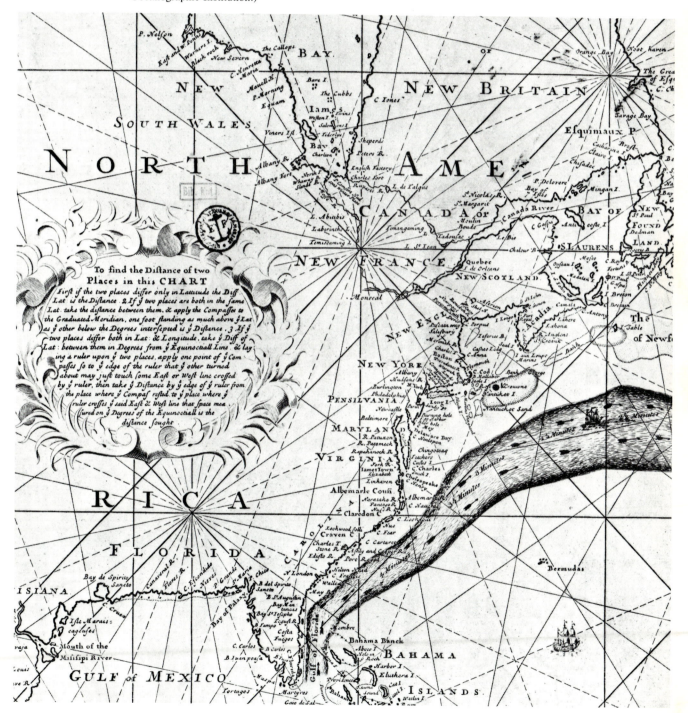

using thermometers to locate its warm waters were used by ship captains to take advantage of the strong currents on their eastbound voyages and to avoid bucking the current on their returns.

Despite the obvious benefits obtained from these advances, there was no systematic exploration or study of the ocean until the mid-1800s. These early explorations were made by government-backed expeditions or for commercial purposes. Universities were not involved in ocean studies until after 1950. During the early centuries of ocean studies, universities were very conservative and not interested in ocean exploration, which was left to sailors and traders.

Primitive early nineteenth-century measuring and sampling techniques caused serious misconceptions about deep-ocean conditions. It was believed, for instance, that the deepest ocean waters did not circulate but lay stagnant at the ocean bottom. Consequently, no incoming oxygen would be available to support life there. Furthermore, the respected naturalist *Edward Forbes* (1815–1854), had collected animals in the Aegean Sea to compare findings with those of Aristotle nearly 2500 years earlier. He noticed that fewer plant and animal species were caught as his nets went deeper. He concluded that most deep-ocean waters were **azoic** or devoid of animal life. This idea gained wide acceptance—so much so that occasional reports of animals taken from below 800 meters (2600 feet) were ignored or discounted.

Forbes's azoic theory was not refuted until the 1860s, when ocean basin topography was studied in connection with the laying of transoceanic telegraph cables. A cable from 2000-meters' depth (6600 feet) in the Mediterranean Sea raised for repairs had on it living animals (corals). This discovery came at a time when interest in deep-sea exploration had been aroused in connection with *Charles Darwin's* (1809–1882) theories about the origin of species through evolution. Scientists became interested in finding stable environments in which conditions had not changed over long time periods. Many scientists thought that the deep-ocean floor might be inhabited by ancient species, the ancestors of modern organisms.

CHALLENGER EXPEDITION

By the 1860s, then, there were compelling scientific reasons for studying the deep ocean. Widespread interest existed in setting up large-scale research projects, which even wealthy scientists could not afford.

The *Royal Society of London* provided financial support. The British Admiralty also provided two ships for North Atlantic deep-sea studies during the late 1860s. As a result of these research efforts, some of the old prejudices and misconceptions disappeared. It was demonstrated that waters move through deep-ocean basins.

Finally, the Royal Society was persuaded to sponsor the most ambitious ocean exploration project that had ever been attempted. This expedition (1872–1876) aboard H.M.S. *Challenger* established a tradition of large-scale research projects that has been an important component of oceanography ever since.

The *Challenger* was a sailing warship with an auxiliary steam engine, its guns replaced by scientific gear (Fig. 1-9). The scientific party under *Sir Wyville Thomson* (1830–1882) had been assigned to investigate "everything about the sea". They planned to study physical and biological conditions in every ocean, recording whatever might influence the distribution of marine organisms. This process included taking water samples and temperature measurements of deep and sur-

FIGURE 1-9
The *Challenger* crew attempting to land on St. Paul's Rocks in the Atlantic during the *Challenger* Expedition.

face waters, recording currents and barometric pressures, and collecting bottom samples in order to study sediments and identify new species. Fishes were caught in nets dragged behind the ship.

Before returning to England (Fig. 1-10), *Challenger* traveled 109,000 kilometers (68,000 miles), collected rocks and sediments from the ocean floor, and took soundings of water depths, using a weighted hemp line. All ocean basins except the Arctic were sampled. Data from the expedition filled 50 large volumes. The reports, written over 23 years were a watershed in the development of ocean science, the beginnings of modern oceanography.

Many misconceptions about the ocean were swept away by the *Challenger* results. The most famous, perhaps, was the debunking of *Bathybius,* a supposed "primordial slime" that had been thought to represent preevolutionary life on the ocean floor. Bathybius had been "discovered" by *Thomas H. Huxley* (1825–1895), a well-known biologist. In 1868 Huxley found a slimy substance in a jar containing a biological specimen several years old. He may have been overly anxious to provide new evidence for his friend Darwin's theory, which was then under attack. Huxley concluded that he had identified the substance from which life originated in the deep ocean. It was a fortuitous idea in that it stimulated interest in marine research. But while at sea, a *Challenger* chemist revealed that the slime was precipitated by mixing alcohol with seawater, a discovery which was very embarrassing to Huxley but did not refute the theory of evolution.

A British government department, the *Challenger* Expedition Commission, was established to analyze and publish the results of the

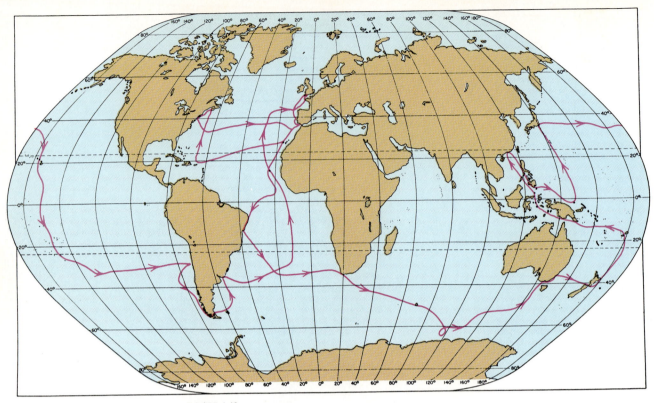

FIGURE 1-10

The track of H.M.S. *Challenger* during the first comprehensive scientific study of the ocean (1872–1876).

voyage. It was headed by *Sir John Murray* (1841–1914), a Canadian-born geographer and naturalist. Murray was a brilliant scientist whose career did not end with publication of the *Challenger* Reports. He is credited with laying the foundations for submarine geology. After persuading the British government to annex uninhabited Christmas Island in the Indian Ocean, Murray made his fortune in phosphate mining. He supported his own research laboratory and funded one oceanographic expedition.

AMERICAN OCEAN EXPLORATION

In the United States, government support for oceanography grew in the nineteenth century, out of a need to solve practical problems. These included ensuring the safety of passengers and goods aboard U.S. ships, maintaining adequate coastal defenses, and protecting fisheries. In 1830 the Navy created a Depot of Charts and Instruments to supervise use of navigational instruments on government vessels. A naval officer, *Matthew Fontaine Maury* (1806–1874), was appointed Superintendent in 1842.

Maury (Fig. 1-11) wanted to make ocean transportation safer and speedier. He had already written a navigation textbook. Maury organized the vast amount of data on wind, current, and seasonal weather stored in ships' logbooks, mostly from whaling ships. To gather more information, Maury furnished ships with blank charts for recording weather and ocean conditions. Wind and current charts compiled in this way revolutionized navigation and cut weeks off transoceanic runs by clipperships, the fastest ships of their time. In addition, Maury's sounding and bottom-sampling projects were used to map the Atlantic Ocean bottom.

In 1853 Maury organized the first international meteorological conference, which led to international cooperation in collecting

FIGURE 1-11
Matthew Fontaine Maury was the first superintendent of the U.S. Navy's Depot of Charts and Instruments. His book *The Physical Geography of the Sea* was the first oceanography book in English. He is often called the father of physical oceanography. (Courtesy of U.S. Navy.)

weather information at sea. His popular and influential book, *The Physical Geography of the Sea* (1855), was the first major oceanographic work in English. He has been called the father of physical oceanography.

To protect its fisheries, the United States established a Fish Commission in 1871 and a research center on Cape Cod. This stimulated development of the *Marine Biological Laboratory* (founded in 1888) and the *Woods Hole Oceanographic Institution* (founded in 1930). The first U.S. ocean research vessel was the *Albatross,* built in 1882.

In addition to support provided by government agencies, wealthy individuals sponsored research in the late nineteenth and early twentieth centuries. For instance, the American scientist *Alexander Agassiz* (1835–1910) maintained a private laboratory in Newport, Rhode Island, to study materials collected on his privately financed expeditions. He also paid for equipment to outfit the *Albatross* to dredge the deep-ocean bottom. On one of Agassiz's expeditions, materials were collected from a depth of 7632 meters (25,200 feet), a record not surpassed until 1951.

GLOBAL OCEAN STUDIES

Virtually all civilian ocean research ceased in 1939 with the outbreak of World War II. Many advances were made in instrumentation, and our understanding of the ocean was improved. For example, advances in predicting wave conditions were essential for the invasion of Europe and the many amphibious assaults on islands in the Pacific. Mapping features of ocean basins, such as the magnetic field, was expanded to improve the ability to detect submarines. These results were used for scientific purposes after the war ended.

The most recent phase of ocean studies began in the 1950s. Government support for ocean research and education greatly increased.

This permitted universities, especially in the United States, to play a major role in ocean studies for the first time.

The new ocean studies required more ships and scientists than any one country could provide. Thus cooperative projects involving ships and other resources from many countries because the norm. The *International Geophysical Year* (IGY) accelerated this trend in 1957–1958. The loosely coordinated IGY projects involved scientists from 67 nations to study earth and ocean phenomena during a short period of intense scientific activity.

The next major event was the *International Decade of Ocean Exploration* (IDOE), led by the United States. The studies during the 1970s involved countries around the world. The intent was to learn more about the ocean as a potential source of food, fuel, and minerals as traditional supplies on land of these commodities were used up. In other words, IDOE projects were intended to solve practical problems.

FIGURE 1-12
Current meters are widely used to measure current speed and direction. In this simple meter, the propeller is turned by the water flowing through it. The number of turns is recorded by the dials. The vane orients the meter in the current. A recording compass (not shown) records the number of turns. (Courtesy Woods Hole Oceanographic Institution.)

For example, one IDOE project studied distributions and characteristics of deep-sea manganese nodules, a potential source of metals such as copper, nickel, and cobalt.

New techniques were developed to map strong but short-lived currents in the ocean, much like storms in the atmosphere. Still other projects developed techniques to improve long-range weather forecasts. Internationally organized studies of the ocean continue to play major roles in improving our understanding of oceanic processes.

STUDYING OCEAN WATERS

Many techniques are used to observe the ocean and to sample its waters. Simplest to understand are those involving direct measurements. For example, *current meters* (Fig. 1-12) measure and record current speeds and directions, much as an anemometer measures wind speed and direction.

Floats are another way to study currents. A float (Fig. 1-13) at the surface can respond by radio to Earth-orbiting satellites reporting its location. Float movements indicate direction and speed of currents. Special floats can remain in a subsurface water mass and report its

FIGURE 1-13
Drifting buoys are used to trace movements of water masses. This buoy is designed to float at the surface. It will be tracked by satellite. Position and water temperatures are reported twice daily. (Photograph by Peter Wiebe, courtesy Woods Hole Oceanographic Institution.)

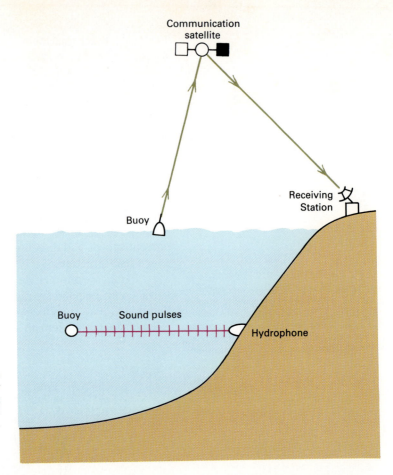

FIGURE 1-14
Buoys are used to follow subsurface water movements and to report their locations using sound pulses. Both the satellite signals and sound pulses are then processed by computers.

location by sending sound pulses to listening stations ashore (Fig. 1-14). Again, changes in float location indicate the presence of currents.

Water samples must be taken for many studies. For instance, microorganisms are filtered from water samples for later study. Specially collected (to avoid contamination) water samples are also taken for chemical analyses. Water-sampling bottles are commonly made of plastic to avoid metal contamination.

Sampling bottles (Fig. 1-15) are lowered from a ship by a metal wire to the desired sampling depth. Once there, an electrical signal closes the valves on the bottle. The wire is hauled up, the bottles recovered, and water samples taken.

Precise temperature measurements are especially important in oceanography. For many decades, precision mercury thermometers mounted on water bottles (Fig. 1-15) measured water temperatures. Now, electrical devices called *thermistors* measure water temperatures and send the measurements back to computers aboard ship. Other electronic instruments mounted on the wire measure electrical conductivity (an indication of salt content) of the water. These measurements too are recorded for later analysis. Temperature and salinity are used to identify water masses.

Remote sensing is also used to study ocean currents. One technique maps subtle changes in ocean surface heights, using Earth-orbiting satellites (Fig. 1-16). Knowing where the ocean surface slopes steeply, oceanographers can map current locations by combining such data with observations made from ships. (We discuss this technique in Chapter 7.)

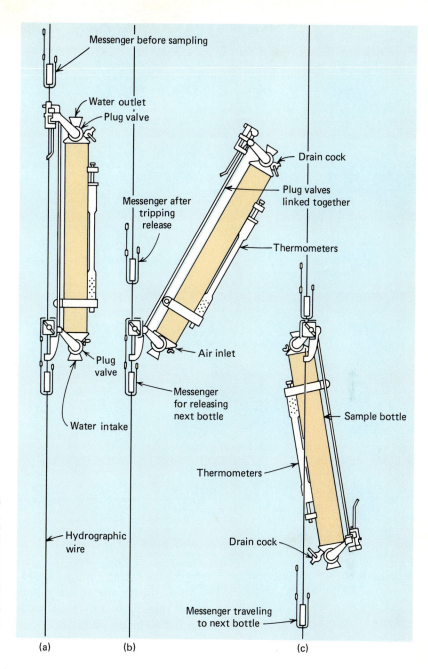

Messenger before sampling

Water outlet
Plug valve

Messenger after tripping release

Drain cock

Plug valves linked together

Thermometers

Air inlet

Plug valve

Messenger for releasing next bottle

Sample bottle

Water intake

Thermometers

Hydrographic wire

Drain cock

Messenger traveling to next bottle

(a)   (b)   (c)

FIGURE 1-15
Fridtjof Nansen, a Norwegian oceanographer, invented a water-sampling bottle. The brass bottles were attached to a wire and lowered from a ship. When the bottle reached the desired depth, a weight, called a *messenger,* was clamped on the line and released. It moved along the wire until it hit the bottle and released a trigger. This allowed the bottle to rotate, closing valves at both ends and collecting a water sample. Thermometers on the bottle recorded water temperatures when the sample was taken.

EXPLORING THE DEEP-OCEAN FLOOR

Systematic study of the deep ocean floor is fairly recent, made possible by new instruments and techniques. Perhaps the most striking advances have been in mapping the ocean bottom. Ocean depths were originally determined by measuring the amount of weighted rope let out before the weight touched the bottom. This slow process produced few measurements, called **soundings.** Many of them were inaccurate. (Mark Twain used the same technique as a Mississippi River pilot to determine where the water was deep enough for a riverboat to pass.)

The first crude maps of the ocean bottom came in the mid-1800s, from data collected while surveying routes for trans-Atlantic subma-

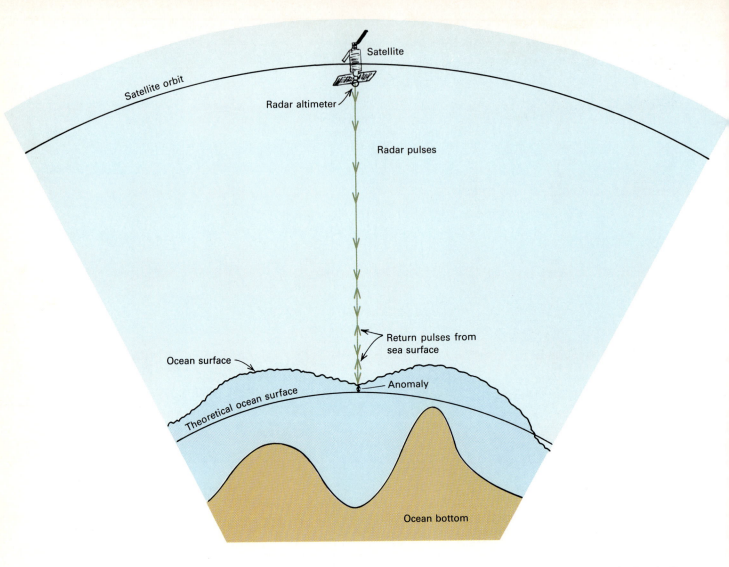

Satellite

Satellite orbit

Radar altimeter

Radar pulses

Return pulses from
sea surface

Ocean surface

Theoretical ocean surface

Anomaly

Ocean bottom

FIGURE 1-16
Instruments on Earth-orbiting satellites use radar
to map the ocean surface topography.

rine cables. Only very large features, such as the relatively shallow Mid-Atlantic Ridge, could be detected.

The use of **echo sounders** (originally developed to detect submarines in World War I) to measure ocean depths was a major advance. This permitted many accurate depth determinations from a moving ship. Thus large areas could be surveyed quickly and cheaply.

Echo sounders work by accurately timing the delay between the transmission of a sound pulse and the arrival of its echo from the bottom. Water depths beneath the ship can be calculated from the known velocity of sound in seawater. Echo sounders were first used by the German *Meteor Expedition* (1925–1927) to map the South Atlantic.

Systematic mapping of deep-ocean basins began in the late 1940s. Scientists at Columbia University's Lamont-Doherty Geological Observatory carried out pioneering surveys, guided by *Maurice Ewing* (1906–1974) (Fig. 1-17). These surveys dramatically improved our knowledge of the sea floor. For example, midocean ridges were mapped in all ocean basins.

Such surveys continue today. Because of the enormous size of the ocean, much of it remains unmapped. Only about 20% has been precisely mapped with modern instruments. Because of concerns about nuclear weapons launched from submarines, much of this information remains classified and unavailable for scientific study.

BOX 1-1

Science and Technology

Science dominates the way that Western society views the world. In this section we consider the scientific method and its use as well as the importance of technological advances in studying the ocean.

Science is based on two premises, namely, that natural phenomena are controlled by relatively simple relationships and that these relationships can be discovered by careful observations or measurements of the processes involved. Observations are analyzed and then organized into statements—called **models**—of how such processes work. These models may permit predictions of future phenomena. Indeed, the ability to predict events or conditions, such as weather or tides, is a major benefit that science offers society. Models are also useful for describing systems and how they work, even without making predictions.

Predictive models are made by developing a **hypothesis**—an idea to be tried. The hypothesis may come from analysis of data or from hunches or even inspired guesses. For example, Sir Isaac Newton hypothesized that ocean tides were caused by the gravitational attraction of the sun and the moon acting on ocean waters. According to legend, Newton's original ideas about the force of gravity came to him when an apple fell on his head while he was sitting under an apple tree. Later he showed how the gravitational attraction of the sun and moon can cause ocean tides. This hypothesis was used to predict tides, a major test of the idea.

Generally, the simplest model that explains and predicts the phenomena is preferred. (One scientist calls this the principle of ''least astonishment.'') Thus the relatively simple model of the earth revolving around the sun was shown to be correct even though ancient astronomers could predict planetary movements using elaborate models involving the sun, planets, and stars revolving around the Earth.

Models can be stated in words (*conceptual models*) or expressed in a mathematical formula (*mathematical models*). Such models must be verified or validated. This is done by testing predictions made using the model against observations, sometimes of past phenomena. A hypothesis confirmed by many such tests is called a **theory.**

Science is also accumulated knowledge gained from applications of scientific methods. It consists of techniques and theories that have been tested over many years. At any time, most scientists are engaged in research on problems posed by the accepted theories but not yet answered. Theories are elaborated to include more points and to make more useful predictions.

Whenever a model or theory fails to predict satisfactorily or to describe the observed phenomena, it must be revised. If the predictions are nearly correct, the model may need only fine tuning. But if predictions made using a model are too much in error, that model is discarded and replaced by another, perhaps one based on a different hypothesis. This change in models is called a *scientific revolution.*

Science makes its biggest advances when new theories are adopted and older, less useful one discarded. A well-known example is Darwin's *theory of evolution,* which has dominated scientific thinking about relationships among species of organisms since its publication in 1857. A more recent example of a scientific revolution that radically altered scientific thinking is the development of plate tectonic theory (discussed in Chapter 3). The utility of any theory is measured by its ability to answer old questions and to pose interesting new problems for study.

Technology—the application of science and knowledge gained through the scientific method and from practical experience—provides instruments needed to observe the ocean and to handle the data obtained from them. For example, much progress in understanding the ocean has come from improvements in our ability to observe oceanic phenomena.

Since the late 1970s, the availability of ocean observations from instruments on Earth-orbiting satellites has changed oceanography. These instruments permit ocean conditions, such as wave heights or sea surface temperatures, to be observed over all the ocean within a few days. The amount of data obtained requires very large computers to process, store, and manipulate it. The rate of advance in understanding the ocean is heavily dependent on the rate at which new instruments and measuring techniques become available.

The ocean bottom can now be observed with other instruments (Fig. 1-18). *Cameras* can photograph the ocean bottom, either on film or electronically. The images are processed and analyzed aboard ship. These techniques are used in large-scale surveys or for searching for submarine cables or downed aircraft. The *Titanic* was found on the North Atlantic floor off Newfoundland in 1985 using such a device.

(a)

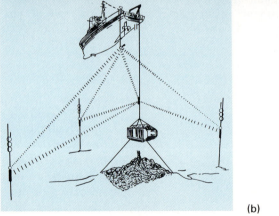

(b)

FIGURE 1-19
Manned submersibles are used to study deep-ocean organisms and processes and to retrieve samples. (Courtesy Harbor Branch Oceanographic Institution.)

Manned *submersibles* permit detailed observations or precise sampling (Fig. 1-19) of ocean-bottom processes.

Specially equipped *drillships* sample rocks and sediment deposits below the ocean bottom. Use of drillships for scientific purposes began in 1967 with the Deep Sea Drilling Program, using the *Glomar Challenger*. The internationally supported Ocean Drilling Program continues such studies using the drillship *Joides Resolution* (Fig. 1-20).

OCEAN EXPLORATION FROM SPACE

Observations made by cameras and other instruments on Earth-orbiting spacecraft (Fig. 1-21) have revolutionized oceanography by dramatically improving our ability to detect rapidly changing ocean features. Frequent observations from satellites permit oceanographers to make weekly or monthly maps of sea surface temperatures or ocean currents, much like daily weather maps. Instead of being average observations of ocean conditions made thousands of kilometers apart and separated by decades in time, these new maps of ocean conditions

FIGURE 1-20
The drillship *Joides Resolution* collects samples of rock and sediment from the ocean bottom for studies of ocean basins. The drill tower is used to lower a pipe with a drill on the end to core rocks and sediments. The ship also has modern laboratory facilities to process and study samples retrieved from the ocean bottom (Courtesy Texas A & M University.)

show events as they happen. Satellites can also observe areas inaccessible to ships, such as the ice-infested polar-ocean waters.

Before *ocean-sensing satellites* were launched, ocean conditions were observed primarily from ships, and only occasionally from aircraft. Ships can cover only about 460 kilometers (250 miles) per day. Thus it was impossible to observe features thousands of miles across that changed within a few days. With ships it was possible to determine only long-term ocean conditions, much like **climate** (average weather conditions) on land. It was impossible to study rapidly changing ocean processes.

The first ocean-sensing satellite was SEASAT, launched in 1978 (Fig. 1-21). Its 3 months of observations showed the utility of satellites in studying ocean conditions. Several ocean-sensing satellites have been launched since then by the United States, Japan, and the European Space Agency.

During a single day orbiting the earth, a satellite obtains millions of ocean observations (Fig. 1-22). Before 1980 we knew relatively little about the ocean; in other words, oceanographers were data poor. Now they receive such quantities of data that the largest and fastest computers are needed to store and manipulate the data.

Satellites and computers have also had a major effect on efforts to use the ocean and to extract its resources. Fishing boats now use satellite observations to locate fish. This saves time and fuel that they would otherwise spend in locating fish.

Ships are now routed to avoid ice and storms. They can also take advantage of favorable currents or avoid unfavorable currents. This

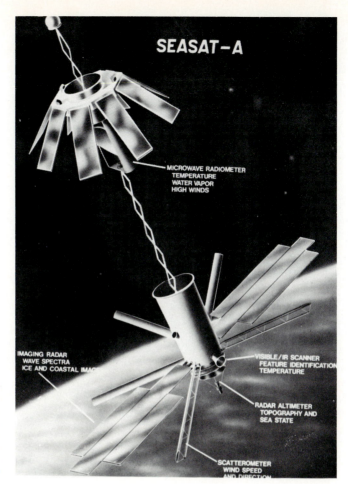

FIGURE 1-21
SEASAT-A—the first ocean-observing satellite—carried instruments to observe the surface and measure sea surface topography, including wave heights. Other instruments measured the visible and infrared spectrum to measure winds over the ocean. (Courtesy NASA.)

FIGURE 1-22
The track covered by a satellite during a 10-day period. Compare this coverage with the track (Fig. 1-8) of the *Challenger* Expedition.

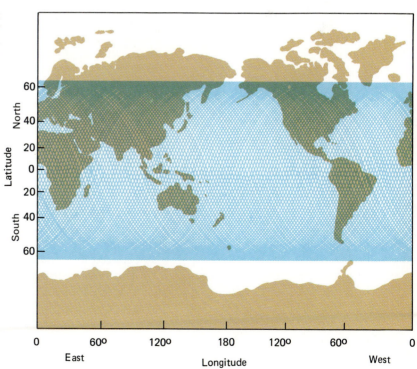

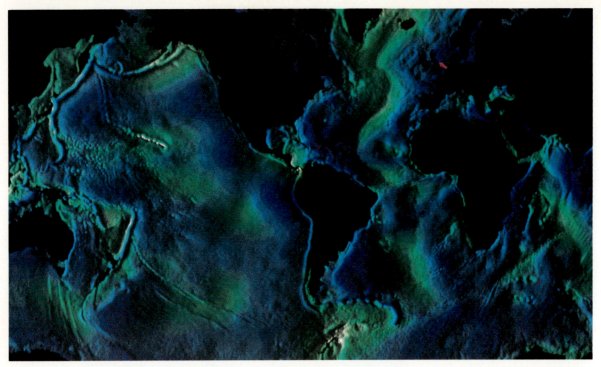

FIGURE 1-23
Images of ocean-surface phenomena are used to plan sampling strategies at sea. (Courtesy NASA.)

shortens the passages and lessens the possibility of storm damage to the ship or its cargo.

Satellites and ships are now used together in studying ocean processes. Images of currents and other features are transmitted to ships. These images (Fig. 1-23) are then used to plan sampling strategies. In this way, ocean processes can be studied as they happen. The results are revolutionizing our view of the ocean and how it works.

LAW OF THE SEA

After European exploration and colonization of the newly discovered continents in the sixteenth century, the "high seas" were free to all nations. *Hugo Grotius* (1583–1645) formulated the legal doctrine of *mare liberum* (freedom of the high seas) in 1635. Until the 1970s the ocean, outside of a narrow territorial sea, extending 3 nautical miles (6 kilometers) from the shore, was freely used by all nations.

By the 1970s growing national interest in the potential of immense quantities of petroleum on the continental margins and metal-rich manganese nodules from the deep-ocean floor challenged this concept. Uncertainties about the legal status of these deposits inhibited investments needed to exploit them.

In 1974 the United Nations convened a *Conference on the Law of the Sea*. The nations of the world negotiated their many conflicting needs in order to develop a consensus, often fuzzy, to divide the ocean among them. The treaty was adopted in 1982. Its implementation markedly changed the legal status of the ocean margin.

The territorial sea was extended to 22 kilometers (12 nautical miles) from the shore. Within this zone each coastal state has rights and responsibilities equal to those it has over the land. Consequently, some major straits through which ships must pass in going from one ocean to another came within the territories of coastal states (Fig. 1-24). Negotiations were necessary to ensure the rights of ships to pass through these areas.

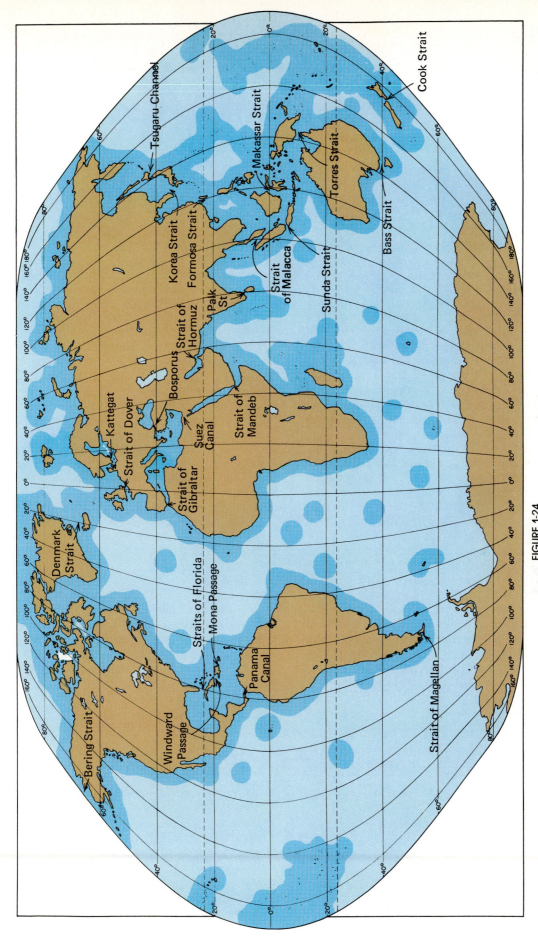

FIGURE 1-24

Exclusive Economic Zones (EEZ) are claimed by coastal states. Several seas, such as the Gulf of Mexico and the Mediterranean and North seas are totally divided among several EEZs. Some strategic straits affected by the EEZs are also shown.

More dramatic was the recognition of a 370-kilometer (200-nauti-cal-mile) *Exclusive Economic Zone*. In this zone coastal states regulate fisheries and resource exploration and exploitation. About one-third of the ocean falls in the Exclusive Economic Zone of some coastal state. The North Sea, Mediterranean, Gulf of Mexico, and Caribbean are totally divided among coastal states.

Thus in the early 1980s, enormous areas of ocean came under national jurisdictions. The United States, for instance, assumed responsibility for an ocean area approximately equal to its land area.

SUMMARY

The ocean provides humans with food, defense, and transportation. More than 1000 years ago Polynesians and Micronesians sailed thousands of kilometers across the Pacific to colonize islands from New Zealand to Hawaii. Chinese ocean voyages extended the influence of the Ming dynasty. None of these voyages contributed to modern ocean science.

European ocean exploration began in the fifteenth century. Ptolemy's maps were extremely influential. The Greek's principal ideas were that the world was a sphere with large ocean areas and that the ocean was navigable.

Portugal led the exploration of the Atlantic. The Portuguese also developed an oceanic trade route connecting Europe with India and Asia, which replaced the overland Silk Route that was closed to European traders. Spain and later other European countries participated in exploration and colonization. European domination of the ocean lasted for more than 400 years. New types of ships carried explorers and, later, traders. Benjamin Franklin mapped the Gulf Stream's location, giving instructions to sailing ships crossing from America to Britain.

Scientific study of the ocean began in the 1800s. Many early measurements were inaccurate, leading to mistaken ideas about the ocean and its circulation. Modern oceanography, or systematic study of the ocean, began with the *Challenger* Expedition in 1872–1876. During the nineteenth and early twentieth centu-ries, government agencies mapped coasts to locate or protect fisheries and to promote shipping.

Technological advances facilitate ocean science. The steam engine powered early oceanographic ships and their sampling gear. Electronic instruments gradually replaced mechanical devices to measure ocean properties at depth. Using ships to study the ocean restricted oceanographers' understanding of the ocean to long-term processes.

Studies of the deep-ocean floor accelerated in the 1950s. Mapping the properties and topography of the ocean bottom led to a better understanding of Earth and its ocean. In addition to studying the ocean bottom from submersibles, ships, and satellites, oceanographers now use drillships to sample the crust beneath the ocean floor to study how basins form and how climate has changed over the past 200 million years.

Earth-orbiting satellites permit observations of rapidly changing features. Large, extremely fast computers are also necessary to store and analyze the data obtained from satellites.

The Law of the Sea Conference, completed in 1982, changed the ocean's legal regime from one of freedom on the high seas to one where coastal states exercise control over the ocean within 200 nautical miles (370 kilometers) of their coast. Exploitation of resources in this part of the ocean under the new regime has been radically changed.

STUDY QUESTIONS

1. Why did ancient sea captains not contribute to ocean science, given their knowledge of the ocean?
2. What did the Polynesians use to navigate in their voyages?
3. What did Prince Henry the Navigator contribute to oceanography?
4. What role did Captain Cook play in exploring the ocean?
5. Explain the importance of the *Challenger* Expedition.
6. List the principal contributions to ocean science of M. F. Maury, Benjamin Franklin, and Maurice Ewing.
7. What has deep-ocean drilling contributed to ocean science?
8. How has the Law of the Sea Conference changed the legal status of the ocean?
9. List some satellite-based techniques now available to observe the ocean, and indicate the contribution that each makes.

Atlas of the Oceans. 1977. New York: Rand McNally. 208 pp.

BOORSTIN, D. J. 1983. *The Discoverers.* New York: Random House. 745 pp. Includes history of early ocean exploration.

DEACON, M. 1971. *Scientists and the Sea. 1650–1900.* London: Academic Press. 445 pp. History of oceanography up to the 20th century.

MAY, W. E., AND L. HOLDER. 1973. *A History of Marine Navigation.* New York: Norton. 280 pp.

MENARD, H. W. 1986. *Islands.* New York: Scientific American Books. 230 pp.

NATIONAL GEOGRAPHIC SOCIETY. 1987. *Into the Unknown: The Story of Exploration.* Washington, D.C. 336 pp.

STOMMEL, H. M. 1984. *Lost Islands: The Story of Islands that have Vanished from Nautical Charts.* Vancouver, B.C. University of British Columbia Press. 146 pp.

The submersible *Alvin* is retrieved by the RV *Atlantis II* after a diving operation. (Courtesy Woods Hole Oceanographic Institution/Photo Researchers.)

Ocean Basins

OBJECTIVES

1. To learn the principal features of the ocean basins;

2. To understand the major features of the ocean floor;

3. To understand the principal differences between ocean basins and continents.

*O*cean basins are a primary feature of the earth's surface. Until recently, the ocean floor was hidden from human observation by miles of ocean water. Now it is possible to map the ocean bottom in detail using satellites and acoustic techniques and to observe ocean-bottom processes from submersibles. With these techniques, major new discoveries come virtually every year. Our improved knowledge of the ocean basins is changing our view of the earth as much as the exploration of the New World changed human perspectives in the fifteenth and sixteenth centuries. We will use maps of the earth's surface to show our understanding of the processes involved.

In this chapter, we examine:

Features of ocean basins;
Distributions of continents and ocean basins; and
Major features of continental margins.

LAND AND OCEAN

From space, we see Earth (Fig. 1-1) nearly covered by ocean waters, accounting for 70.8% of its surface. Land accounts for the remaining 29.2%, most of it in the Northern Hemisphere (Fig. 2-1).

If we look at the earth's surface from the South Pole, the world ocean appears as three large basins extending northward from Antarctica, where they connect (Fig. 2-2). The land areas separate the ocean basins.

The Pacific dominates the earth's surface and holds 52% of its ocean waters (Fig. 2-3). The Atlantic and Indian oceans account for the rest, except for the 2% of the water that occurs as ice, primarily in Antarctica and in the Greenland ice caps.

If we could drain the ocean basins and view Earth from space (Fig. 2-4), the most conspicuous features of its surface would be the enormous mountain ranges in the middle of the Atlantic Ocean, the eastern South Pacific, and in the western Indian Ocean as well as circling Antarctica. We call this set of globe-circling mountain ranges **midocean ridges** even though not all are in the middle of their respec-

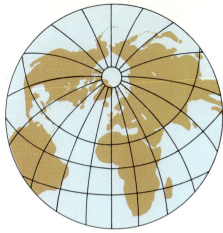

Land hemisphere

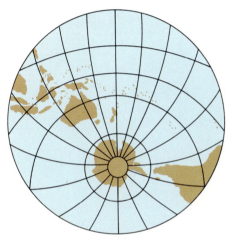

Water hemisphere

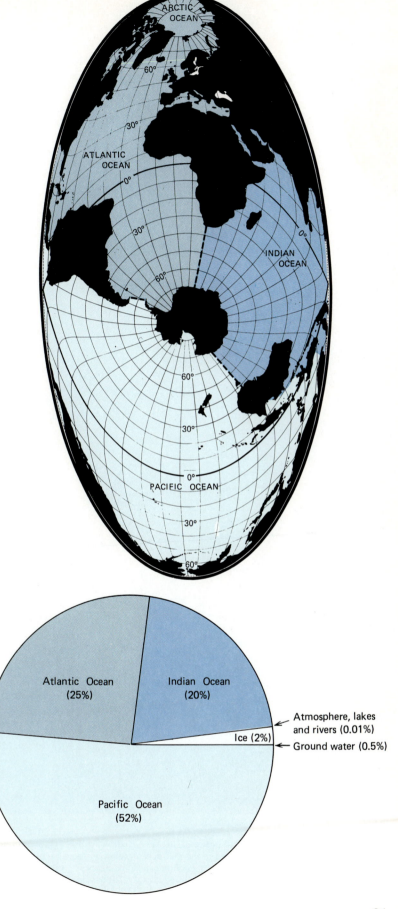

FIGURE 2-1
The land hemisphere is centered on western Europe and has roughly equal amounts of land and water. The water hemisphere is centered on New Zealand and water dominates. (After Judson, Kauffman, and Leet, 1987.)

FIGURE 2-2
The ocean is centered on Antarctica. For convenience, it is divided into three oceans (upper right).

FIGURE 2-3
Distribution (in percent) of water on Earth's surface. Only a minute fraction occurs in lakes, rivers, or the atmosphere (lower right).

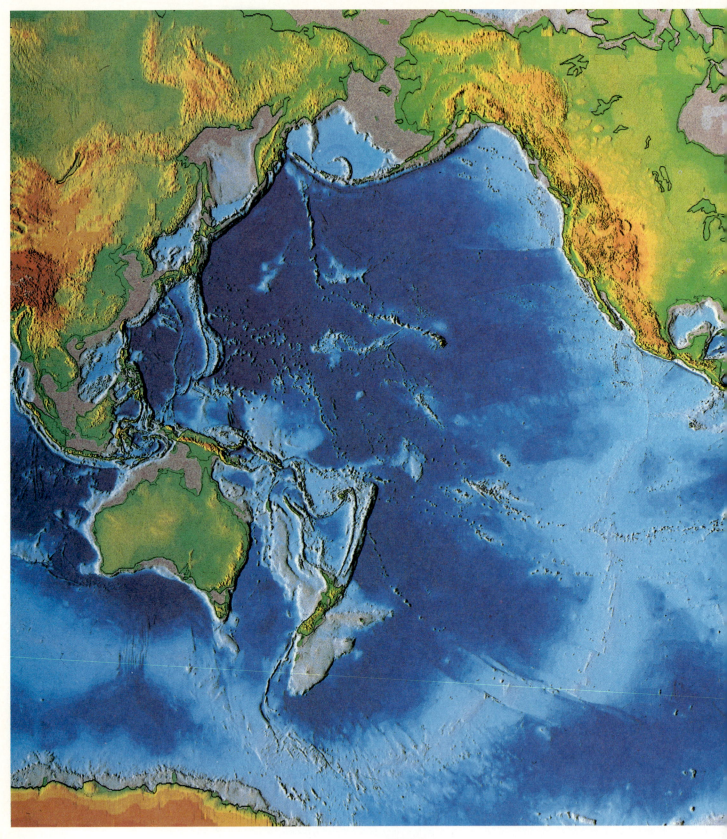

FIGURE 2-4
Topography of the continents and ocean basins. The shallowest parts (shown in light
blues) are near the continents and in the mid-ocean mountain ranges which is most
conspicuous in the Atlantic but occurs in all oceans. The deepest parts of the ocean
basins (shown in dark blues) are in long, curved arcs, called trenches, which are most

CHAPTER TWO   OCEAN BASINS

conspicuous near the western margin of the Pacific and south of Alaska and the Aleutian Islands. Note that the high mountain ranges (shown in reds) form long belts that border the Pacific and cut across Asia. The deep trenches and the mountains are formed by movements of Earth's outermost layer, called the lithosphere. (Data from the National Geophysical Data Center, NOAA.)

tive basin. (The name comes from the Mid-Atlantic Ridge, which is indeed in the middle of its basin.)

The other principal feature of the ocean basin is the deepening of the ocean bottom (shown in Fig. 2-4 by the darker shades of blue) away from the midocean ridges. Around the margins of the Pacific are the deepest parts of the ocean, the **trenches.**

On land, the most conspicuous features are mountain ranges (higher elevations shown in red browns). They form elongate north-south belts along the western margin of the Americas and an east-west belt across southern Asia. In short, the orientation of mountain ranges on land parallels the orientation of the midocean ridges. There are two principal directions for each: north-south and east-west.

Also conspicuous on a waterless Earth would be the margins of the land masses, the **continental margins.** Thus on a water-free Earth, we could see the two principal levels of its surface, land and ocean bottom and the connection between them. We discuss these different surfaces in the next section.

To show these relationships, we will use maps of the earth's surface.

BOX 2-1

Maps

Maps are flat representations of the earth's surface. Since Earth is a sphere, each map (a flat surface) involves distortions of the features shown. No map can accurately represent size, shape, distance and direction; thus each map is a compromise. Only a globe shows correct sizes, shapes, distances, and directions. Globes are obviously unsuited for use in books or for navigation, so we are stuck with maps and their attendant problems.

Maps are used extensively in studying the ocean to show distributions of features and processess. Some of these features involve the continents. Others involve the ocean basins. Because of their various limitations, different maps are used in this book.

In making a map, the first step is devising the grid system. Each point on the surface of a sphere has a **latitude**—angular distance north or south of the equator—and a **longitude**—angular distance east or west of a designated starting point. The most commonly used starting point for measuring longitude is the **prime meridian,** a north-south line running through Greenwich, England, near London. Each map projection takes the system of latitude and longi-

tude lines from the sphere and transfers them to a flat sheet. In the process, various aspects of the earth's surface are distorted.

The most common maps are **Mercator projections** (Fig. B2-1-1). This projection is readily identified by the rectangular grid of straight lines representing latitude and longitude. Mercator projections are most accurate near the equator and most distorted in the high latitudes. It is impossible to show the polar regions on such a projection. Such maps are commonly used for navigation because they preserve directions.

A special projection—the *interrupted homolosine* projection—is used to show features of ocean basins. It preserves the areas of the various parts of the basins.

Finally a special projection—the *Hoelzel planisphere*—is used to show land areas. It shows coastal ocean areas well but interrupts the Pacific Ocean along the *International Date Line*. Note that this projection shows the polar regions much larger than they actually are. The lines of longitude should converge to a point rather than a line.

FIGURE B2-1-1 (right)
Various map projections are used in the text to display different aspects of the Earth's surface. (a) The Mercator Projection is a familiar one. It is used in the text for general-purpose displays of data. It is widely used for navigation. Note the distortions at the high latitudes. (b) The Interrupted Homolosine Projection is somewhat less familiar. It is used in the text to show ocean features, since it shows all three ocean basins without interruption. It is also an equal-area projection, so that all parts of the ocean are shown without areal distortions. The Hoelzel Planisphere projection (c) is used to show land features. It shows the shapes of the continents with much less distortion in the high latitudes than the Mercator projection.

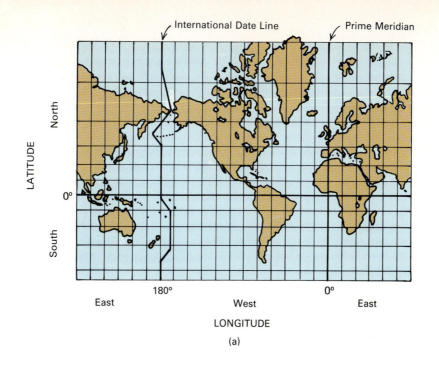

(a)

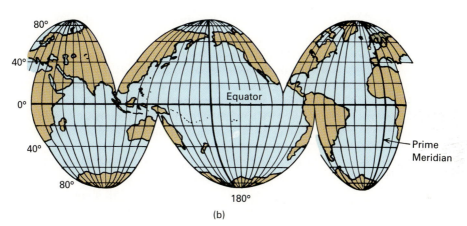

(b)

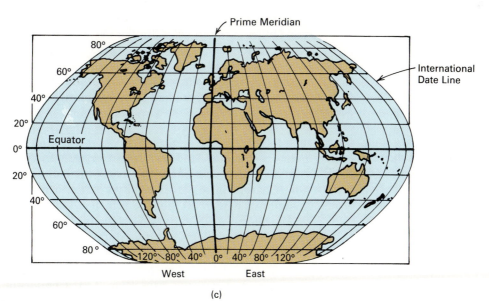

(c)

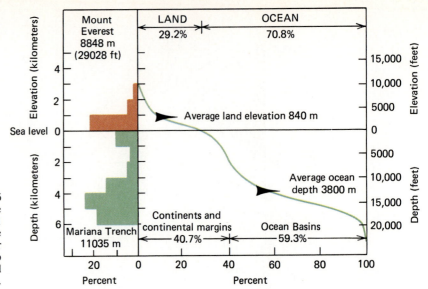

FIGURE 2-5
Hypsometric diagram shows the fraction of the Earth's surface in elevation or depth zones of 1 kilometer (left side). The right side shows the cumulative fraction of Earth's surface shallower than a given depth. Note that Earth has two dominant levels—one corresponding to the land and the other to the ocean bottom.

HYPSOGRAPHIC CURVE

Earth's surface has two distinctly different levels. The land stands above sea level, about 840 meters, while the ocean bottom averages about 3800 meters below sea level. This relationship between the height of the land and the depths of the ocean basins is shown in a **hypsographic curve** (Fig. 2-5), which plots the amount of the Earth's surface at each elevation or depth.

Compared with human dimensions, the extreme heights and depths of the earth's surface are enormous. Mount Everest in the Himalaya Mountains of northern India stands 8848 meters above sea level, while the deepest part of the Mariana Trench is 11,035 meters below the sea surface. Even if we cut down the land and dumped the material in the ocean to make Earth a smooth sphere, it would still be covered by water 2430 meters deep.

Compared with Earth's radius (6370 kilometers), these heights and depressions are minute. In fact, the heights of the mountains and the depths of the trenches on the earth's surface are smaller than the imperfections on the best globe available.

In summary, Earth is essentially a smooth sphere. The ocean basins are very shallow depressions. And the ocean is a thin film of water, interrupted here and there by continents.

CONTINENTS AND CONTINENTAL MARGINS

Four continental land masses—Eurasia-Africa, North and South America, Antarctica, and Australia—interrupt the ocean basins (Fig. 2-6). Changes in sea level modify coastal locations and features (discussed in Chapter 10), but the principal boundary between continent and ocean basin is submerged below the sea surface. This is the continental margin, which separates continents and ocean basins—a fundamental feature of our planet (see Fig. 2-7).

Continental margins consist of continental shelves, continental slopes, and continental rises (Fig. 2-8). **Continental shelves** are the submerged upper parts of continental margins. From the shoreline, shelves slope toward the shelf break (average depth 130 meters). There they join the steeper **continental slope,** which extends down to depths

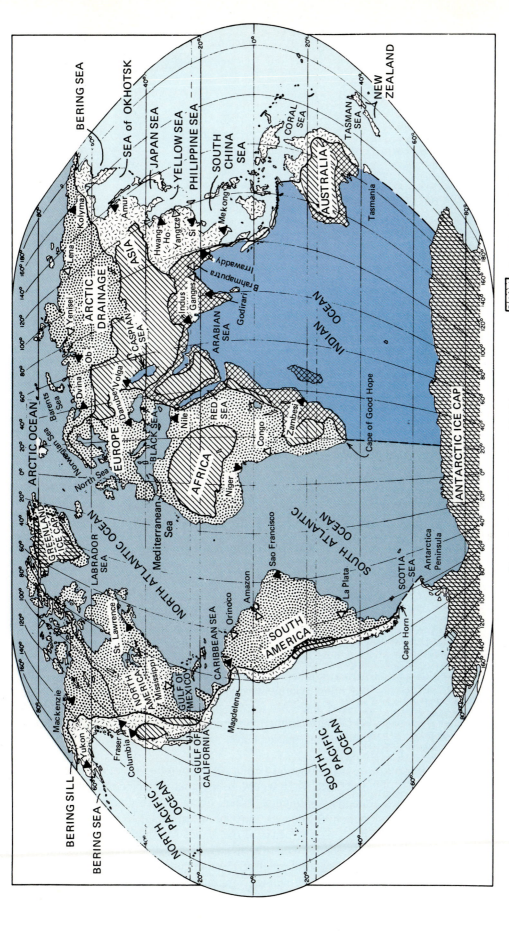

FIGURE 2-6

Map of Earth's surface showing the boundaries between major ocean basins. Areas of the continents whose rivers drain into each ocean basin are indicated, as well as the mouths of some major rivers. Note the large land area that drains into the Atlantic and Arctic oceans.

▽ Rivers discharging more than 15,000 cubic meters (525,000 cubic feet) per second

▼ Rivers discharging more than 3000 cubic meters (100,000 cubic feet) per second

Runoff to Atlantic Ocean and Arctic Sea

Runoff to Pacific Ocean

Runoff to Indian Ocean

No runoff to oceans

Ice

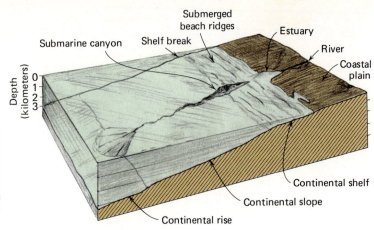

FIGURE 2-7
Schematic representation of the continental margin, showing the continental shelf, the coastal plain, the continental slope, and the continental rise. Note that a submarine canyon cuts across the shelf break and extends to the base of the continental slope, where it flows out onto the continental rise.

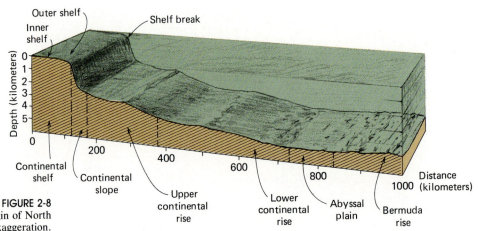

FIGURE 2-8
Diagram of the North Atlantic margin of North America. Note the vertical exaggeration.

of 2 to 3 kilometers. Off Antarctica, the continental shelf breaks are about 500 meters deep because the weight of the ice on the land has depressed the continent and its continental margins. The shelf break is unusually deep around the Arctic Ocean for similar reasons.

Around much of the Pacific where there is active mountain building, continental shelves are only a few tens of kilometers wide. Wide continental shelves occur where continental margins have not experienced mountain building for many millions of years. The shelf around Africa is unusually narrow because the continent stands higher than the other continents. (We discuss the reason in the next chapter.) **Coastal plains** are former continental shelves, now exposed above water. Where mountains occur along the coast, as in Southern California, continental margins are rugged, broken by submerged ridges and basins.

Continental slopes are the outer edges of continental blocks. On a dry Earth, they would be its most conspicuous surface feature. Spectacular continental slopes occur where coastal mountains parallel a trench, the deepest parts of ocean basins. Near the west coast of South America, the peaks of the Andes Mountains reach elevations around 7 kilometers, while the nearby Peru-Chile Trench is 8 kilometers deep. This is a total relief of 15 kilometers within a few hundred kilometers. Such steeply sloping, mountainous margins are common around the Pacific.

Submarine canyons (Fig. 2-7) cut through continental shelves and slopes. They look much like river valleys on land. Some submarine canyons are as large as the Grand Canyon of the Colorado River. A few are associated with major rivers—for instance, Hudson Canyon or the Congo Canyon. In these cases, the upper parts of the canyons probably formed when sea level was much lower than its present level. Many submarine canyons have no obvious connection with rivers.

Continental rises occur at the base of continental slopes. They slope gently seaward and connect with the deep-sea floor.

MIDOCEAN RIDGES

We begin our study of the deep-ocean basins (Fig. 2-9) with midocean ridges, which stand above the deep-ocean floor. They are the most conspicuous features of the ocean basins (Fig. 2-4). They are 60,000 kilometers long and cover 23% of the earth's surface.

The rugged *Mid-Atlantic Ridge* stands 1 to 3 kilometers above the deep-ocean floor. Its most prominent feature is its steep-sided central valley (Fig. 2-10), called a **rift valley;** it is 25 to 50 kilometers wide and 1 to 2 kilometers deep. The rift valley of the Mid-Atlantic Ridge is bordered by rugged mountains whose tallest peaks come to within 2 kilometers of the sea surface (Fig. 2-11). The *East Pacific Rise* is much broader and less rugged (Fig. 2-12). Much of the East Pacific Rise has no rift valley. It intersects North America in the Gulf of California.

In the Indian Ocean, the *Mid-Indian Ridge* (similar to the Mid-Atlantic Ridge) intersects Africa-Asia in the Red Sea (Fig. 2-13). To the south, the Mid-Indian Ridge joins the *Pacific-Antarctic Ridge,* which circles Antarctica (see Fig. 2-9).

Small earthquakes occur frequently on crests of midocean ridges (Fig. 2-14). These earthquakes coincide with the location of the central rift valleys. Indeed, earthquakes have been used to locate active ridges in little-known ocean areas such as the Arctic Ocean.

DEEP-OCEAN FLOOR

The *deep-ocean floor* (deeper than 4 kilometers) occupies about 30% of the earth's surface. Low **abyssal hills** (less than 1 kilometer above the surrounding ocean bottom) cover about 80% of the Pacific floor and about half of the Atlantic. They are also abundant in the Indian Ocean. These hills are thus among the most common features on the earth's surface. They are typically about 200 meters high. Many appear to be extinct volcanoes.

Immense areas of exceedingly flat ocean bottom, called *abyssal plains,* lie near the continents (Fig. 2-10). These are among the flattest parts of the earth's surface. They commonly occur at the seaward margins of the continental rises. Most abyssal plains appear to be covered with thick deposits of sediments that likely came from the nearby land.

FRACTURE ZONES

Hundreds of **fracture zones** cut the ocean floor and offset midocean ridges (Fig. 2-10). They are narrow (10- to 100-kilometer) elongated belts of rugged topography with steep ridges and valleys. Some are 4000 kilometers long (Fig. 2-9). They commonly parallel each other across large parts of ocean basins; this is especially obvious in the North Pacific (Fig. 2-4). Some of the deepest parts of the ocean floor

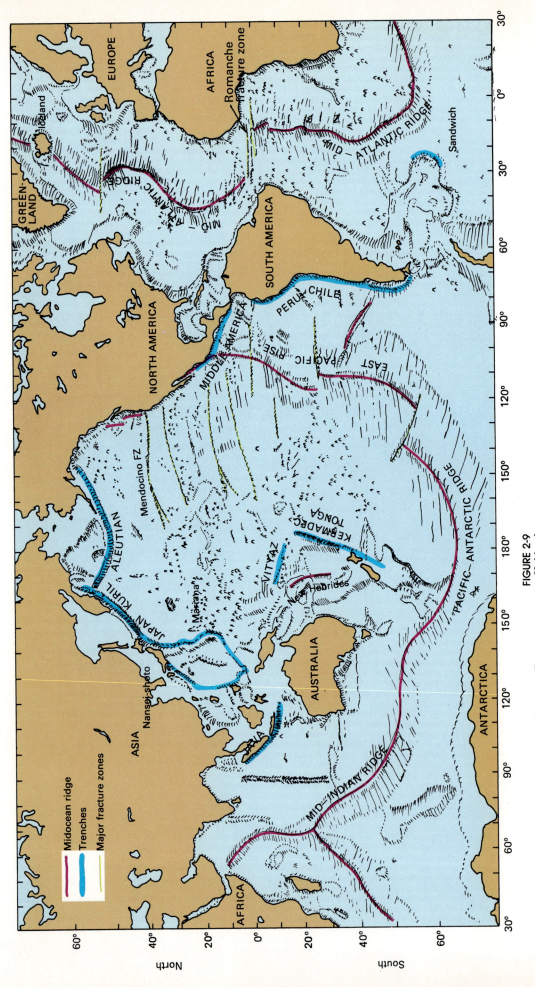

FIGURE 2-9

Map of the ocean bottom, showing major features. (Base map courtesy Hubbard Scientific Company.)

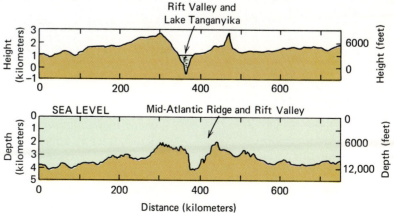

FIGURE 2-10
The North Atlantic Ocean, depicted without water. (Depths are in feet.) Note the conspicuous Mid-Atlantic Ridge and its central rift valley in the middle of the ocean basin. Also note the many offsets in the Mid-Atlantic Ridge. These are fracture zones. (Painting by Heinrich Berann, courtesy Aluminum Corporation of America.)

FIGURE 2-11
Profiles of the rift valley of the Mid-Atlantic Ridge and the Tanganyika rift in East Africa. Note the similarity in form and size of the two features.

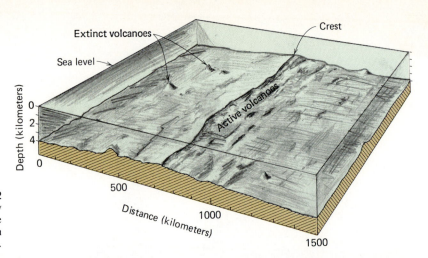

FIGURE 2-12
Block diagram showing ocean-bottom topography along the East Pacific Rise. Unlike the Mid-Atlantic Ridge, rift valleys are less common on the East Pacific Rise.

FIGURE 2-13
The Red Sea occupies the rift formed by the Mid-Indian Ridge intersecting Africa-Asia. (After Judson, Kauffman, and Leet, 1987. Photograph courtesy NASA.)

occur in fracture zones. Sea-floor depths usually change markedly across fracture zones.

Large fracture zones continue into the continents. The *Mendocino Fracture Zone* comes onto land at Cape Mendocino, California. From there, it continues as the San Andreas Fault to connect with the East Pacific Rise in the Gulf of California.

CHAPTER TWO   OCEAN BASINS

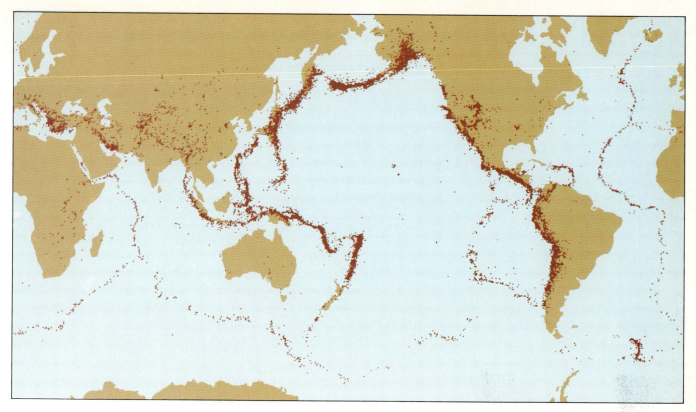

FIGURE 2-14
Earthquakes mark the locations of midocean ridges and trenches. (After Judson, Kauffman, and Leet, 1987.)

One of the largest fracture zones is the *Romanche Fracture Zone* (Fig. 2-10, lower right), which offsets the Mid-Atlantic Ridge by 950 kilometers near the equator. It is part of a group of large transform faults. The entire transform zone consists of deep valleys about 100 kilometers wide, separated by ridges, one of which comes to the sea surface as a small island. The Romanche Fracture Zone contains the deepest part of the Atlantic, 7960 meters. (Because of its great depth, it was included with the trenches for many years.) This deep gap in the Mid-Atlantic Ridge permits bottom waters from the western Atlantic to flow into the deep basins of the eastern Atlantic.

The *Puerto Rico Trench* (Fig. 2-10) is part of another large fracture zone that runs east-west, separating the Gulf of Mexico and the Caribbean basin. It may at one time have been a trench, like the ones we discuss next, but is no longer active.

TRENCHES

Trenches are the deepest parts of the ocean floor. They are typically 3 to 4 kilometers deeper than the surrounding ocean floor (Fig. 2-15). They are relatively narrow features, only a few tens of kilometers wide, but many thousands of kilometers long. Most occur in the Pacific, especially the western Pacific. In fact, most of the Pacific is bordered by trenches. The greatest depth anywhere in the ocean is 11,035 meters, in the *Mariana Trench,* southeast of Japan.

There is also a trench in the South Atlantic (*South Sandwich Trench*) and one (*Java Trench*) in the Indian Ocean (Fig. 2-16).

Trenches are associated with active volcanoes and earthquakes (Fig. 2-16). Many are near chains of volcanic islands (called island arcs), which we discuss later when we consider volcanoes.

VOLCANOES

Volcanoes and volcanic islands are common in the ocean (Fig. 2-17); there are many tens of thousands of extinct volcanoes. They usually stand 1 kilometer or more above the surrounding ocean floor. Volcanic activity on land occurs primarily near the edges of ocean basins, especially around the Pacific.

Most volcanic eruptions occur—usually unnoticed—on mid-ocean ridges. There are two modes of oceanic eruptions. Some lavas form tranquil flows in which the surface layer cools first. This forms an insulated cover for the still-molten lavas in the interior. This type of lava flow forms **pillow lavas** (Fig. 2-18), rounded masses of volcanic rock. These are typical of deep-ocean eruptions.

The other type of volcanic eruption is more common in relatively shallow waters. These are explosive eruptions, producing lots of volcanic ash (fragments of lava rock) when lavas are quenched by cold seawater (Fig. 2-19).

Curved chains of volcanic islands, called *island arcs,* are associated with mountain building and trenches. For instance, the Aleutian Islands, southwest of Alaska, are a single chain of volcanoes with a trench on the seaward side, as shown in Fig. 2-15).

The **Indonesian-Philippine** area has several trenches, including two large ones (Fig. 2-16) with many active volcanoes. Large, destructive eruptions occur in Indonesia, the most volcanic region on earth. In 1815, Tambora on the island of Sumbawa erupted, killing 92,000 persons. The large amount of ash it injected into the atmosphere caused the summer of 1816 to be unusually cold throughout the Northern Hemisphere. In 1883, *Krakatoa*, a volcano between Java and Sumatra, erupted, killing 36,000 and generating destructive sea waves.

FIGURE 2-15

The Aleutian Islands, a simple volcanic island-arc system with a trench. Many of the volcanoes are active. Trench depths greater than 6 kilometers are shown in black.

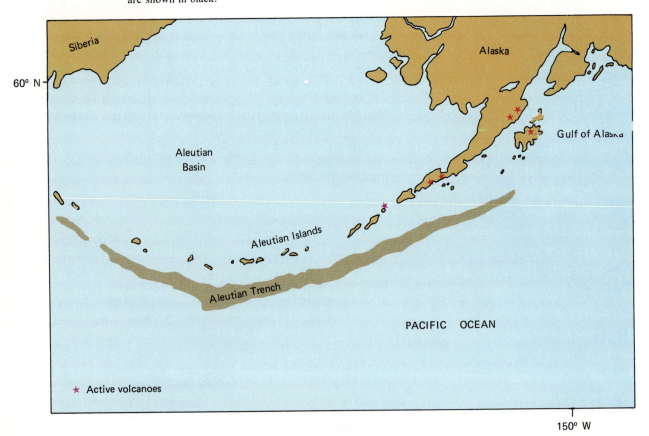

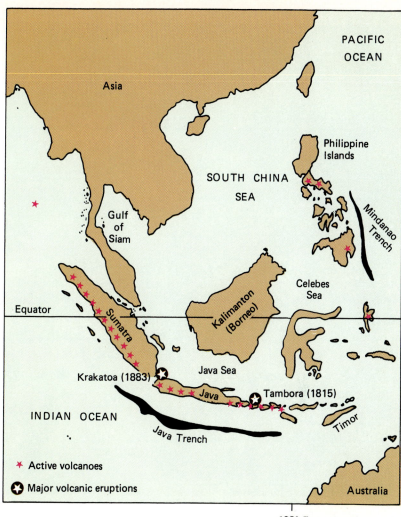

FIGURE 2-16
Indonesia, a complex island-arc system with several trenches, lying between Australia and Asia, has the largest number of active volcanoes in the world. Trench depths greater than 6 kilometers are shown in black.

FIGURE 2-17
Locations of some active volcanoes on land. (After Judson, Kauffman, and Leet, 1987.)

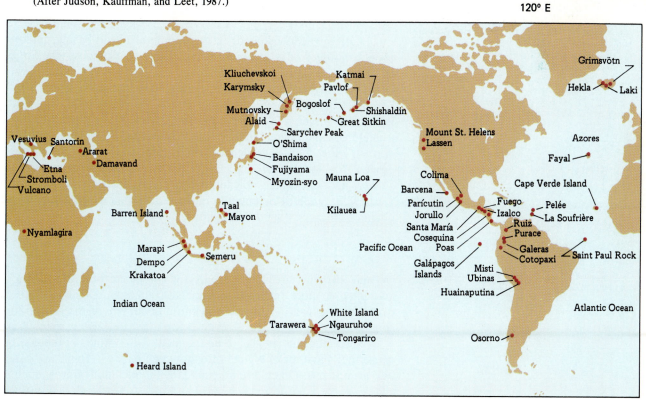

FIGURE 2-18
Toothpastelike structures formed during submarine volcanic eruptions in the axial valley of the Mid-Atlantic Ridge. These rocks are probably less than 10,000 years old. The arm of the submersible (*Alvin*) is picking up a rock. (Courtesy Woods Hole Oceanographic Institution.)

Active volcanoes occur in the middle of ocean basins, far from midocean ridges or island arcs. Some of these volcanoes—the largest on earth—are called *shield volcanoes* (Fig. 2-20). They stand several kilometers higher than the deep-ocean floor and form high islands. Examples of such active volcanoes are on the island of Hawaii in the Pacific and Reunion Island in the Indian Ocean. While it is active, such a shield volcano builds an enormous cone. After the volcano becomes dormant, the cone is eroded by wind, rain, and waves, often being eroded to a submarine bank. Once it is submerged, erosion is much slower, and subsurface volcanoes can persist for tens of millions of years. The Emperor Seamounts in the North Pacific are a submerged chain of volcanoes, 40 to 60 million years old.

OCEANIC PLATEAUS

Oceanic plateaus are isolated parts of the ocean floor that stand a kilometer or more higher than their surroundings. Oceanic plateaus occur in all ocean basins and constitute about 3% of the ocean floor (Fig. 2-21).

Some plateaus appear to be fragments of continental blocks, because they have rocks typical of continental masses. These are called **microcontinents.** Others are associated with volcanic features, such as

(a)

(b)

FIGURE 2-19
Kovachi, an active submarine volcano in the Solomon Islands, South Pacific. (a) When it erupted in October 1968, explosions ejected water and steam 60 to 90 meters into the air and discolored surface waters for 130 kilometers (80 miles). (b) An island formed by the 1961 eruption was eroded away, leaving a submerged bank. (Courtesy Smithsonian Institution Center for Short-Lived Phenomena.)

large volcanic islands. One of these is the Walvis Ridge in the South Atlantic, which is connected to the active volcano on the island of Tristan da Cunha. Another is the Ninetyeast Ridge in the Indian Ocean, whose origin is related to the large volcano on the island of Reunion in the southern Indian Ocean.

We now consider these features in each of the major ocean basins.

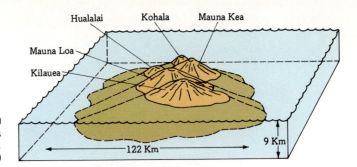

FIGURE 2-20
Schematic diagram showing the five volcanoes that have coalesced to form the island of Hawaii. (After Judson, Kauffman, and Leet, 1987.)

PACIFIC OCEAN

The *Pacific Ocean* is the deepest and largest basin (Fig. 2-3), occupying more than one-third of the earth's surface (Fig. 2-6). Mountain building dominates the Pacific basin margins, with the young mountains creating barriers between continents and ocean. Along the Pacific's eastern margin (the western coasts of the Americas from Alaska to Peru), rugged mountains parallel the coastlines. Submerged continental margins are narrow because the mountain slopes continue below sea level. These mountains block the flow of rivers along most of its margins, especially on the Pacific coast of the Americas. Therefore, the Pacific is less affected by continents than either the Atlantic or Indian oceans.

Islands are abundant in the Pacific, especially in the southern and western portions. Many of the islands are volcanoes, some still active. The *Hawaiian Islands* are good examples. They are the exposed tops of a chain of enormous volcanoes 3000 kilometers long that extends from Hawaii, with its active volcanoes, to Kure Island, a small sand island on top of a deeply submerged, eroded volcano more than 20 million years old. The chain changes direction and goes almost due north to the Aleutian Trench as a chain of submerged volcanoes called the Emperor Seamounts. Island arcs are common in the western Pacific, extending from the Aleutians on the north to New Zealand on the south.

ATLANTIC OCEAN

The *Atlantic Ocean* is a relatively narrow basin connecting the Arctic and Antarctic oceans (Fig. 2-2). The boundary between the Atlantic and Indian oceans runs from South Africa's Cape of Good Hope along longitude 20°E to Antarctica. The boundary between the Atlantic and Pacific oceans is drawn between the Antarctic Peninsula and Cape Horn, South America. The Arctic Ocean is considered part of the Atlantic, so the boundary with the Pacific is the *Bering Strait* between Alaska and Siberia.

The Atlantic is relatively shallow, averaging 3310 meters deep. It has large continental margins, several shallow marginal seas, and the Mid-Atlantic Ridge. There are relatively few islands in the Atlantic. *Greenland,* the world's largest island, is a piece of the North American continent.

The Atlantic Ocean receives large amounts of water and sediment from rivers. The Amazon and Congo rivers flow into the equatorial Atlantic. Together they discharge about one-quarter of the world's river flow to the ocean. Other large rivers flow into marginal seas and into the Arctic Ocean.

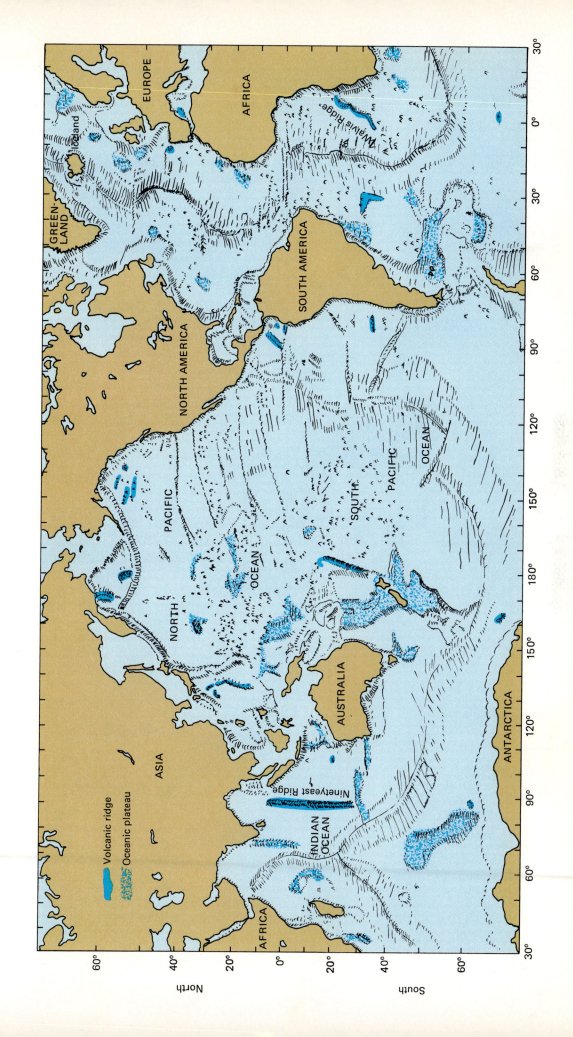

FIGURE 2-21
Plateaus on the ocean floor.

Volcanic ridge
Oceanic plateau

The *Arctic Ocean,* an arm of the Atlantic (Fig. 2-2), is almost completely surrounded by land. During most of the year, sea ice covers its surface. In the central Arctic the sea-ice cover is permanent. Continental margins form one-third of the basin.

INDIAN OCEAN

The Indian Ocean lies primarily in the Southern Hemisphere (Fig. 2-6). Its Pacific boundary runs through Indonesia and extends from Australia southward to Antarctica along longitude 150°E. It is the smallest of the three major ocean basins. Continental shelves around the Indian Ocean are relatively narrow, especially around Africa.

Three of the world's largest rivers (Ganges, Brahmaputra, Indus) discharge into the northern Indian Ocean (Fig. 2-6). Thus this region has an abundance of both fresh water and sediment from the discharge of these rivers.

The northern Indian Ocean also has two major sources of warm saline waters. The *Red Sea* and the *Arabian Gulf* have little river discharge, since they are surrounded by desert. Because of the intense evaporation, they produce warm saline subsurface waters that can be traced for hundreds of kilometers below the ocean surface.

There are few islands in the Indian Ocean. Madagascar, the largest, was apparently once part of Africa. There are a few volcanic islands and several groups of low-lying carbonate islands.

MARGINAL OCEAN BASINS

Marginal ocean basins are large ocean-bottom depressions near continents. They are separated from the open ocean, usually by submarine ridges or islands. They are typically more than 2 kilometers deep and thus are partially isolated from the nearby open ocean. These areas often occur in areas of active mountain building.

There are three different types of marginal basins. Most common are basins associated with island arcs. Here submarine ridges separate them from the open ocean and may completely isolate the deeper waters in the basin. There are several examples of such basins in the Indonesia area.

A second type of marginal basin lies between continents. The *Mediterranean Sea* lies between Europe and Africa, and the *Gulf of Mexico–Caribbean Sea* lie between North and South America.

The third type is the long, narrow, marginal sea surrounded by continent. The *Red Sea* (Fig. 2-13) and the *Gulf of California* are examples of such basins.

SUMMARY

Oceans dominate the earth. Earth's surface has two levels: Land stands about 840 meters above sea level; the ocean bottom is about 3800 meters below sea level. If Earth were a smooth ball, it would be covered by ocean 2430 meters deep. *Continents* interrupt ocean basins, forming north-south barriers. Ocean waters cover about one-third of the continents. Continental margins are the boundary between continents and ocean bottom.

The **midocean ridge** is the world's largest mountain range. The Mid-Atlantic Ridge is a rugged mountain chain in the center of the basin. Similar ridges occur in all ocean basins. Midocean ridges form about 23% of the earth's surface (about 32% of the ocean bottom). They are the most active volcanic regions on earth; they have many shallow earthquakes.

Trenches are the deepest parts of the ocean basins—maximum depths are about 11 kilometers. They occur primarily at the margins of the ocean basins, usually near continents.

Fracture zones are elongate regions of mountainous topography. They connect offset segments of midocean ridges.

The *deep-ocean floor* is nearly flat or gently rolling. It is mostly covered by sediment deposits, especially near the continents.

The *Pacific* is the largest ocean basin, containing about half the earth's free water. It is nearly surrounded by active mountain building with active volcanoes and many earthquake belts.

The *Atlantic* is a relatively narrow, S-shaped basin. It connects the Arctic and Antarctic polar regions. The Arctic Sea is nearly landlocked and partially isolated by submarine ridges from the rest of the Atlantic Ocean. The *Indian Ocean* is primarily a Southern Hemisphere ocean. Small, partially isolated *marginal ocean basins* occur on the edges of the major basins, especially in areas of active mountain building.

STUDY QUESTIONS

1. Describe the two different levels of the earth's surface.

2. On an outline map, sketch the locations of earthquake belts, active volcanoes, midocean ridges, and subduction zones.

3. Give the characteristic features of trenches, midocean ridges, and fracture zones.

4. Draw a cross section of a rapidly spreading midocean ridge. Indicate where volcanic activity and hydrothermal vents are most likely to occur. Label significant features.

5. Describe the three major ocean basins. Indicate where they connect and what separates them.

6. Discuss the various types of volcanoes and where they are found.

7. Describe the three different types of marginal ocean basins. Where do they occur?

SELECTED REFERENCES

BALLARD, R. D. 1983. *Exploring Our Living Planet*. National Geographic Society, Washington, D.C. 366 pp. Illustrated presentation of ocean features and their origins.

FRIEDMAN, H. 1985. *Sun and Earth*. New York: Scientific American Library. 251 pp.

JUDSON, S., M. E. KAUFFMAN, AND L. D. LEET. 1987. *Physical Geology, 7th ed*. Englewood Cliffs, N.J.: Prentice-Hall. 484 pp. Standard geology text.

KENNETT, J. P. 1982. *Marine Geology*. Englewood Cliffs, N.J.: Prentice-Hall. 813 pp. Comprehensive treatment of ocean basin geology.

PRESS, F., AND R. SIEVER. 1986. *Earth*, 4th ed. New York: W. H. Freeman and Co. 656 p. Standard text on Earth sciences

WEINER, J. 1986. *Planet Earth*. Toronto: Bantam Books. 370 pp. Survey of ocean, Earth, atmospheric and astronomic sciences.

Mt. St. Helens, Washington erupted in 1980 discharging large quantities of ash and volcanic gases into the atmosphere. The volcano is associated with the subduction zone on the Oregon-Washington coast. (Courtesy Krafft-Explorer/Science Source/Photo Researchers.)

Plate Tectonics

OBJECTIVES

1. To understand the principal features of the earth;

2. To grasp the principal features of plate tectonic theory;

3. To understand how the evolution of the earth has affected ocean basins, seawater, and atmosphere.

*E*arth is a dynamic planet. Its surface—continents and ocean basins—slowly but constantly changes. New oceanic crust is formed by volcanic eruptions along midocean ridges and moves slowly away as a rigid plate. Over millions of years oceanic crust cools and becomes denser while it moves toward the trenches, where it sinks under the adjacent plate and is assimilated into the underlying mantle. In the process, continents are moved and ocean basins change their shape. Thus the distribution of ocean basins and continents described in Chapter 2 is only a snapshot of the earth's slowly changing surface. In this chapter we examine:

> Structure of the earth;
>
> Plate tectonics and crustal movements;
>
> Changes in the earth's magnetic field and how these changes are used to determine ages of oceanic crust;
>
> Effects of seawater circulating through newly formed oceanic crust;
>
> Oceanic plateaus, continental margins, and mountain building;
>
> Formation and destruction of ocean basins; and
>
> History of ocean basins.

ORIGIN OF EARTH

Earth formed about 4.6 billion years ago. It began in a giant disk of dust and gases orbiting the sun, an average star in the solar system, part of an ordinary galaxy of stars. The grains came together, forming comets, asteroids, and planets, which continue to orbit the sun. Planets nearest the Sun (Mercury, Venus, Earth, and Mars) consist largely of rocks. Planets farther from the sun consist primarily of gases and ice.

The rocky planets melted soon after formation. The densest constituents sank toward the center, while the less dense materials remained at the surface. Also early in their history (before 4.0 billion years ago), the planets were bombarded by meteors that extensively cratered their surfaces. These craters are still seen on the Moon, Mercury, and Mars, but not on Earth. Venus is hidden by clouds, so we know little about its surface.

Each planet has had a different history, primarily controlled by its size. Small planets cool rapidly and do not sustain prolonged volcanic activity. Thus they retain their cratered surfaces. The moon is a familiar example. Large planets cool slowly, so volcanic activity continues for long periods. Such activity continually renews and reshapes Earth and maybe Venus.

Volcanic activity is the source of Earth's ocean and atmosphere. Water and various gases are released from molten rocks during eruptions. On the moon, any gases released would have escaped, due to its small size. Earth is large enough to retain water and other gases. Water and gases are still released, but the quantity is small compared to the ocean and atmosphere.

EARTH STRUCTURE

Earth now consists of several concentric spheres (Fig. 3-1). The **core** is very dense because it contains iron and nickel. The inner core is solid. Movements of the liquid, outer core create the earth's magnetic field. The **mantle** consists of dense rock and is mostly solid. Part of the upper mantle, called the **asthenosphere,** is nearly molten and can flow very slowly.

The **lithosphere** is the rigid outer layer of the earth and is 50 to 100 kilometers thick. It floats in the asthenosphere. The upper part of the lithosphere is called the **crust**; it is rigid because of its low temperatures. There are two types of crust. Continents are made of **continental crust,** consisting of thick accumulations (30 to 40 kilometers) of many different kinds of rocks, generally rich in aluminum and silica. Ocean basins are underlain by thinner (5 to 7 kilometers thick) **oceanic crust,** primarily basaltic rocks, which are rich in magnesium and iron.

The **hydrosphere** is the free water (not combined in rocks and minerals) on earth. The ocean contains 98% of the free water; the remaining 2% is frozen in glaciers, primarily on Greenland and Antarctica (Fig. 2-3). The **atmosphere** is the gaseous outer envelope of the earth. It consists principally of nitrogen, oxygen, and variable amounts of water vapor, carbon dioxide, and dust.

FIGURE 3-1
Earth's internal structure, showing the thickness of each layer. (After Judson, Kauffman, and Leet, 1987.)

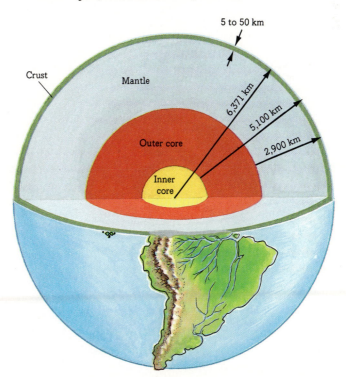

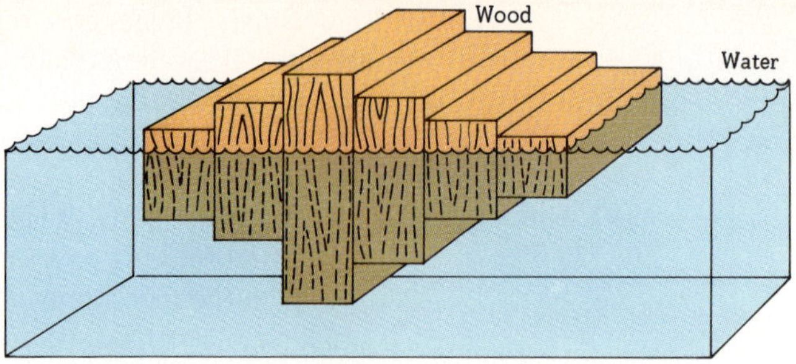

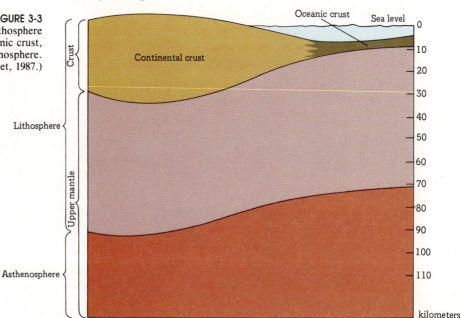

FIGURE 3-2
Isostasy is a condition of floating equilibrium, represented here by wooden blocks floating in water. The tops of the thickest blocks stand highest above the water surface. (After Judson, Kauffman, and Leet, 1987.)

ISOSTASY

Lithospheric plates float in the asthenosphere in a balance called **isostasy**. This accounts for the two levels of the earth's crust, which we discussed earlier. Let us see how this happens.

The simplest way to visualize isostasy is to consider wood blocks of different thickness floating in water (Fig. 3-2). The thickest block stands higher above the water surface than the thinnest one. Thus if oceanic and continental crust had the same density (as the wood blocks do), we would expect the tops of the thick continental crust to stand higher than the thin oceanic crust. There is still another factor. Continental rocks are less dense than the oceanic basalts, so the continents stand higher than the ocean floor (Fig. 3-3).

Lithospheric plates also move vertically as loads are placed on them or removed, such as volcanoes or glaciers. When the Hawaiian Island volcanoes formed on the Pacific plate, it was then about 80 million years old and fairly rigid. The volcanoes' weight caused the plate to bend. This bending created a moat around the base of the islands about 500 meters deeper than the general level of the ocean in that area. It also formed a broad arch now located about 250 to 300 kilometers from the island.

Large glaciers are loads that form during ice ages and then melt when the climate warms. Plates are depressed by the glacier's weight. This accounts for the deeply submerged continental margins around Antarctica and in the Arctic Ocean. After glaciers melt, the plates rise. Scandinavia is now rising due to the disappearance of its glaciers about 11,000 years ago.

FIGURE 3-3
Schematic representation of the lithosphere (consisting of continental crust, oceanic crust, and upper mantle) floating on the asthenosphere. (After Judson, Kauffman, and Leet, 1987.)

CHAPTER THREE   PLATE TECTONICS

BOX 3-1

Plate Tectonic Theory—History of An Idea

Plate tectonic theory has dramatically changed how scientists view the earth and its history. It revolutionized the way that scientists think about a host of earth processes, ranging from origins of continents to the history of ocean circulation. Acceptance of this simple, but elegant, idea took decades because of the reluctance of many scientists to accept it. It is informative to see how a scientific revolution occurs by reviewing the development and acceptance of this idea.

Alfred Wegner (1880–1930), German meteorologist, balloonist, and polar explorer, first advanced an idea he called *continental drift*. He noticed many puzzling aspects of continents. For instance, the shorelines of South America and Africa matched closely. Mountain ranges in Africa and South America stopped at the shore but matched mountains on the other side. Fossils in rocks on each side of the Atlantic as well as the characteristic features left by ancient glaciers also matched. To account for all these findings, Wegner theorized that all continents were once part of a supercontinent, which he called **Pangaea.**

While recovering from wounds received during World War I, Wegner published his ideas in 1915. Continental drift found many early supporters in Australia, South Africa, and to some extent in Europe. But his theory was not accepted in North America. The principal objections were that his evidence (fossils, paleoclimates) was not convincing. In particular, there were objections to the mechanisms he proposed for the movement of continents. Wegner proposed that continents were rigid bodies moving through the ocean basins. The driving forces from these movements were the variations in the gravitational attraction of the earth's equatorial bulge and the westward drift due to the attractions of the sun and moon. For several decades there were heated arguments but no new evidence.

Until the 1960s, most North American geologists thought that continents and ocean basins were permanent features of the earth's surface. These ideas were increasingly challenged, however, by data gathered in the 1950s as ocean basins were mapped in detail and their geophysical properties measured.

Based on his studies of these newly available data, *Harry H. Hess* (1906–1969) (Fig. B3-1-1) of Princeton University theorized in 1962 that the earth's outer shell moved. He speculated that these movements caused continents to break apart, move, and form new ocean basins. Hess further hypothesized that oceanic crust formed through volcanic ac-

tivity at midocean ridges and moved toward trenches, where they were destroyed (Fig. B3-1-2). Because of the strong feelings on the subject among his colleagues, he labeled his paper "geo-poetry." The American geophysicist *R. Dietz* called this idea **sea-floor spreading.** The Canadian geologist *J. Tuzo Wilson* provided many of the basic concepts, including the role of transform faults.

Magnetic anomalies under the ocean basins had also been mapped in great detail because of the possibility of locating submarines through their disturbances of the earth's magnetic field. The striped patterns of the magnetic anomalies intrigued scientists for many years.

Finally, *F. Vine* and *D. Matthews* of Cambridge University proposed that these patterns of magnetic anomalies recorded ancient reversals in the earth's magnetic field. In their model, the newly formed crust acted like a gigantic tape recorder, indicating the orientation of the magnetic field at the time it formed. Each stripe was a piece of oceanic crust formed while the field was in one orientation, before it reversed. Thus magnetic anomaly patterns made it possible to date the age of ocean-floor formation. This technique is still widely used.

In 1965, the British geophysicist Sir *Edward Bullard* (1907–1987) showed that the continents could be fitted together along their 100- and 1000-meter contours. A few areas of overlap corresponded to recently formed river deltas or coral reefs (Fig. B3-1-3). This convinced many geologists that the continents could be fitted together in a reasonable way and thus that it was feasible to consider that they could all be pieces of a single supercontinent, just as Wegner had proposed.

Another development supporting the idea of

FIGURE B3-1-1
Professor Harry Hess of Princeton University postulated that the oceanic crust was formed at midocean ridges and moved toward the trenches, where it was destroyed.

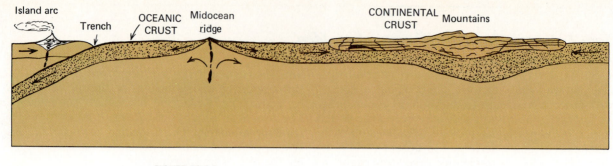

FIGURE B3-1-2
Major features of sea-floor spreading as postulated by Hess.

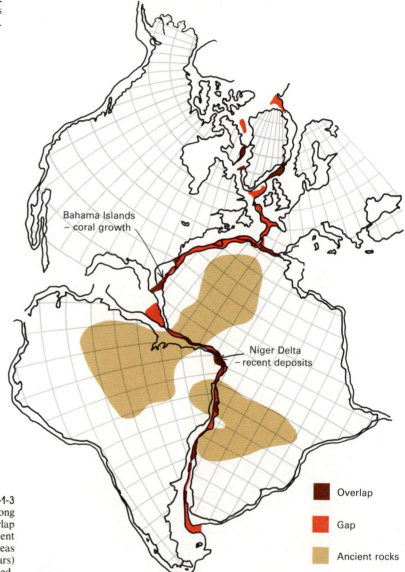

Bahama Islands
– coral growth

Niger Delta
– recent deposits

Overlap

Gap

Ancient rocks

FIGURE B3-1-3
Continents on both sides of the Atlantic fit along the 500-meter depth contour. Areas of overlap are sites of coral reef growth or sediment deposition since the Atlantic basin formed. Areas of ancient rock (older than 500 million years) were split apart when the Atlantic opened.

sea-floor spreading came from drilling the South Atlantic basin floor. The samples from beneath the sediment cover were progressively older moving away from the Mid-Atlantic Ridge. For most geologists, this was conclusive evidence.

Finally, other scientists put together the evidence from studies of earthquakes in subduction zones, along the transform faults between midocean ridges, and along fracture zones. They formulated the concept of plate tectonics in its present form.

As we have seen, the crust and the uppermost part of the mantle make up the lithosphere (Fig. 3-3), which floats on the asthenosphere. The lithosphere consists of several large and many small plates that move (Fig. 3-4) because heat from the core causes massive movements in the mantle rocks.

Lithospheric plates have three different kinds of boundaries. At midocean ridges, plates move away from each other. This is also called a *spreading center*. At **trenches,** plates converge. And at **transform faults,** plates slide past each other (Fig. 3-4).

Crust forms when molten rock (called **magma**) from the mantle erupts or fills fissures at a ridge crest and solidifies. New crust forms at the rate of about 20 cubic kilometers per year. Newly formed crust slowly moves away from the relatively shallow midocean ridge. As it ages, the crust cools, becomes denser and therefore floats lower in the asthenosphere. The youngest crust lies about 2500 meters below the sea surface (Fig. 3-5, p. 60). The oldest oceanic crust is about 6000 meters deep. The lithosphere also thickens as it cools, increasing from a few kilometers in thickness when newly formed to about 150 kilometers in thickness.

Lithosphere is destroyed at the same rate as it forms. This occurs at trenches where dense lithosphere sinks into the mantle and is resorbed, a process called **subduction.** In brief, you can view the process as a gravity-driven conveyer belt, moving lithospheric plates from shallow midocean ridges, where the plates form, to the deep trenches, where they are destroyed.

So far, we have treated plate movements as occurring on a flat surface, but they are actually on a sphere. Plates on a sphere rotate about fixed poles (Fig. 3-6, p. 60). One consequence of this rotation is that transform faults form *small circles*—that is, circles perpendicular to the pole of rotation. They help in reconstructing plate movements through time.

FIGURE 3-4

Schematic representation of the major features of plate tectonics. Convergent and divergent plate boundaries are shown. Arrows show directions of plate movements. (After Judson, Kauffman, and Leet, 1987.)

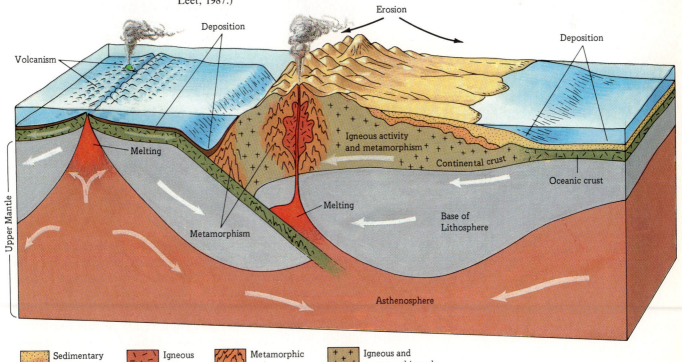

 Sedimentary rock

 Igneous rock

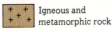 Metamorphic rock

Igneous and metamorphic rock

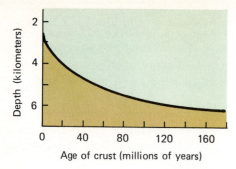

FIGURE 3-5
Ocean-bottom depths increase with age of the crust. The shallowest ocean bottom is the midocean ridges. The deepest and oldest crust occurs near land. (After J. G. Sclater, R. N. Anderson, and M. L. Bell. 1971. The elevation of ridges and evolution of the central eastern Pacific. *J. Geophys. Res.* 76:7888–7915.)

The position of plate movements on a sphere determines relative rates of crustal formation. The lowest rates of crustal formation occur near poles of rotation; highest rates occur halfway between them. Poles of rotation for the Pacific and Atlantic ocean plates are near the earth's magnetic poles. The corresponding pole for the Indian Ocean is not well known.

Movements of plates (Fig. 3-7) on the earth's surface are interconnected. Changing the direction and rate of movement for one plate involves the others. Plate movements may change for many reasons, such as a continent-continent collision.

Three plate boundaries intersect to form **triple junctions.** There are two principal types. Three intersecting subduction zones form an *unstable triple junction*. Such a triple junction moves as the plates are consumed in the subduction zones. Such a triple junction occurs east of Japan (Fig. 3-7).

Three intersecting ridges form a *stable triple junction,* which remains relatively fixed in location. Such a triple junction occurs on the East Pacific Rise south of Easter Island (Fig. 3-7). Another is in the South Atlantic on the Mid-Atlantic Ridge.

FIGURE 3-6
Representation of three rigid plates moving on the surface of a sphere. The relative rates of motion are least near the pole of rotation and greatest midway between the two poles. Relative differences in motion are taken up along transform faults. As plate B rotates through angle *a* (shown at top), new surface is added at the ridge crest. An equal amount is destroyed at the subduction zone. (After Kennett, 1982.)

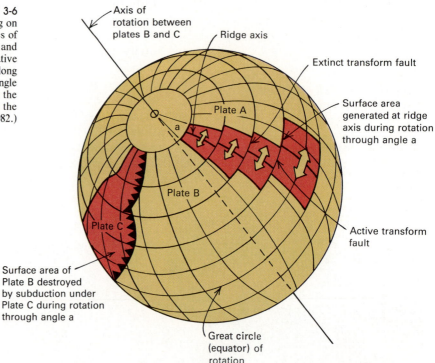

HOT SPOTS

Plumes of magma that rise from deep within the mantle erupt and cause **hot spots,** centers of prolonged volcanic activity (Fig. 3-7). Many hot spots occur on or near midocean ridges; Iceland is an example. Others, such as Hawaii, occur in lithospheric plates, far from the edges where most active volcanoes are situated.

Locations of long-lived hot spots apparently remain fixed for up to 100 million years. As plates move across such hot spots, they form chains of volcanoes, which record directions of plate movements (illus-

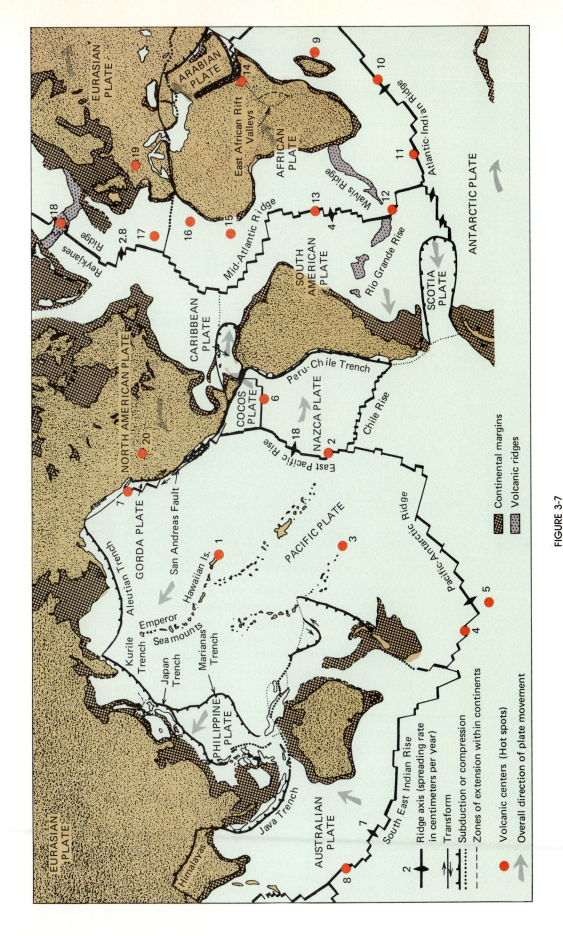

FIGURE 3-7

Major lithospheric plates and their boundaries, showing principal hot spots. Legend: 1. Hawaii. 2. Easter Island. 3. Macdonald Seamount. 4. Bellany Island. 5. Mt. Erebus. 6. Galapagos Islands. 7. Cobb Seamount. 8. Amsterdam Island. 9. Reunion Island. 10. Prince Edward Island. 11. Bouvet Island. 12. Tristan da Cunha. 13. St. Helena. 14. Afar. 15. Cape Verde Islands. 16. Canary Islands. 17. Azores. 18. Iceland. 19. Eifel. 20. Yellowstone. [After D. L. Anderson. 1971. The San Andreas Fault. *Scientific American* 225(5):52–71.]

FIGURE 3-8
The Hawaiian Islands and the Emperor
Seamounts formed as the Pacific plate moved
over a hot spot. The lines of volcanoes record
the change in direction of plate movements.

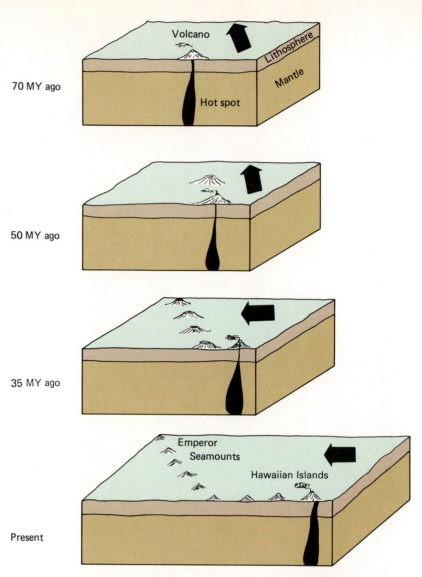

70 MY ago

50 MY ago

35 MY ago

Present

trated in Fig. 3-8). Volcanoes on the youngest islands remain active
while they are near the hot spot. After the crust has moved the volcano
away from its source of magma, it becomes extinct. Eventually, extinct
volcanoes are eroded by waves and rivers, first becoming small islands,
then submerged banks, and finally seamounts.

The Hawaiian Islands are an example of such an island chain
formed over a hot spot (Fig. 3-8). Volcanoes on the island of *Hawaii*
are still active because they are still near the hot spot. Indeed, there is
volcanic activity at a submerged site southeast of Hawaii, called Loihi.
A new island will eventually form there. Moving northwest, the volcanoes are extinct and the islands more deeply eroded. The oldest island
is Kure, a low carbonate-sand island built on the eroded and deeply
submerged volcano. A chain of ancient submerged volcanoes (Emperor Seamounts) continues almost due north, marking a change in
direction for the plate about 40 million years ago (Fig. 3-9).

MAGNETIC ANOMALIES

So far, we have used ocean-bottom features primarily for data about
plate movements. Much evidence supporting plate tectonics came
from observations of sea-floor properties. In fact, studies of the earth's

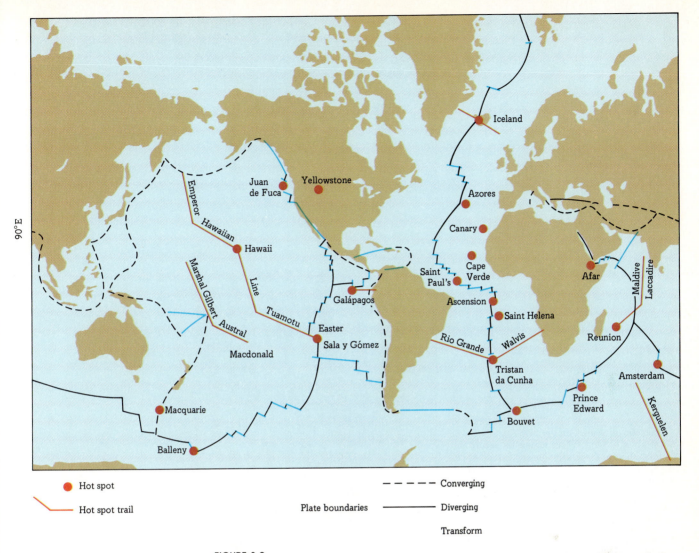

● Hot spot

／ Hot spot trail

Plate boundaries

- - - - Converging

——— Diverging

Transform

FIGURE 3-9

Some of the major hotspots. The thin black lines represent the volcanic ridges or lines of volcanoes formed as the plates moved over the hotspot locations. (After Judson, Kauffman, and Leet, 1987.)

magnetic field provided some of the most compelling evidence of plate movements. These anomalies are now used to determine when parts of the oceanic crust formed. Let's see how.

The magnetic field over the ocean exhibits long, irregular bands of deviations (either stronger or weaker) from the predicted magnetic field; these are called *magnetic anomalies* (Fig. 3-10). For many years, these magnetic patterns were a great scientific mystery.

We now know that minerals record the orientation of Earth's magnetic field at the time when the rock cooled. The anomalies result from frequent reversals of the magnetic field every hundred thousand years or so. During times of normal magnetic orientation, the north and south magnetic poles occupy their present positions. At times of reversed orientation, the pole locations are reversed. (No one yet knows what causes the poles to reverse.)

The magnetic field is slightly weaker over crust formed when the poles were reversed than it is over rocks formed during times of normal orientation (the present orientation). Sensitive instruments, called *magnetometers*, towed behind aircraft or ships can map these variations.

Studies of magnetic orientation of minerals in volcanic rocks of known ages provide a time scale to interpret anomaly patterns. The relative lengths of time in each orientation provide a scale for determin-

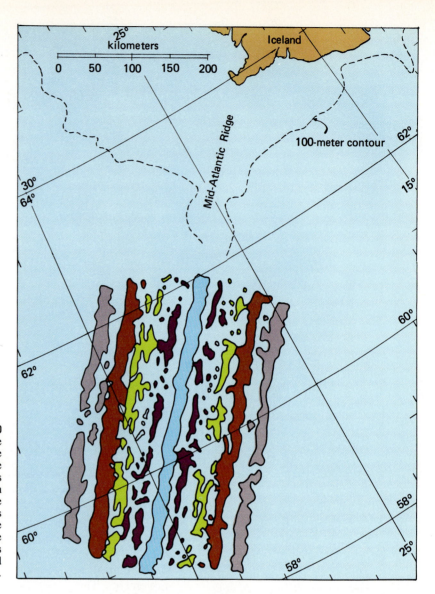

FIGURE 3-10
Magnetic anomalies on the Mid-Atlantic Ridge south of Iceland. These anomalies record the orientation of the Earth's magnetic field at the time when the crust formed. Shaded areas indicate where the magnetic field is stronger than the average magnetic field of the Earth. These crustal segments formed when the Earth's magnetic field had the same orientation as the present field. Light areas indicate where the magnetic field is weaker. These crustal segments formed when the magnetic field was reversed from its present orientation.

ing ages of ocean-floor rocks. (A similar record is also recorded in strata recovered from the ocean bottom.)

Figure 3-11 shows oceanic crust forming at a midocean ridge over a period of time when Earth's magnetic field reversed several times. The crust records these reversals like a gigantic tape recorder. Such anomaly patterns have been mapped over most of the ocean, so we now know when the ocean floor formed (Fig. 3-12, p. 66). As we would expect, the youngest crust occurs in a band centered on the midocean ridges. The faster the spreading rate, the wider the band of young crust. The oldest oceanic crust (up to 190 million years old) occurs in the North Pacific near Asia and along the margins of the North and South Atlantic. About half the ocean bottom formed in the past 80 million years.

Some parts of the oceanic crust do not exhibit magnetic anomalies. This crust formed during a time (80 to 120 million years ago) when Earth's magnetic field apparently did not reverse. Thus at this time we have no way of determining the ages of these crustal segments. There

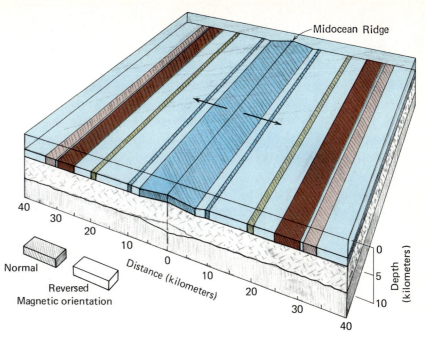

FIGURE 3-11
Striped pattern of magnetic anomalies results when minerals in oceanic crust record the direction of the Earth's magnetic field at the time of crustal formation.

is also evidence that deeply buried oceanic crust may lose its magnetic record because the minerals are heated due to the thick sediment deposits overlying them. Since the oldest crust usually lies near the continents, this process may destroy the magnetic evidence of the oldest rocks in the ocean basins.

HYDROTHERMAL CIRCULATION

Much of the heat from newly formed, hot oceanic crust is removed by seawater flowing through the rocks. Seawaters enter through open fractures and permeable zones, penetrating several kilometers into recently solidified volcanic rocks. They pick up heat along the way. The hot waters flow out through crevices and vents on the ocean floor and quickly mix with overlying waters. The total flow is immense, estimated at about 0.5% of the annual river flow. Another way to look at this circulation is to realize that all the water in the ocean circulates through newly formed crust every 5 to 10 million years. (Thus the entire ocean has flowed through recently formed ocean floor many hundreds of times during the history of the ocean.)

These discharges occur in areas of recent volcanic eruptions. Only limited parts of any midocean ridge are active at any time. Volcanic eruptions at any location are separated by thousands to tens of thousands of years, depending on local spreading rates. Volcanic eruptions occur more frequently on rapidly spreading segments than on slow-spreading ones.

Three types of vents have been observed by scientists in submersibles. Most spectacular are *black smokers* (Fig. 3-13, p. 68). They discharge superheated waters (300 to 400°C) at high flow rates, much like a fire hose. Because of their high temperatures, these waters are less dense than seawater; thus they rise, forming large plumes. These plumes are black because of chemical reactions that occur in the waters, forming sulfur-bearing minerals. Black smokers form large,

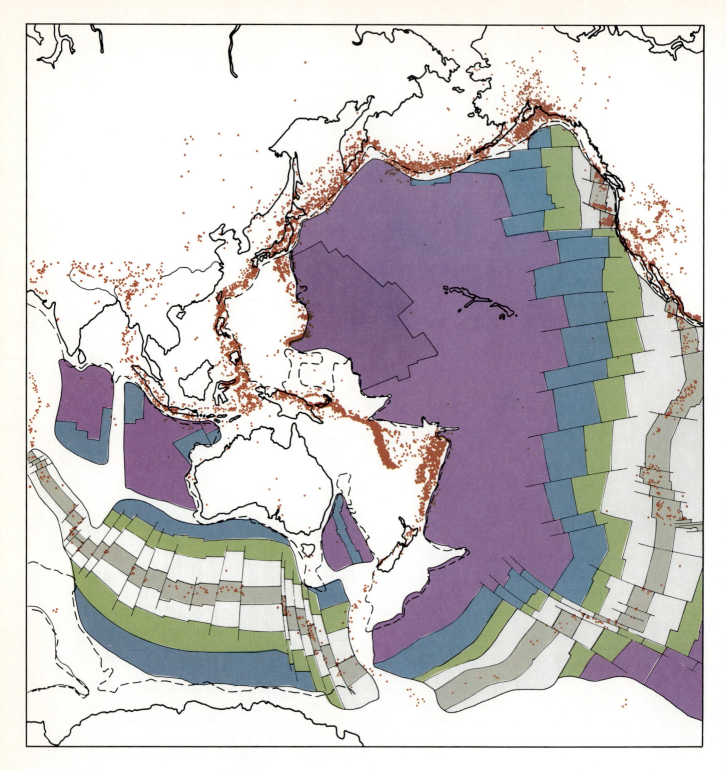

FIGURE 3-12
Ages of oceanic crust as determined from magnetic anomalies. Note the symmetrical patterns of crustal ages in the Atlantic and Indian oceans. Large areas of relatively young crust have been subducted in the Pacific, east of the East Pacific Rise. (After Judson, Kauffman, and Leet, 1987.)

fragile, chimneylike mounds up to 10 meters high made of porous silica (a glasslike substance), native sulfur, and sulfur-bearing minerals. The sulfur-bearing minerals color the mounds with yellows and blacks, like a Halloween decoration.

Cooler vents (called *white smokers*) are also common. These vents discharge waters between 25 and 250°C because the circulating fluids have mixed with cold ocean waters. Least spectacular is a third

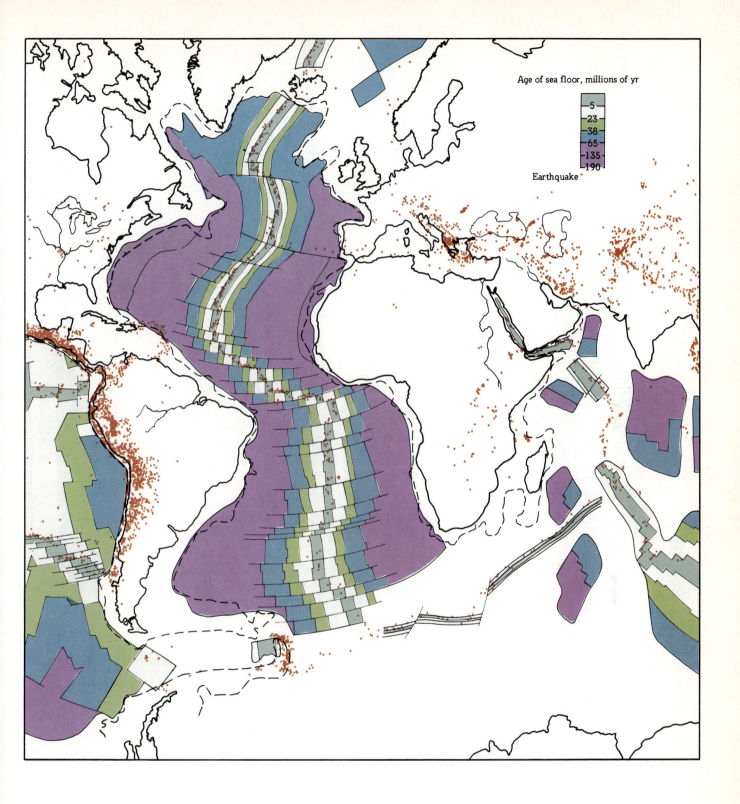

Age of sea floor, millions of yr

5
23
38
65
135
190

Earthquake

type of discharge where relatively cool (5° to 25°C) waters flow out through cracks and fissures in the ocean floor. These waters are only a few degrees warmer than surrounding ocean waters. All three types of discharges support abundant growths of bottom-dwelling organisms. (We discuss these in Chapter 14.)

Hydrothermal circulation continues for millions of years as the rocks cool. Eventually, fractures fill with mineral deposits and can no

FIGURE 3-13
A black smoker on the East Pacific Rise.
(Photograph by Dudley Foster, courtesy Woods
Hole Oceanographic Institution.)

longer pass fluids. Furthermore, sediments accumulate on the ocean floor, covering areas where seawater flowed into the rocks. These processes eventually seal the rocks and stop the flows of hydrothermal fluids.

Comparable processes occur around active volcanoes on land. In some areas, such hydrothermal circulation systems have been tapped to obtain energy.

BOX 3-2

Geothermal Power

Hot rocks formed by volcanic activity can be used to provide power for human uses. Waters flowing through them typically reach temperatures of 35° to 400°C. Some of the techniques used in the past have been quite simple. For example, Polynesians in Hawaii used hot waters from volcanic springs for cooking as well as for religious purposes. Iceland still makes extensive use of its geothermal resources. There warm waters are used to heat houses and greenhouses.

There are also schemes to develop the geothermal resources of Hawaii and Iceland to generate

large amounts of electrical power. Since the small populations on the islands cannot use all the power, it would be exported. In Hawaii, this involves building large geothermal plants on the island of Hawaii. A submarine cable would connect to the islands with larger populations, providing them with power.

A more ambitious scheme has been proposed for Iceland. It calls for a submarine cable lying on the submerged ridge between Iceland and the Faeroe Islands, connecting to Scotland. Still another option is to use the electricity to generate hydrogen (or some other fuel), which could be used elsewhere.

ACTIVE MARGINS

Convergent (active) plate boundaries, also called *Pacific-type margins,* are marked by active volcanoes, many earthquakes, and young mountains. Oceanic crust is destroyed there as lithospheric plates are pulled down into the mantle. Chains of volcanic islands, called *island arcs,* are formed by the many volcanic eruptions. Through many millions of years, continents are formed at convergent margins.

Earthquakes are common near Pacific-type margins. Deep- and intermediate-focus earthquakes (100 to 700 kilometers below the surface) are characteristic of these margins. Earthquakes at midocean ridges, while very common, rarely occur deeper than about 70 kilometers. Thus deep earthquakes indicate subduction.

In subduction zones, plates move as large slabs. In so doing, they drag against the rocks above and below. It is in these zones at the top and bottom of slabs that most earthquakes occur. Earthquakes result from strains in the rock which accumulate as the plates move and deform the rocks along their margins. The strains accumulate until the rocks fail. At that time pieces of crust move past each other, and large amounts of energy are released. After the stored energy is released, the rocks stop moving and begin to accumulate more strain as plate movements continue.

The sinking plate and the materials on it melt as it sinks into the hot mantle. Some of the molten rock rises to form the volcanic islands that make up island arcs. In some circumstances, subduction can cause formation of new oceanic crust as a result of the volcanic activity.

When the subducting plate is old and dense, it sinks into the mantle as a steeply dipping slab (Fig. 3-14). Such subduction occurs widely in the western Pacific basin, where the oldest oceanic crust occurs. In several of these subduction zones volcanic activity occurs in *back-arc spreading centers,* where new oceanic crust is formed. The Mariana and Scotia (South Atlantic) island arcs both have back-arc spreading centers and have actively forming small basins behind the island arcs.

Where the subducting crust is young and therefore still warm and relatively buoyant, the slab dips at a shallow angle (Fig. 3-15). This occurs along the eastern margin of the Pacific, near the East Pacific Rise, where the American plate is overriding recently formed crust. Volcanoes occur on land, and there are many earthquakes along the Pacific coast of North and South America for this reason.

The land near the subduction zone is elevated because it is under-

FIGURE 3-14
Subduction involving old, dense oceanic crust that dips steeply (Mariana-type subduction zone). Note the back-arc spreading where young oceanic crust forms in a narrow basin between the active island arc and the older, extinct arc.

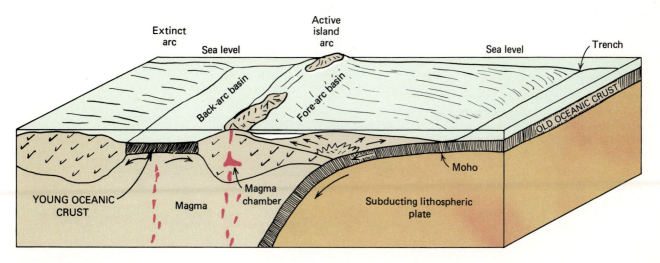

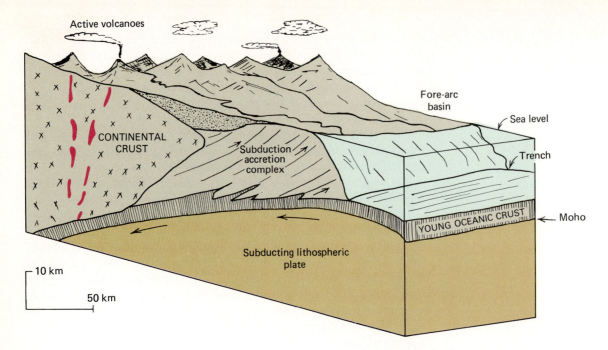

Active volcanoes

CONTINENTAL CRUST

Subduction accretion complex

Fore-arc basin

Sea level

Trench

YOUNG OCEANIC CRUST

Moho

Subducting lithospheric plate

10 km

50 km

FIGURE 3-15
A shallowly dipping plate of young, buoyant ocean crust is subducted beneath continental crust. This is typical of subduction zones of the Pacific coasts of North and South America. Note the thick wedge of deformed sediment deposits scraped off the subducted plate.

lain by the subducted Pacific plate. In addition to the volcanoes, young mountain ranges form from materials scraped off the subducting plate (Fig. 3-15). Both the chains of volcanoes and the young mountain ranges parallel the coast line.

Off Central America and southern Mexico, a trench marks the location of the subduction zone (Fig. 2-9). But off the United States and Canada, the trench is filled with sediments derived from the land, so the subduction zone is not visible.

When two continents collide, the processes are different. Since the rocks on both sides of the convergence zone are relatively light (low density), they are not easily subducted. Instead, the plates respond by folding and by thrusting slices of one plate on top of the other. This overriding and uplift is now happening in India. The other possibility is folding and thickening of the crust. This is happening in Turkey and nearby countries where Africa and Asia are now colliding.

Where there are small volcanoes, they may be broken up into smaller blocks and drawn down into the subduction (Fig. 3-17). Larger volcanoes, volcanic ridges, and oceanic plateaus are often too large for this to occur. They are instead welded to the continent, forming an **exotic terrain.** Many such blocks of former oceanic crust and oceanic plateaus have been identified in North America from Alaska to Mexico. Other Pacific-type margins doubtlessly have similar blocks, but they have not yet been studied. Indeed, most continents consist of many such blocks, large and small, that have been welded together at ancient subduction zones.

FIGURE 3-16
Mountain chains form as continents collide. Such processes usually involve volcanic activity as well.

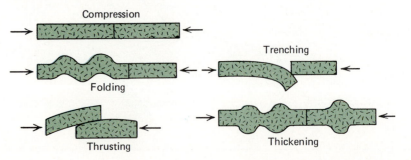

Compression

Folding

Thrusting

Trenching

Thickening

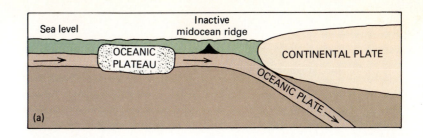

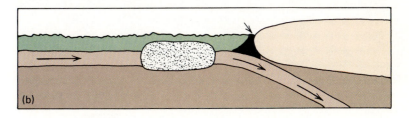

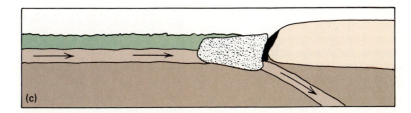

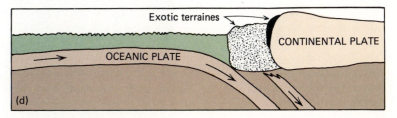

FIGURE 3-17
Oceanic plateaus and volcanic ridges are scraped off oceanic crust as it is subducted. These oceanic structures are welded to the continents, forming new continental crust.

Where pieces of ocean floor have been welded onto continents, they have often incorporated deposits of sulfide minerals formed at spreading centers. These have been mined for copper, lead, zinc, and silver since antiquity. Indeed, such deposits on the Arabian Peninsula provided King Solomon's wealth in biblical times. Similar deposits are still mined for copper on the island of Cyprus in the Mediterranean.

When very large continents collide, they may cause a reorientation in local spreading. This apparently happened when India collided with Asia about 40 million years ago.

PASSIVE MARGINS

Passive margins, also called *Atlantic-type margins,* have no mountain building. They occur in the interior of lithospheric plates and are usually formed by splitting of continents early in the spreading cycle. Thus any mountains associated with passive margins are usually very old, derived from previous spreading cycles.

Thick sediment deposits and old oceanic crust are typical of passive margins (Fig. 3-18). These deposits usually cover the boundary between oceanic and continental crust. They also record the early history of the formation of the basin and various changes in sea level.

Economically, passive margins are important because of the accumulations of oil and gas that they often contain. Most of the world's giant oil and gas fields occur in such deposits.

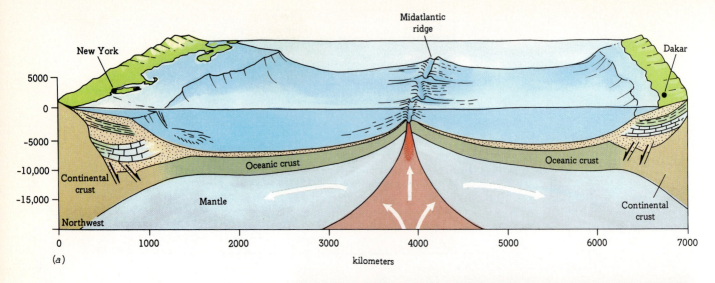

(a)

FIGURE 3-18
A passive or Atlantic-type margin. Note the thick sediment deposits shown in the cross section. The shoreline of New York (Long Island) and New Jersey is typical of a passive margin with flooded river valleys and barrier beaches along the shorelines. We discuss these features in Chapter 10. (After Judson, Kauffman, and Leet, 1987.)

(b)

MANTLE CONVECTION

Movements of lithospheric plates are caused by *mantle convection,* which extends deep within the earth's interior (Fig. 3-19). The mantle convects, or overturns, because it is warmed at the core-mantle boundary and then rises to the surface, where it cools and then sinks again. Oceanic crust is simply the upper surface of the mantle and therefore moves with it, along with the continents. The energy causing mantle convection comes from the core (probably the inner core), which is still cooling.

Hot-spot volcanoes originate at unusually hot areas of the core-mantle boundary. There the overlying mantle melts, forming plumes of magma that rise and penetrate the crust as volcanoes. These spots do not move with the overlying mantle and thus remain fixed in location for tens of millions of years. (We will use the chains of volcanoes formed as plates move over hot spots to help us reconstruct plate movements.)

The mantle is cooled at the top by volcanic eruptions at midocean ridges and hot spots, by seawater circulation through newly formed crust, and by heat conduction through the ocean bottom. (Heat conduction is the same process that warms a pot on a hot burner.) The lithospheric plates thicken as the crust and upper mantle cool. When the cool, dense lithospheric plates sink at trenches, this too cools the mantle. These plates sink through the mantle and eventually come to rest on the core-mantle interface, which is a major change in density. The plates remain there for many millions of years as they warm and their density decreases enough to rise and begin the cycle again.

These plumes of rising magma and the sinking plates can be imaged by *seismic tomography*. This technique uses seismic waves from earthquakes that pass through the earth to be detected by hundreds of seismographs around the world. After analyzing travel times between

FIGURE 3-19
Schematic representation of mantle convection, showing its relationship to the movements of major lithospheric plates.

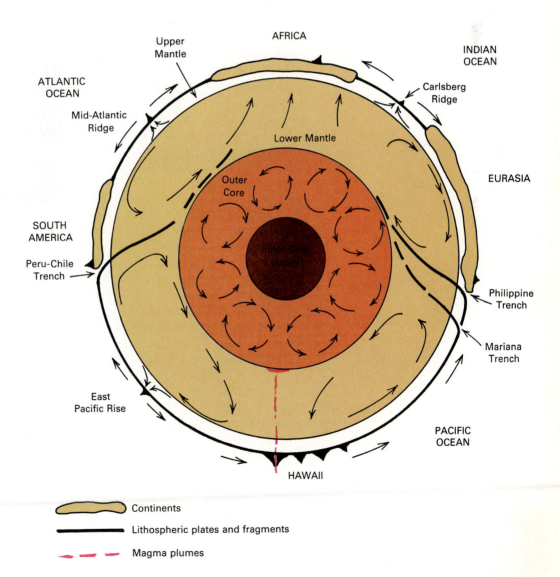

earthquakes and seismographs, computers map the regions of warm rock (having slower wave speeds) and cold rock (higher wave speeds). These maps show the rising and sinking portions of the mantle. This technique is similar to the x-ray technique called computerized axial tomography or CAT used by doctors to image organs in the human body.

Movements in the mantle also appear to be coupled to the source of the earth's magnetic field. The fluid outer core also convects as a result of heating at its lower boundary (with the inner core) and cooling at the core-mantle boundary. Thus sinking of dense lithospheric plates cools the outer core. With increased ability to map the core-mantle boundary, it may soon be possible to determine what causes the magnetic reversals.

FORMATION AND DESTRUCTION OF OCEAN BASINS

Ocean basins form through the breakup of continents (Fig. 3-20). This process begins when a continent remains in one location for several hundred millions years. Thick continental crust impedes heat flow from the earth's interior. Eventually the underlying mantle heats, expanding and uplifting the overlying lithosphere (Fig. 3-20). This is happening now in Africa, which has apparently remained in its present location for 100 to 300 million years. It now stands several hundred meters higher than the other continents, which have moved over this period.

Continued uplift stretches the overlying continental crust, as it is now doing in the East African rift valleys. The crust eventually breaks, forming narrow, fault-bounded valleys. In humid climates these valleys fill with fresh water, forming deep lakes, such as the East African lakes.

As valley floors subside, they fill with sediment, often rich in organic matter. Eventually the valleys subside to sea level and connect with the ocean. At this point, freshwater lakes become long, narrow oceans as the continental fragments separate. Modern analogues are the Red Sea and the Gulf of California.

The long, narrow seaway impedes ocean circulation and favors evaporating conditions in the basin. High mountains on both sides of the basin cause winds to blow along the length of the basin. The uplifted basin sides divert rivers away from the embayment, so little or no fresh water reaches the newly formed sea. Thick salt deposits form if the basin is partially cut off from the sea.

As the oceanic crust ages, it cools and grows denser. Eventually it becomes so dense that it begins to sink into the underlying mantle, starting a new cycle of subduction, as is now occurring in the West Indies. There Atlantic crust is subducting under the Caribbean plate. Similar processes are occurring in the South Atlantic in the South Sandwich Trench.

Through continued subduction, the ocean basin narrows, and eventually the two sides collide. In the process, the ocean disappears, and a mountain range forms where the sediment deposits are squeezed between the two continental blocks. During the collision, sediments and large pieces of ocean floor are attached to the continental crust. In this way, continents grow by incorporating large pieces of ocean floor, oceanic plateaus, and fragments of former continents. Each such cycle takes about 400 million years from start to finish. Thus there have been about 10 such cycles since the earth's surface solidified. We know most about the cycle we are in, so we will briefly discuss its history and the effects on each ocean basin.

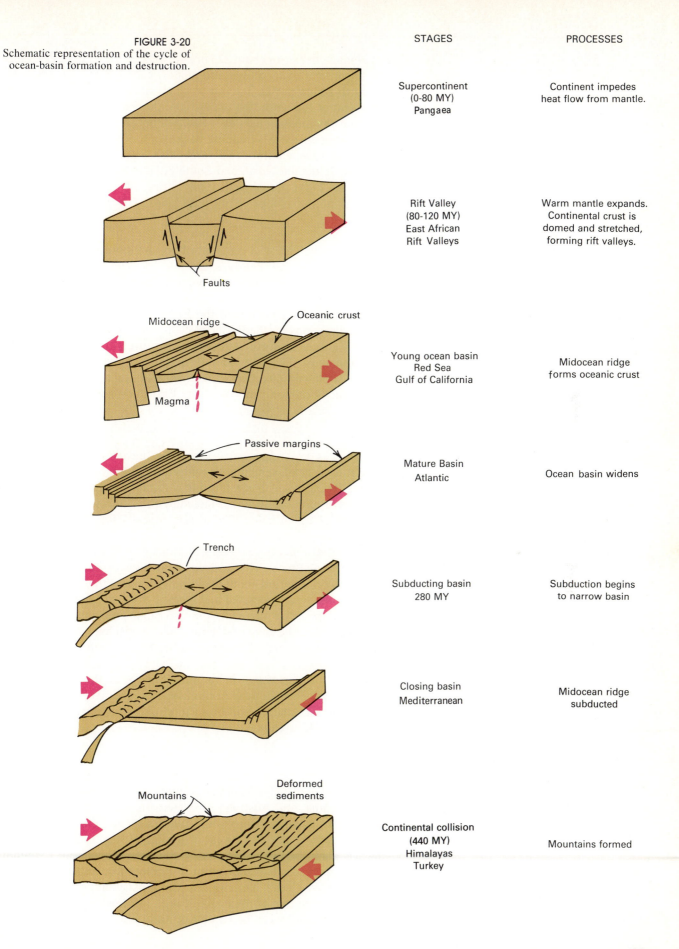

FIGURE 3-20
Schematic representation of the cycle of
ocean-basin formation and destruction.

STAGES	PROCESSES
Supercontinent (0-80 MY) Pangaea	Continent impedes heat flow from mantle.
Rift Valley (80-120 MY) East African Rift Valleys	Warm mantle expands. Continental crust is domed and stretched, forming rift valleys.
Young ocean basin Red Sea Gulf of California	Midocean ridge forms oceanic crust
Mature Basin Atlantic	Ocean basin widens
Subducting basin 280 MY	Subduction begins to narrow basin
Closing basin Mediterranean	Midocean ridge subducted
Continental collision (440 MY) Himalayas Turkey	Mountains formed

Faults

Midocean ridge — Oceanic crust

Magma

Passive margins

Trench

Mountains — Deformed sediments

PRESENT SPREADING CYCLE

The present cycle began about 225 million years ago. At that time there was a single land mass or supercontinent, called *Pangaea*, after the Greek earth-goddess [Fig. 3-21(a)]. A single ocean, *Panthalassia*, covered the rest of the earth, the ancestor of the Pacific.

About 180 million years ago, Pangaea began to break up, initially forming a long east-west–trending basin, called *Tethys* [Fig. 3-21(b)]. Tethys split Pangaea into two large subcontinents, *Laurasia* and *Gondwana*. Later the North and South Atlantic basins formed as the Mid-Atlantic Ridge developed. This rifting of Pangaea eventually formed the three existing ocean basins [Fig. 3-21(c)].

HISTORY OF THE PACIFIC OCEAN

The Pacific is the oldest, a remnant of previous crustal spreading cycles. The basin is now nearly surrounded by subduction zones and much decreased in size since the present cycle of seafloor spreading began.

The Hawaiian Islands and the Emperor Seamounts record two spreading directions for the Pacific plate as it moved over the Hawaiian hot spot. An ancient north-south spreading direction formed the Emperor Seamounts. The Hawaiian Islands formed during the present northwest-southeast direction. The change in spreading direction occurred about 40 million years ago. The East Pacific Rise in its present location is a relatively young feature.

HISTORY OF THE ATLANTIC OCEAN

The Atlantic is expanding east-west under the influence of the Mid-Atlantic Ridge. Thick sediment deposits around the margins have buried the early history of basin formation. Thick salt deposits occur in many areas around the Atlantic, including the Gulf of Mexico and the eastern side of the Atlantic. These salt deposits formed early in the basin's history.

About 100 million years ago, bottom waters in the basin were stagnant several times. This caused deposition of carbon-rich sediments. For a brief period about 6 million years ago, the Mediterranean Sea was almost totally isolated and formed thick salt deposits.

The Atlantic now connects the two polar ocean areas. About 6 million years ago, the Iceland-Faeroe Ridge subsided. This permitted cold, dense waters from the Norwegian Sea to flow south into the open Atlantic basin.

Subduction has begun in the oldest parts of the Atlantic—the South Sandwich Trench, near Antarctica, and in the West Indies. This suggests that the Atlantic will begin to close again as subduction becomes more widespread.

HISTORY OF THE INDIAN OCEAN

The Indian Ocean is the youngest ocean basin. It began with the breakup of Gondwanaland about 125 million years ago when Africa, Antarctica, India, and Australia separated. India moved north to collide with Asia, leaving conspicuous tracks of its movement on the sea floor. The basin is the shallowest of the major basins because of its young age. It reached its present shape and circulation about 15 million years ago, when Australia moved to its present location.

The Indian Ocean basin is also the most complex. It has four major ridges, with two of them still active. The north-south spreading direction associated with the Circum-Antarctic Ridge dominates.

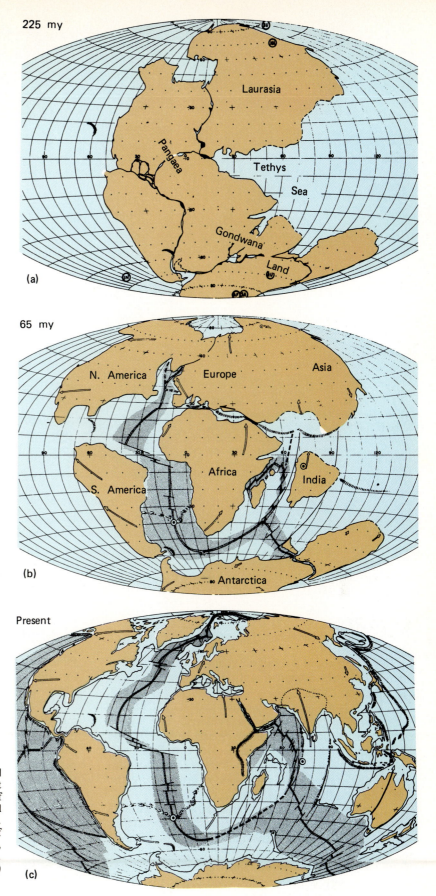

225 my

Laurasia

Pangaea

Tethys

Sea

Gondwana

Land

(a)

65 my

N. America

Europe

Asia

S. America

Africa

India

Antarctica

(b)

Present

FIGURE 3-21
Movements of lithospheric plates during the past 225 million years following the breakup of Pangaea (a), formation of the Atlantic (b), and opening of the Indian Ocean (c). (After R. S. Dietz and J. C. Holden. 1970. Reconstruction of Pangaea: breakup and dispersion of continents, Permian to present. *J. Geophys. Res.* 75:4939–56.)

(c)

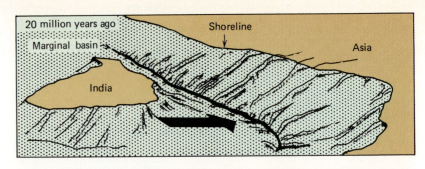

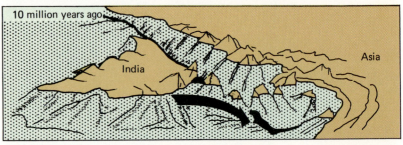

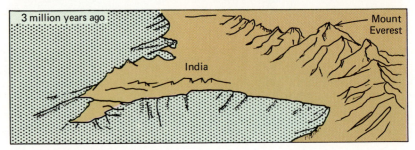

FIGURE 3-22
About 50 million years ago, the northern edge of the Indian subcontinent collided with Asia and began to underthrust it, eventually forming the Himalaya Mountains. (After J. Weisberg and H. Parish. 1974. *Introductory Oceanography.* New York: McGraw-Hill.)

The Himalaya Mountains are another result of the opening of the Indian Ocean (Fig. 3-22). They began to form about 50 million years ago when India, moving northward, underthrust Asia. This collision eventually formed continental crust about 70 kilometers thick. The former continental slope, now the edge of the Himalaya Plateau, rises sharply above the low-lying region to the south.

SUMMARY

Earth formed about 4.6 billion years ago and is one of the inner planets. Because of its size, volcanic activity continues on earth.

Earth is made of concentric spheres. The core is rich in iron and nickel and is very dense. The mantle consists of dense rock. The lower part is rigid but the upper mantle is nearly molten and can flow very slowly. The crust is the rigid outer shell and includes continental and oceanic crust. The hydrosphere is the free water on the Earth's surface.

Continents are thick accumulations of granitic rocks; one-third of continental crust lies under oceans.

The crust and underlying upper mantle constitute the lithosphere, the rigid outer shell of the earth, which floats on the asthenosphere, a process called isostasy. The lithosphere is broken into rigid plates that move more or less independently of each other. Plate movements form and destroy ocean basins and move continents as well as add material to continental blocks.

Plate boundaries are midocean ridges (divergences), trenches (convergences), and transform faults, where plates slide past each other.

Oceanic crust forms at midocean ridges and moves slowly away as it cools, becomes denser, and gradually sinks lower in the asthenosphere. Midocean ridges are offset by transform faults, which parallel plate movements. Plates move toward trenches, where they are subducted into the mantle and destroyed.

Convection in the mantle supplies molten rock (magma) at midocean ridges. The youngest parts of the plate are elevated at the midocean ridges due to their high temperature and resulting low density. Oldest crust is densest and sinks at the trenches. Hot spots are persistent plumes of magma that rise at volcanic centers, causing long-continued volcanic activity. Plates moving across hot spots form ridges and chains of volcanic islands.

Earth's magnetic field reverses itself at irregular

intervals. Rocks record the magnetic field's orientation when they form at midocean ridges. They retain this record, much like a gigantic tape recorder. Oceanic crust consists of alternating bands of rocks with different magnetic orientations, which create a striped pattern in the magnetic field. These patterns of alternating magnetic orientation can be used to date the age of formation of the crustal plates.

Seawater circulates through newly formed, hot crustal rocks, removing heat and altering the rocks by chemical reactions between the rocks and seawater. These hot waters discharge in vents in the rift valleys, forming deposits of sulfide minerals. The sulfides also support the growth of bacteria which feed abundant growths of bottom-dwelling organisms.

Fracture zones are bands of irregular topography marking the extensions of the inactive portions of transform faults. Fracture zones separate different sea-floor provinces.

Pacific-type margins are areas of active mountain-building at convergent plate boundaries. They are usually marked by trenches with many earthquakes and volcanic activity. Many, especially in the Pacific, have island arcs. Continental crust is also formed at convergent margins by accreting oceanic plateaus.

Atlantic-type margins of continents are called passive margins because they lie in the interior of the plates and have no volcanic or earthquake activity.

The boundaries between continental and oceanic crust at these old margins are covered by thick sediment deposits, which are often rich in oil and natural gas.

Ocean basin formation begins when a continent remains in one position for about 100 million years. Thick continental crust impedes heat flow, causing the underlying mantle to heat up and to expand. This expansion stretches the continental crust and causes long, narrow rift valleys to form, as is happening now in East Africa. As spreading continues, the basins widen, forming narrow ocean basins, such as the Red Sea or the Gulf of California. A midocean ridge occurs in the middle, as in the present Atlantic.

Newly formed crust is hot and therefore relatively buoyant. As crust ages, it cools and becomes denser. Eventually, dense crust begins to sink into the mantle, initiating subduction, which results in a narrowing of the ocean basin. Finally, when the basin closes, mountains are built, which mark the location of the former ocean basin. The Appalachian Mountains in the eastern United States mark the location of an ancestral Atlantic which closed about 400 million years ago.

The Pacific is the oldest basin—a remnant of the previous spreading cycle. The Atlantic and Indian oceans are younger, both having formed in the present cycle.

STUDY QUESTIONS

1. Describe how Earth formed.
2. Explain why Earth has a layered internal structure.
3. Explain how reversals in the earth's magnetic field are used to determine the ages of specific parts of the ocean floor.
4. Explain the principal features of plate tectonic theory.
5. What makes lithospheric plates move?
6. Illustrate the three types of crustal plate boundaries.
7. Explain how a volcanic center ("hot spot") under an ocean basin forms a chain of volcanic islands.
8. Draw a diagram of an active (Pacific-type) continental margin, labeling the most significant features.
9. Draw a diagram of a passive (Atlantic-type) continental margin, labeling the most significant features.
10. Discuss the ways in which hydrothermal circulation affects oceanic crust.
11. Explain isostasy.

SELECTED REFERENCES

ALLEGRE, C. 1988. *The Behavior of the Earth*. Cambridge, Mass.: Harvard University Press. 272 pp. Discusses plate tectonics.

BALLARD, R. D. 1983. *Exploring Our Living Planet*. National Geographic Society, Washington, D.C. 366 pp. Well illustrated.

JUDSON, S., M. E. KAUFFMAN, and L. D. LEET. 1987. *Physical Geology, 7th ed.* Englewood Cliffs, N.J.: Prentice-Hall. 484 pp. Standard geology text.

KENNETT, JAMES P. 1982. *Marine Geology*. Englewood Cliffs, N.J.: Prentice-Hall. 813 pp. Ocean basin geology.

SULLIVAN, W. 1974. *Continents in Motion. The New Earth Debate*. New York: McGraw-Hill. 399 pp. Well-written treatment of plate tectonic theory and how it developed.

WEGENER, A. 1924. *The Origin of the Continents and Oceans*. (trans. from the 3d German ed. by J. G. A. Skerl), London: Methuen & Co. Original statement of continental drift.

WOOD, R. M. 1985. *The Dark Side of the Earth: The Battle for the Earth Sciences, 1800–1980*. Winchester, Mass.: Allen & Unwin. Describes the scientific revolution brought about by plate tectonics.

WYLLIE, P. J. 1976. *The Way the Earth Works: An Introduction to the New Global Geology and its Revolutionary Development*. New York: John Wiley. 296 pp. Introduction to plate tectonics.

A black smoker on the East Pacific Rise. These discharges of hot waters and sulfide minerals are part of the system that controls the salinity of ocean waters. (Photograph courtesy Dudley Foster, Woods Hole Oceanographic Institution/Photo Researchers.)

Seawater

OBJECTIVES

1. To understand the importance of the thermal and physical properties of water in ocean processes;

2. To understand the processes controlling the composition of sea salts;

3. To differentiate conservative and nonconservative behavior of seawater constituents.

Water is the most common substance on the surface of our planet. Most of it—98%, to be exact—occurs as seawater filling the ocean basins. Seawater is a complex mixture: about 96.5% water, 3.5% salt, and minute amounts of other substances, such as dirt and dissolved organic matter. The water, salt, and the atmosphere above it all came from rocks in the earth's interior, extracted by volcanic eruptions over the 4.5 billion years of Earth history.

In this chapter, we study how the unusual properties of water influence the physical behavior of seawater and how sea salts are added to and removed from seawater. In the next chapter, we learn about interactions between the ocean and atmosphere. In Chapter 11, we learn about the influence of marine organisms on seawater and its constituents.

This chapter discusses:

Origin of ocean and atmosphere;

Factors controlling the chemical and physical properties of water;

States of matter;

Water's capacity to store heat and the effect of heat on water's properties;

Salinity and its effect on seawater properties;

Conservative and nonconservative properties;

Dissolved gases;

Acidity and alkalinity;

Carbon dioxide and carbonate cycles; and

Processes controlling sea-salt composition.

ORIGIN OF THE OCEAN AND ATMOSPHERE

Volcanic activity is the source of Earth's ocean and atmosphere. Water and various gases are released during volcanic activity from the molten rocks that erupt from volcanoes. (Some small amount of gases may have been captured early in Earth's history.) The water released has collected in the ocean basins to form the ocean.

Earth is large enough to retain water and atmospheric gases, so it

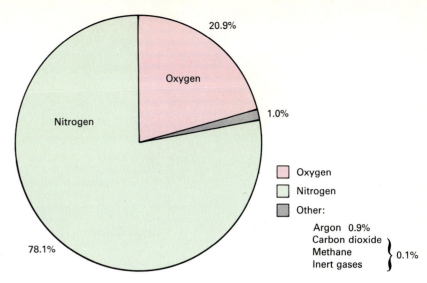

FIGURE 4-1
Relative abundance (by volume) of gases in a dry atmosphere.

has both an ocean and large amounts of fresh water (together called the *hydrosphere*) as well as an atmosphere. The Moon and Mercury lost water and most gases as they were released from rocks through volcanic eruptions. The surface of Mars exhibits signs of running water, and at some time in its history it may have had a shallow ocean. Where the water went and why the planet is so much colder now is not known. Venus may still have an ocean, but its opaque atmosphere prevents our seeing it. Earth's ocean makes it unique in the solar system.

The larger planets, such as Earth, retain their atmospheres. Most atmospheric gases (Fig. 4-1)—water, carbon dioxide, nitrogen—can be detected in volcanic eruptions on Earth. The principal difference between Earth's atmosphere and that of nearby planets is the scarcity of carbon dioxide and the abundance of oxygen and nitrogen on Earth. These are due to the presence of life on Earth. During photosynthesis by plants, oxygen is released, and water and carbon dioxide are removed from the atmosphere and combined to form carbon compounds. Most of the carbon is buried in sediment deposits, keeping it out of the ocean and atmosphere.

MOLECULAR STRUCTURE OF WATER

Water has many unusual chemical properties (Table 4-1). These properties influence the ocean and marine organisms. So we begin our discussion of seawater by examining the reasons for water's unusual properties. We start with the molecular structure of water—what holds it together.

Attractive forces between atoms hold molecules together. For example, strong bonds called *covalent bonds* form when each atom retains its electrons but shares them with adjacent atoms. When molecules separate into their constituents, each retains its own electrons.

Ionic bonds are strong bonds between molecules or atoms that have lost or gained electrons. One example of ionic bonding is found in common table salt (NaCl), which is formed by combining sodium (Na^+) and chlorine (Cl^-). Ionic bonds result from mutual attractions between opposite electrical charges. When the atoms in such a molecule separate, each will gain a positive charge (having lost an electron) or a negative charge (having gained an electron). *Van der Waals forces* are very short range electrostatic attractions between chemical species.

TABLE 4-1

Anomalous Physical Properties of Water and Their Effects on Seawater and the Ocean*

PROPERTY	COMPARISON WITH OTHER SUBSTANCES	IMPORTANCE IN OCEAN
Heat capacity	Highest of all solids and liquids except liquid ammonia	Prevents extreme ranges in ocean temperature Heat transfer by currents is large
Latent heat of fusion	Highest except ammonia	Acts as thermostat at freezing point owing to uptake or release of latent heat
Latent heat of evaporation	Highest of all substances	Important in heat and water transfers to atmosphere
Thermal expansion	Temperature of maximum density decreases with increasing salinity. For pure water, it is at 4°C.	Freshwater and dilute seawaters reach maximum density at temperatures above freezing point
Surface tension	Highest of all liquids	Controls drop formation and behavior; also surface phenomena, such as capillary waves
Dissolving power	Dissolves more substances and in greater quantities than any other liquid	
Transparency	Relatively great	Absorption of radiant energy is large in infrared and ultraviolet. In visible portion of energy spectrum there is relatively little selective absorption—hence is "colorless"

* Modified after Sverdrup et al., 1942.

They are relatively weak compared to bonding forces and are easily formed and easily broken.

Hydrogen bonds determine many physical and chemical properties of water. These bonds result from the small size of hydrogen atoms compared to the large oxygen atoms which make up most of the volume of water molecules. Each oxygen atom has three electron pairs in its electron cloud. It shares one pair with two hydrogen atoms in forming covalent bonds in the water molecule.

Water's unusual properties arise primarily from its asymmetrical molecular structure (Fig. 4-2). On an atomic scale, this asymmetrical structure makes the oxygen side of the molecule negative because of the two unshared electron pairs. The two hydrogen atoms on the other side make it positive. The net charge on a molecule is zero (the negative and positive charges are equal). However, the molecular structure causes it to behave as if it had a positive side and negative side. Such molecules are called *polar molecules*.

The two hydrogen atoms form weak hydrogen bonds with oxygens of adjacent water molecules. These bonds are responsible for water's ability to store large amounts of heat with small temperature changes. (See Table 4-1 for the unusual properties of water, mostly arising from hydrogen bonding.)

To obtain an idea of the relative strength of these bonds, consider the amounts of energy (expressed in kilocalories, the amount of heat required to raise the temperature of 1000 grams of water by 1°C) required to break them in 18 grams of water:

Covalent bonds:	100s of kilocalories
Ionic bonds:	10s of kilocalories
Hydrogen bonds:	4.5 kilocalories
van der Waals bonds:	0.6 kilocalorie

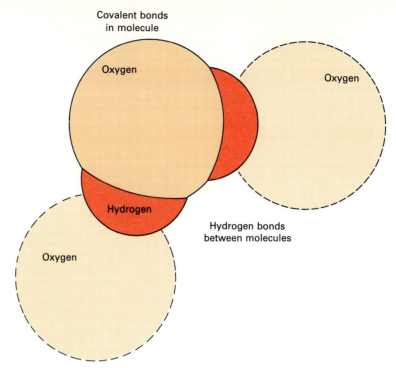

Covalent bonds
in molecule

Oxygen

Oxygen

Hydrogen

Hydrogen bonds
between molecules

Oxygen

FIGURE 4-2
Schematic representation of a water molecule
showing the bonds between hydrogen atoms and
adjacent oxygen atoms in water molecules on the
right and lower left. Note the separation between
the side of the molecule having two hydrogen
atoms and the side having no hydrogen atoms.

The sun's heating of the ocean surface provides enough energy to
break hydrogen and van der Waals bonds but not enough to break ionic
and covalent bonds. In fact, energy absorbed through evaporating water (breaking hydrogen bonds) and released through condensing water
(re-forming hydrogen bonds) is important in exchanging heat between
ocean and atmosphere.

STATES OF MATTER

All substances on Earth exist in one of three states of matter: gas,
solid, or liquid. Each has distinctive properties arising from the structure of the molecules.

Gases are the simplest to understand, because the molecules
move independently. Gases can thus fill any container and have neither
size nor shape. Molecules striking the sides of a container exert pressure which can be changed in two ways. We can add (or remove) gas,
thus changing the rate at which molecules strike the sides. Alternatively, we can increase (or decrease) the temperature of the gas. This
makes molecules move faster (or slower) and thereby changes the rate
at which they hit the sides. In each case, gas pressure changes.

Solids are the opposite of gases, having definite size and shape.
Furthermore, they break or bend when enough force is applied. Most
solids have fixed internal structures in which atoms (or molecules)
cannot readily move from their positions, although they can vibrate.
Some movement is possible, however, if there are defects, such as
holes, in the structure. Vibrations (and movements) increase as temperatures rise.

Liquids are intermediate between solids and gases. They have
definite volumes (size), but they flow to conform to the shape of any
container in which they are placed. In liquids, atoms and molecules are
loosely bonded to each other; otherwise they would behave like gases.
But the bonds are easily broken, allowing them to flow.

As we shall see, water occurs in all three states on Earth, making
it unique in the solar system.

STRUCTURES OF ICE AND WATER

Hydrogen bonding between water molecules affects the structures of ice and water, giving them unusual properties. Water vapor, on the other hand, behaves like a normal gas because there is no bonding between molecules.

Water molecules in ice are held together by hydrogen bonds, as shown in Fig. 4-3. The oxygen atoms form six-sided puckered rings arranged in layers, each layer a mirror image of the adjacent ones. The result is a fairly open network of atoms, giving ice a lower density (0.92 gram per cubic centimeter) than that of water (1 gram per cubic centimeter) because molecules in ice are not as tightly packed as they are in water.

Despite the somewhat open structure of ice, impurities, such as sea salt, do not easily fit into the structure. So salt and dissolved gases are excluded from sea ice as it freezes. These sea salts remain in pockets of unfrozen liquid (brine) which migrate out of the ice. (We learn more about sea ice later.)

Liquid water has properties intermediate between those of solids and gases (Table 4-1). Water apparently consists of two different types of molecular aggregates. The amount of each type is determined by the temperature, pressure, and chemical composition of any dissolved salts.

One type of molecular aggregate consists of clusters of hydrogen-bonded water molecules, sometimes called the structured part of water (Fig. 4-4). These clusters form and re-form rapidly—10 to 100 times during one-millionth of a microsecond. The lifetime of any cluster is

FIGURE 4-3

(a) Crystal structure of ice showing the six-sided rings formed by 24 water molecules. (b) In the same volume of liquid water, there would be 27 water molecules. (c) The ice lattice is also shown to give an idea of its three-dimensional character.

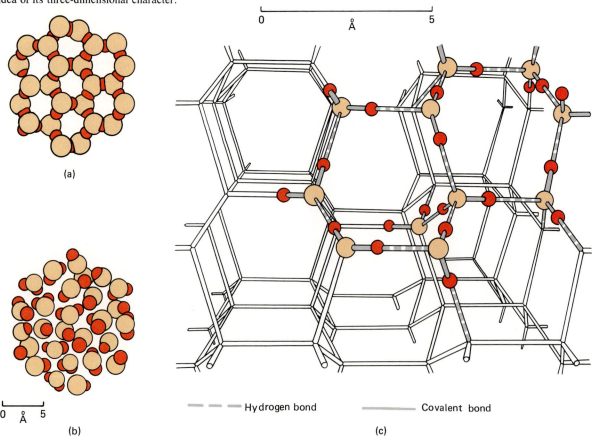

(a)

(b)

(c)

Hydrogen bond   Covalent bond

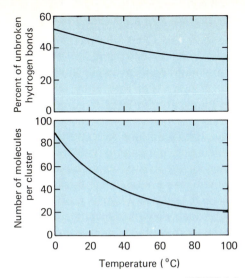

FIGURE 4-4
Effects of temperature on the relative abundances of unbroken hydrogen bonds and the number of molecules per cluster. (Data from G. Nemethy and H. A. Scheraga. 1962. Structure of water and hydrophobic bonding in protein. *J. Chem. Phys.* 36:3382–3400.)

exceedingly short, but clusters persist long enough to influence water's physical behavior. Clusters are smaller and have fewer unbroken hydrogen bonds at higher temperatures.

These structured portions of water are less dense than the unstructured portions. When pressure exceeds about 1000 atmospheres (corresponding to depths greater than 10,000 meters in the ocean) these clusters disappear. They also break up when water evaporates.

The other constituent of water is unstructured. It consists of closely packed, "free" water molecules that surround the structured portions. These molecules move and rotate without restriction. Interactions with adjacent water molecules are much weaker than in the structured portion. Because of the close packing of the molecules, the density of the unstructured portion is greater than the structured portion.

Because of its unusual structure, liquid water differs from hydrogen compounds of other elements similar to oxygen, as indicated in Fig. 4-5. Based on the behavior of other compounds, water's melting point and boiling point are 90 and 170°C higher, respectively, than would have been predicted. In other words, if water were a "normal" compound with no hydrogen bonding, it would occur only as a gas at Earth's surface temperatures and pressures.

TEMPERATURE EFFECTS ON WATER

FIGURE 4-5
Melting and boiling points of water and chemically similar compounds. (After R. A. Horne. 1969. *Marine Chemistry: The Structure of Water and the Chemistry of the Hydrosphere.* New York: Wiley-Interscience. 568 pp.)

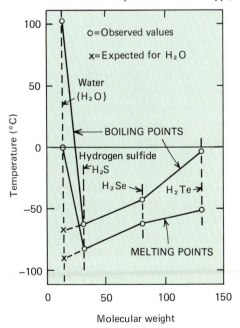

Temperature changes alter water's internal structure and its properties. Much of the heat absorbed by water goes into changing its internal structure, breaking hydrogen bonds, for instance. Thus the temperature of water rises less than other substances after absorbing a given amount of heat. For this reason, we say water has a high **heat capacity.**

Water on Earth's surface prevents wide variations in its surface temperatures. By comparison, surface temperatures on the moon, where there is no water, range from about +135°C at noon to about −155°C at night. The highest temperature recorded on Earth was 57°C at Death Valley, California, and the lowest was −68°C at Verkhoyansk, Russia, both far from the ocean.

Much of Earth's incoming solar radiation goes into evaporating water and melting ice. In the ocean, water temperatures rarely exceed 30°C or go below −2°C, obviously a much smaller temperature range than on land. Let us see what happens to water during these changes and how it affects the earth's heat balance.

First, we define a measure of heat, the **calorie**—the amount of heat required to raise the temperature of 1 gram of liquid water by 1 degree Celsius (1°C). This means that we can change the temperature of 1 gram of water by 50°C by supplying 50 calories of heat. Alternatively, we can change the temperature of 50 grams of water by 1°C with the same amount of heat.

To break bonds in ice and liquid water requires energy, usually heat. Conversely, the bonds release energy when they re-form. To illustrate this process, consider what happens when 1 gram of ice just below the freezing point is slowly heated. As we add heat to ice, its temperature increases about 2°C for each calorie of heat we add (Fig. 4-6). When the ice reaches its melting point (0°C), the temperature remains constant. Heat added goes into breaking hydrogen and van der Waals bonds. After we have added about 80 calories, enough hydrogen bonds have broken, and the last bit of ice melts, leaving only liquid water.

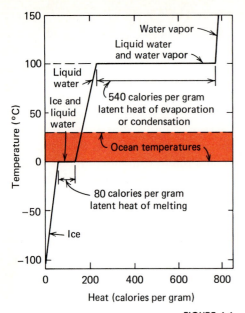

FIGURE 4-6
Temperature changes when heat is added or removed from ice, liquid water, or water vapor.
Note that temperatures do not change when mixtures of ice and liquid water or liquid water and water vapor occur together. (Redrawn from M. G. Gross. 1985. *Oceanography*. Columbus Ohio. Charles E. Merrill Publishing Company.)

As we continue to supply heat, the water temperature rises but at a lower rate, 1°C per calorie of added heat. This rate of temperature change remains nearly constant between 0°C and 100°C. At 100°C (the boiling point of liquid water), water vapor begins to form. The temperature remains constant while both water vapor and liquid water are present. After we have added about 540 calories per gram, all the remaining hydrogen bonds are broken. Thus each water molecule exists alone, in other words, as a gas. If we continue to heat the gas, we find that the temperature rises again at 2°C per calorie of heat added. If we started with water vapor and cooled it, we would reverse this process and gain heat each time there was a change in state.

Heat added to water is taken up in two ways. One is **sensible heat,** heat that we detect either through touch or with thermometers. This temperature change results from the increased vibrations of molecules in solids or liquids or from motions in the gaseous state.

The other form is called **latent heat,** which is the energy required to break bonds. As we have seen, at the melting point and the boiling point of water there was no change in temperature as long as two states of matter coexisted; the added energy went into breaking bonds in the disappearing form of matter. This is called latent heat because we get back all the heat when the process is reversed. Condensing water vapor to form liquid water releases 540 calories per gram at 100°C. Freezing water at 0°C releases 80 calories per gram. The difference between the latent heats of melting and evaporation arises from the fact that only a small fraction of the hydrogen bonds are broken when ice melts, while all the hydrogen bonds are broken when water evaporates.

Although water freezes at 0°C and boils at 100°C, water molecules can go from vapor to solid or liquid at other temperatures. The processes involved are similar to those described earlier, but the amount of heat released (or taken up) is different. The latent heat of evaporation changes as follows:

TEMPERATURE (°C)	LATENT HEAT OF EVAPORATION (calories per gram)
0	595
20	585
100	539

VISCOSITY

Viscosity—internal friction or resistance of a fluid to flow—is highly sensitive to the molecular structure of water. We can think of viscosity as the resistance of molecules—or groups of molecules—to moving past each other. Water's viscosity decreases with increasing pressure, because higher pressures favor the unstructured portions of water. With less structured water at higher pressures, the free water molecules can move past each other more easily.

In near-surface waters, where viscosity is most important (for floating organisms), viscosity is affected by salinity and temperature. Water's viscosity increases with increased salinity, because water molecules form hydration shells around each ion. Thus when there are more ions in the water (higher salinity), there are more water molecules involved in hydration sheaths. Water in such sheaths does not move as readily as free water molecules, so the viscosity of seawater increases with salinity. Temperature effects are more important. Viscosity increases with lower temperatures, again, because lower temperatures favor the structured portions of water.

Viscosity is extremely important to minute organisms. (We discuss these in Chapter 12.) On such a small scale, the viscosity of water is analogous to molasses. Thus tiny organisms must cope with swimming, sinking, and capturing food in a highly viscous fluid. Viscosity is not as important to processes involving humans and at large scales. Thus many factors affecting minute organisms are difficult for us to comprehend.

SALINITY

The quantity of salt in a volume of seawater varies due to additions or removals of water. To indicate how much salt is in seawater, oceanographers use **salinity** (abbreviated **S**), which is defined as the amount of salt (in grams) dissolved in 1 kilogram of seawater. Salinity is expressed in parts per thousand, written ‰. (In other words, 1% is 10‰.)

When salts dissolve in water, bonds between atoms and some molecules break because of interactions between crystal lattices and water molecules. Thus salts dissolved in seawater form electrically charged atoms or molecules, called **ions.**

The six most abundant ions constitute 99% by weight of sea salts, as shown in the table. Note that the four most abundant constituents amount to more than 97% by weight.

	PERCENT BY WEIGHT
Chloride (Cl^-)	55.07
Sodium (Na^+)	30.62
Sulfate (SO_4^{2-})	7.72
Magnesium (Mg^{2+})	3.68
Calcium (Ca^{2+})	1.17
Potassium (K^+)	1.10
TOTAL	99.36

Salinity is determined by measuring seawater's **conductivity,** its ability to conduct an electrical current through the water. The more salt in the water, the greater the conductivity.

DENSITY

The density of a substance controls whether it sinks or floats in a liquid. **Density** is the ratio of mass to unit volume (usually expressed in grams per cubic centimeter). A *stable density-stratified* system is one in which less dense fluids float on heavier ones. An example is gasoline floating on water. (A stable system is one that returns to its original state after being disturbed. An unstable system returns to some other state, not the original one.)

In an *unstable density-stratified* system, a parcel rises or sinks until it reaches a level of comparable density. A substance (or liquid) less dense than the liquid around it floats upward. Conversely, if its density is greater than its surroundings, it displaces less mass than its own and sinks.

A mass sinks through fluids less dense than itself. Eventually it reaches a level where the fluid below is denser, and the fluid above is less dense. At this point, there is no force acting on the parcel and it will remain at that level. Furthermore, vertical displacement of the fluid will be countered by forces keeping it at the same relative position. A fluid parcel may spread out at the appropriate density level, forming a layer. Such behavior is common in the ocean. Indeed, such

BOX 4-1

Resources from Seawater

Salt is the most common material taken from seawater. *Evaporating basins,* located in many dry coastal regions, use solar energy to evaporate seawater. Evaporation of brine is carefully controlled so that only sodium chloride or another desirable component of sea salt is removed at a single stage. Salts recovered from seawater must otherwise be treated to remove magnesium sulfate (Epsom salt) and calcium carbonate. Evaporating ponds for extracting sea salts operate around southern San Francisco Bay (Fig. B4-1-1). Sea salt is used by the chemical industry, as is magnesium, a lightweight metal. Bromine extracted from seawater is used as a component of antiknock compounds in gasoline.

Despite the variety and large amount of valuable materials dissolved in ocean water, the extremely dilute nature of seawater makes it prohibitively expensive to produce most of them. Gold is an example. Its concentration in seawater is extremely low, about 4 grams per 10^{12} grams of seawater. This amounts to about 5×10^{12} grams of gold in the ocean. The cost of energy for pumping seawater, added to the cost of chemical treatment, far exceeds the value of the gold recovered. High energy costs further discourage recovery of metals dissolved in seawater. Another problem in many areas is the difficulty of obtaining undiluted, uncontaminated seawater.

Growth of cities in arid regions makes water from the ocean an increasingly important resource. One simple way to use water from seawater is to use a greenhouse. Evaporation from pans of seawater occurs when the sun is shining, and condensation takes place at night on the cool interior surfaces. This creates a humid, nearly tropical environment where plants can be grown. Such artificially made oases are useful for small-scale production of high-value products like fresh vegetables or fruit.

Fresh water is easily extracted from seawater by using a simple still (Fig. B4-1-2). Seawater is fed into a closed container and heated. Water vapor, free of salt, is condensed. The salt remaining behind and most of the original seawater (typically 90%) are discharged as a hot brine.

Recovery of fresh water from a single-stage evaporation process like that in Fig. B4-1-2 is impracticable. Too much heat is lost in the cooling water, and high energy costs restrict this process to small-scale or emergency uses.

One way to increase the efficiency is to use the seawater for cooling before it is fed into the boiler. In

FIGURE B4-1-1
Seawater flows into enclosed ponds on southern San Francisco Bay. There heat from the sun evaporates the water, leaving behind salt deposits when the brines are evaporated to dryness. The red color of the brines is due to growths of one-celled algae. (Photograph copyright Hans Halberstadt, Photo Researchers.)

this way, energy consumption is reduced. It is also possible to reduce the boiling point by carrying out the process under a slight vacuum, a technique known as *flash evaporation*.

Another way of desalinating seawater is to duplicate the natural freezing cycle to separate water and sea salts. In this process, a coolant is added to seawater, which evaporates and removes heat from the liquid surface, causing it to freeze. The ice crystals are removed and washed free of brine and salt and can then be melted to obtain fresh water. The coolant is recycled and brines are discharged with approximately twice the salinity of the original seawater. Such a process is particularly suited to application in northern areas where cold ocean waters are available, thereby reducing costs.

FIGURE B4-1-2

Simple single-stage evaporation-condensation apparatus to recover fresh water from seawater.

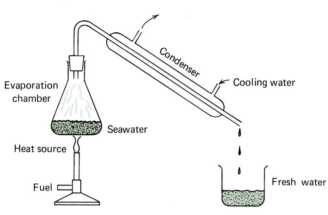

density differences drive the winds and deep-ocean currents, as we see in later chapters.

Next we discuss how increasing salinity or pressure causes density to increase, while increasing water temperature causes its density to decrease. (Remember that density is a ratio; it decreases because of either a decrease in the numerator or an increase in the denominator.) Warming water increases the vibrations of atoms and molecules. They occupy more volume, thus reducing density. For a constant mass, density decreases as volume increases. Increased pressure causes density to increase because water is slightly compressible. (We usually ignore water's compressibility in discussing ocean processes.)

Ice, water vapor, and seawater, like most materials, become less dense with increasing temperature. Pure water, on the other hand, is unusual. It has a density maximum at 3.98°C (Fig. 4-7). When liquid water warms from 0°C, its density increases slightly until it reaches a maximum at 3.98°C. Upon further warming, its density decreases.

The density maximum for pure water is especially important in lakes. In winter, surface waters cool at the surface until they reach 4°C. At that temperature, surface water is denser than the waters below, so it sinks. The process is called *overturning* or *convective sinking*. Overturning continues until the entire lake mixes, and water temperatures are uniformly about 4°C. At that point, overturning stops, and bottom waters are protected against further cooling even though surface ice may form. Overturning of lake waters protects the lake bottom from freezing. It also supplies dissolved oxygen to the deeper portions of the lake. If water behaved like other liquids, entire lakes would freeze.

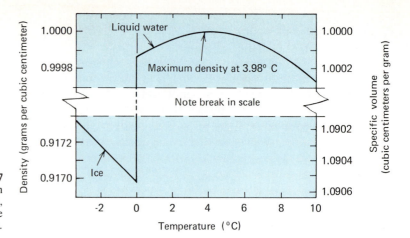

FIGURE 4-7
Effect of temperature on density, expressed in grams per cubic centimeter, and specific volume, expressed in cubic centimeters per gram, for ice and pure liquid water.

FIGURE 4-8
Variation in seawater density (in grams per cubic centimeter) as affected by temperature and salinity. Note the changes in the temperature of maximum density and the point of initial freezing caused by changed salinity. (The dashed line is discussed in the text.)

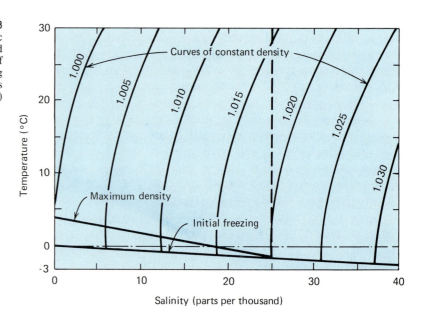

Dissolved salt eliminates water's density maximum. The temperatures of maximum density and of freezing both decrease as salinity increases (Fig. 4-8), but the temperature of maximum density drops more rapidly than the freezing temperature. (The two curves intersect at a salinity of 24.7‰). Thus seawater of average salinity (35‰) begins to freeze before it reaches the temperature of maximum density.

As previously discussed, temperature markedly influences the relative proportions of structured and unstructured portions of liquid water. Because these two constituents have different densities, water temperature strongly affects its density.

The anomalous density maximum in liquid water is explained by the increase in the structured portion of water below about 4°C. Water becomes less dense because the denser unstructured portion of the mixture is diminished. Addition of salt inhibits formation of the structured portion. This causes the temperature of maximum density to decrease sharply as salinity increases.

Like most substances, ice becomes less dense with increasing temperature (Fig. 4-7). At 0°C, it has a density of 0.917 gram per cubic centimeter. It is therefore about 8% less dense than liquid water at the same temperature and floats on water.

BOX 4-2

Mediterranean Salinity Crisis

Between 5 and 6 million years ago, the Mediterranean basin was partially cut off from the Atlantic. During that time, a worldwide drop in sea level of about 40 to 50 meters further isolated the basin. As a result, thick salt deposits accumulated in the basin. Such events of salt deposition have happened many times in the history of the ocean, but few have been so well studied.

The Mediterranean formed when Africa moved north and collided with Europe at Gibraltar. Now the Mediterranean is connected with the Atlantic by the narrow Strait of Gibraltar. But when sea level dropped, the basin was further isolated. A gigantic waterfall may have formed as seawaters flowed into the basin from the Atlantic. In any case, the reduced flows could not keep up with evaporation, and water levels dropped. The Nile and Rhone rivers flowing into the basin cut deep gorges.

As waters in the basin evaporated, their salinity increased. Eventually, concentrations reached levels where salt crystallized and accumulated on the bottom. If the present Mediterranean dried out, it would take about 1000 years to evaporate the water, forming a salt deposit 70 meters thick. Such deposits in the basin are now 2 to 3 kilometers thick. Thus the volume of salt there is equivalent to filling the basin with seawater and then drying it out 30 to 40 times.

It is likely, however, that seawater continued to flow into the basin. Probably, the basin never totally dried out, although isolated areas may have been saline lakes or even deserts for short periods. When sea level rose again about 5 million years ago, the improved connection with the Atlantic permitted salinities to drop, so that salt deposits no longer formed.

Removal of so much salt lowered seawater salinity worldwide, about 2‰. This caused a salinity crisis which affected the evolution and distribution of animals around the Mediterranean. Similar events must have occurred many times, as indicated by the large salt deposits found now in the Gulf of Mexico and around the margins of the Atlantic.

CONSERVATIVE AND NONCONSERVATIVE PROPERTIES

Salinity is changed by adding fresh water through rain or snow, river discharge, and thawing ice, or by removing it through evaporation and freezing of ice. These are all physical processes. They leave the relative proportions of the major ions in seawater unchanged. This is an example of salinity as a **conservative property.**

Conservative properties are used to trace movements of water masses after they leave the sea surface and move through the deep ocean. After leaving the ocean surface, both water mass temperature and salinity are conservative properties. That is, they change only through mixing with other water masses with different temperatures and salinities. (We discuss the use of tracers in studying deep-ocean waters in following chapters.)

Salinity is most variable near the air-sea interface, at current boundaries where water masses of different salinities mix, and in the coastal ocean. Even when salinity changes are relatively small, such changes are useful in tracing water mass movements, because salinity is measured very precisely.

The relative abundance of many constituents in seawater is changed as they are taken up by organisms as they grow or are released through decomposition after death. Such constituents are **nonconservative properties.** Examples of nonconservative properties are concentrations of dissolved oxygen or materials used by plant growth, such as nitrogen and phosphorus. Other nonconservative constituents participate in chemical reactions between seawater and sediment particles. For example, aluminum interacts with sediment particles, removing it from seawater. In general, the major constituents in seawater are conservative properties, while most of the minor constituents are nonconservative.

In later sections of this chapter we discuss processes affecting major elements. In Chapter 11, we discuss biological and chemical processes affecting nonconservative properties.

SALINITY EFFECTS

Natural waters usually contain some dissolved salts. River waters contain, on the average, about 0.01% (0.1‰) dissolved salts; average seawater contains about 3.5% (35‰) of various salts (Fig. 4-9). Even rainwater contains small amounts of salts and gases that it picked up while falling through the atmosphere.

FIGURE 4-9
Major and minor constituents in seawater.

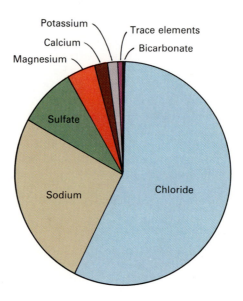

FIGURE 4-9
Major and minor constituents in seawater.

Dissolved salt affects the physical properties of water in several ways. We have already discussed the effect on temperature of maximum density. Other properties affected by salts are as follows.

1. *Temperature of initial freezing*—lowered by increased salinity.
2. *Vapor pressure*—decreased by increased salinity.
3. *Osmotic pressure*—increased by increased salinity (water molecules move through a semipermeable membrane from a less saline solution to a more saline solution; the opposing pressure necessary to stop this movement is called the osmotic pressure). This is important for marine organisms.

Consider the temperature of initial freezing. Pure water freezes at 0°C and the temperature of the water-ice mixture remains fixed until only ice remains. In seawater the temperature of initial freezing is lowered by increased salinity. Furthermore, salts excluded from the ice remain in the liquid, causing the brine to become still saltier. So the temperature of freezing for the remaining liquid is still lower. Then temperatures must drop before additional ice forms. So seawater has a lower temperature of initial freezing than pure water and no fixed freezing point.

Salts affect the internal structure of water. Some dissolved ions, such as sodium and potassium, shift the equilibria toward water's unstructured phase. Other ions, such as magnesium, "favor" the structured portions. Changes in the relative proportions of these constituents affect such properties as *viscosity,* (Fig. 4-10).

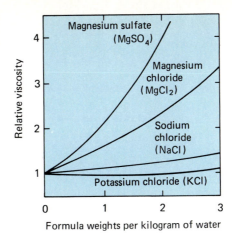

Effects of various salts on viscosity of water at 25°C. (After R. A. Horne. 1969.)

DISSOLVED GASES

Atmospheric gases dissolve in seawater at the air-water interface. Conversely, gases also pass through the interface in the opposite direction—back into the atmosphere. Water can dissolve only a limited amount of any substance at a given temperature and pressure. When that limiting value is reached, the amount of gas going into solution equals the amount going out. At that point, the water is *saturated* with the gas, which is then said to be present in *equilibrium concentration*.

Temperature, salinity, and pressure affect saturation concentrations for a gas. In the normal range of oceanic salinity, temperature is dominant, as indicated in Fig. 4-11. Generally, such gases as nitrogen and oxygen or rare (inert) gases (helium, neon), which do not react chemically with water, become less soluble in seawater as temperature or salinity increases. Seawater at all depths is saturated with most atmospheric gases. Exceptions are dissolved oxygen and carbon dioxide, which are involved in life processes. Dissolved oxygen concentrations in the deep ocean are controlled primarily by biological processes. These are discussed in Chapter 11.

Nitrogen, the most abundant atmospheric gas, is also common in

Saturation concentrations of oxygen and nitrogen in pure water and seawater. Note that both gases are soluble in pure water and that their solubilities decrease with rising temperatures.

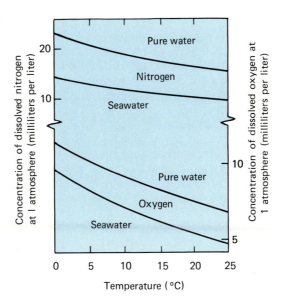

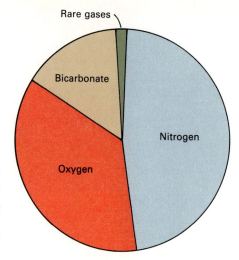

FIGURE 4-12
Relative abundances (by weight) of gases dissolved in surface seawater (S = 36°/oo, T = 20°C). Note the relatively high abundances of bicarbonate (HCO_3^-) and the rare gases, compared with their low concentrations in the atmosphere.

seawater (Fig. 4-12). It is not involved in most biological processes; thus it remains near saturation throughout the ocean.

Helium (He), a rare gas, is produced by radioactive decay in rocks and sediments. Helium from mantle rocks is also released by hydrothermal vents. Otherwise, rare gases—argon, krypton, and xenon—behave like nitrogen and their concentrations vary little in seawater.

ACIDITY AND ALKALINITY

A property of seawater that we must now consider is its acidity and alkalinity. An **acid** is a compound containing hydrogen that releases its hydrogen ion (H^+) when dissolved in water. A strong acid is one that readily releases its hydrogen ions. A weak acid does not release its hydrogen ions so easily. Another type of compound, called a **base,** releases hydroxyl ions (OH^-). Like acids, some bases are strong in that they readily release their hydroxyls. Others are weak.

BOX 4-3

Gas Hydrates

Sediments on continental slopes and rises contain water in an unusual form. At the high pressures and relatively low temperatures of the ocean bottom, water forms an icelike solid called a *gas hydrate*. These solids differ from ordinary ice in that they contain natural gas, principally methane. The gas molecules stabilize the structure. Methane comes from decomposition of organic matter buried in the sediments. Deeper below the sediment surface, heat from the earth's interior warms the deposits, so that gas hydrates are no longer stable. Thus at greater depths, natural gas occurs in voids between grains.

Gas hydrates have been detected in deep-ocean sediments for many years. Acoustic techniques used to study the subbottom strata frequently showed reflectors that closely paralleled the ocean bottom. Drilling of deep-ocean sediments recovered gas hydrates which were known from permanently frozen soils of polar regions.

There are still many unanswered questions about gas hydrates. Do they trap natural gas deposits that might someday be exploited as another energy source? Will deep-ocean waters warm enough that the gas hydrates will decompose and release their gases to the ocean? This would increase the amount of methane in the atmosphere. Methane is a greenhouse gas that helps the atmosphere to hold heat. An increase in heat trapping by the atmosphere might further add to the warming of the earth's surface.

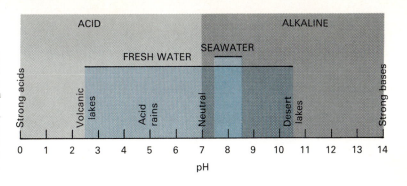

FIGURE 4-13
pH is a measure of the acidity or alkalinity of a solution. The pH of seawater has a narrow range because it is buffered by the carbonate–carbon dioxide system in the ocean. Freshwater systems are not so well buffered, so they have a wider range of pH values. Volcanic lakes are more acidic, while desert lakes tend to be more alkaline. Acid rains contain acids made by gases released during the burning of fossil fuels.

Both hydrogen and hydroxyl ions occur normally in seawater as water molecules break up (dissociate) and re-form. This reaction can be written as:

$$H_2O \rightleftarrows H^+ + OH^-$$

Water itself is neutral. That is, the concentrations of hydrogen and hydroxyl ions are equal. If we add an acid and increase the amount of hydrogen ions, the water becomes acid. Conversely, if we add a compound that increases the amount of hydroxyl ions, the water becomes alkaline.

The acidity or alkalinity of water is expressed using a *pH scale* (Fig. 4-13), based on the abundance of hydrogen ions. If the solution is neutral, it has a pH of 7. The most acid solution has a pH of 0 and the most alkaline one has a pH of 14. Seawater is normally mildly alkaline, with a pH of 8.1, because of the carbon dioxide dissolved in it.

CARBON DIOXIDE AND CARBONATE CYCLES

Carbon dioxide has the most complicated behavior of all the gases in seawater. It occurs as a dissolved gas and in several other chemical forms (Fig. 4-14). It is highly soluble in seawater, entering the sea from the atmosphere. Respiration of plants and animals also produces carbon dioxide throughout the ocean. When carbon dioxide (gas) dissolves, it forms a weak acid (H_2CO_3), which acts as a buffer to stabilize the acidity of seawater. (Acidity is another way of expressing the abundance of hydrogen ions in water.)

FIGURE 4-14
Carbonate–carbon dioxide cycle in the ocean.

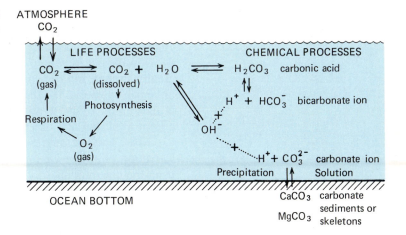

Addition of acid (H^+) creates more bicarbonate, which combines with acids to keep the acidity of the mixture unchanged. Thus respiration and decomposition processes producing carbon dioxide hardly affect the acidity of seawater, nor does removal of carbon dioxide during photosynthesis. Carbonate and bicarbonate freely give up and accept hydrogen ions in seawater, thus creating a buffer against sharp changes in acidity.

Because of its complex chemical behavior, carbon dioxide is involved in many processes in the ocean. Formation and destruction of carbonate shells affect the abundance of carbonate in seawater. If carbonate is used in plant production, the chemical reactions (Fig. 4-14) cause bicarbonate (HCO_3^-) to dissociate. Carbonate precipitation as calcium carbonate ($CaCO_3$) in plant and animal tissues causes a net loss of carbon dioxide (CO_2).

PROCESSES CONTROLLING SEA-SALT COMPOSITION

Some constituents dissolved in seawater come from gases (chlorine, sulfur dioxide) released to the atmosphere through volcanic eruptions. Volcanoes continue to release gases during eruptions. Most of these gases are quickly removed by rainfall and go into the ocean.

Volcanic eruptions on the sea floor supply calcium and other chemically related elements. When seawater reacts with the still-hot volcanic rocks, these elements are extracted from the rocks and carried away by the circulating fluids. The importance of this process is still unknown.

The third major source of materials dissolved in seawater is decomposition (called **weathering**) of rocks exposed on the land. Rocks formed at high temperatures and pressures in the earth's interior usually form in the absence of free oxygen. When they are exposed to atmospheric oxygen and rain, they slowly break down. (The effects of this process can be seen in the gradual obliteration of inscriptions on tombstones.) The insoluble mineral grains released first become part of the soils. They eventually enter the ocean as particles after being transported by winds, rivers, or by ice. The soluble portions released during rock weathering are removed by running water and carried to the ocean by rivers. For instance calcium and magnesium in the ocean come from the alteration of sea-floor volcanic rocks and from decomposition of rocks on land.

There is good evidence that sea-salt composition and probably seawater salinities have not varied much during the past 1.5 billion years. Since large amounts of dissolved constituents are supplied each year, some processes must be removing them at roughly the same rate. Otherwise, salinity of ocean waters would rise and probably sea-salt composition would change. The types and abundances of organisms living in the ocean would very likely have changed if salinity had been greatly altered. But there is no evidence in the fossils of ancient marine organisms preserved in sedimentary rocks that this has occurred. (This is an example of the steady-state condition of the ocean. It is changing little if at all through time.)

The simplest process for removing dissolved constituents is the formation of *evaporite deposits*. If an arm of the sea is nearly cut off from the ocean in an arid climate, evaporation of seawater may leave behind the dissolved sea salts. Limestones and related rocks form first, followed by deposits of more soluble sea salts—sodium, potassium, chloride, and sulfate. As we have already learned, such deposits are likely to be formed in the history of a new ocean basin.

Biological processes also remove dissolved constituents from seawater. Perhaps the most conspicuous example is the removal of calcium by shell-forming organisms, such as clams and oysters and the less familiar one-celled organisms living in near-surface ocean waters.

Finally, chemical reactions between cold seawater and hot, newly formed volcanic rocks remove constituents from seawater. This process is especially important for removing magnesium and sulfate.

RESIDENCE TIMES

A useful concept for characterizing substances in seawater is **residence time,** the time required to replace the amount of a given substance in the ocean completely. This concept can work in two ways, using either the rate of addition or the rate of removal of elements incorporated in sediments depositing on the ocean bottom. Using the second option, we can define residence time:

$$\text{Residence time} \atop \text{(in years)} = \frac{\text{Amount present}}{\text{Removal rate}}$$

Sodium's calculated residence time in the ocean of 68 million years is one of the longest residence times for an element. Residence times of some other constituents are shown in Table 4-2.

An element's residence time in the sea is related to its chemical behavior. Elements like sodium are little affected by sedimentary or biological processes. They have residence times of millions of years. Elements used by organisms or readily incorporated in sediments, such as aluminum or iron, have much shorter residence times, ranging from a few hundred to a few thousand years.

We can also calculate residence times for water. There is a net removal, due to evaporation (which does not fall back on the ocean surface as rain) of a layer of water about 10 centimeters thick from the ocean surface each year. This water falls on the land and returns to the ocean through river discharge. Recall that the ocean has an average depth of about 4000 meters. From these figures we find a residence time of 40,000 years for water.

TABLE 4-2
Estimates of Residence Times for Some Elements in Seawater

ELEMENT	MILLIONS OF YEARS	REMOVAL PROCESS
Sodium	68	Evaporite deposition
Chloride	100	Evaporite deposition
Potassium	7	Reactions with clay
Calcium	1	Shell formation
Lead	0.0004	Removal by particles
Aluminum	0.0001	Absorption on clays

SUMMARY

Earth's ocean and atmosphere came from volcanic activity. Seawater consists of 96.5% water, 3.5% sea salts, some dissolved organic matter, a few particles, and some dissolved gases. Many of the distinctive aspects of seawater are due to the physical and chemical properties of water itself. Some properties are altered by adding salt. Many features of water are due to its molecular structure (a polar molecule), which favors formation of bonds among molecules.

Water can occur as a gas, liquid, or solid (the three states of matter), depending on its temperature. In water vapor, molecules move freely and expand to fill any volume. As a solid, water molecules are bonded together, forming a lattice which is less dense than liquid water. Voids in the ice lattice are too small to accommodate gases or salts in the structure, so they are excluded as water freezes. Liquid water is an unusual substance in that it contains both structured and unstructured regions; changing temperature, pressure, or salt content alters the relative abundances of the two. Increased pressure favors the denser, unstructured phase.

Heat is absorbed as bonds are broken or released when bonds form. Because of the large amount of bonding in liquid water, it has an unusually high heat capacity, the ability to absorb heat with relatively little temperature increase. The size of the structured clusters of water molecules decreases as water temperature rises. Heat is also absorbed (or released) when matter changes from one state to another.

Viscosity—internal resistance to flow—is sensitive to the molecular structure of water. Water's viscosity increases with higher salinities, lower temperatures, and lower pressures, all favoring the more structured component.

Sinking (or floating) is determined by relative density (mass per volume); less dense objects (or liquids) float in more dense ones. Seawater density is controlled by temperature, salt content, and pressure. Water density decreases with increasing temperature because its volume expands. Freshwater has a density maximum at 3.98°C. Seawater begins to freeze before it reaches its density maximum.

Six constituents account for 99% of sea salts: chloride, sodium, sulfate, magnesium, calcium, and potassium. They occur as ions (electrically charged atoms or groups of atoms) because salts dissolve in water, leaving the ions separate. Salinity, the total amount of dissolved salts, expressed in parts per thousand, is used to indicate the amount of salt dissolved in seawater; it is usually measured electrically.

The major constituents in seawater occur in constant proportions; these are called conservative substances. Substances occurring in low concentrations which vary as a result of chemical or, usually, biological processes are called nonconservative substances.

The amount of gas that can be dissolved in seawater is controlled by temperature and salinity; the lower the temperature and salinity, the greater the amount of gas that can be dissolved. Surface seawater exchanges gas with the atmosphere. When the amount taken up equals the amount lost, seawater is said to be saturated with the gas. Nitrogen is near saturation levels throughout the ocean. Oxygen is taken up from the atmosphere at the surface and is also released by plants; it is near saturation in surface waters and depleted in deep-ocean waters.

Seawater's acidity-alkalinity balance is controlled by the carbon dioxide–carbonate cycle in the ocean.

Carbon dioxide has a complicated chemical behavior in seawater. It occurs as a dissolved gas and combines with water in the form of bicarbonate and carbonate. Its abundance is controlled by photosynthesis in near-surface waters and by the formation and destruction of carbonate shells and the destruction of organic matter in the deep ocean.

Sea-salt composition is controlled by chemical reactions between seawater and hot, newly formed crust at midocean ridges. Some constituents are removed by biological processes; others, by evaporation. Sodium and chloride have the longest residence times.

STUDY QUESTIONS

1. Explain the origins of the ocean and atmosphere.
2. List the principal gases in the atmosphere.
3. Draw a diagram of a water molecule. Explain why water is a polar molecule.
4. Describe and contrast the molecular structures of ice, liquid water, and water vapor. Discuss the significance of the different structures on the amount of energy required (or given off) by changing from one structure to another.
5. List the six major constituents of seawater and indicate the percentage of each.
6. Define residence time. Discuss its significance and give examples of elements having long residence times; short residence times.
7. Explain the carbon dioxide–carbonate cycle in seawater without living organisms or carbonate solids.
8. Discuss why seawater rarely gets colder than −1.8°C.
9. Describe why fresh water in the bottom of large lakes does not get colder than 4°C.
10. Discuss the difference in chemical behavior between conservative and nonconservative constituents.
11. Discuss the processes controlling the composition of sea salts.
12. Why is the concentration of carbon dioxide dissolved in seawater greater than in the atmosphere?
13. What resources are extracted from seawater?

BERNER, E. A., AND R. A. BERNER. 1987. *The Global Water Cycles*. Englewood Cliffs, N.J.: Prentice-Hall. 397 pp.

BROECKER, W. S. 1974. *Chemical Oceanography*. New York: Harcourt Brace Jovanovich. 214 pp. Discussion of processes affecting chemical composition of sea water.

DEMING, H. G. 1975. *Water: The Foundation of Opportunity*. New York: Oxford University Press. 342 pp. General discussion of water.

HORNE, R. A. 1969. *Marine Chemistry: The Structure of Water and the Chemistry of the Hydrosphere*. New York: Wiley-Interscience. 568 pp. Advanced.

HSU, KENNETH S. 1983. *The Mediterranean Was a Desert: A Voyage of the Glomar Challenger*. Princeton, N.J.: Princeton University Press. Lively account of the discovery of the Mediterranean salinity crisis.

Hurricane Gladys was photographed about 250 kilometers southwest of Tampa, Florida on October 18, 1968 from *Apollo 7*. Note the well developed spiral structure of the storm. (Courtesy NASA/Photo Researchers.)

CHAPTER 5

Atmosphere

OBJECTIVES

1. To understand the basic features of the atmosphere and their causes;

2. To grasp how ocean and atmosphere interact;

3. To understand the effects of the Earth's rotation on winds;

4. To explain daily and seasonal effects on weather;

5. To differentiate weather and climate and the processes that affect them;

6. To present the history of Earth's climate and its changes.

*T*he ocean and atmosphere act together as the earth's fluid outer layer. We now examine the atmosphere, the processes that control its movements, and its interactions with the ocean. In this chapter, we discuss:

The atmosphere and the processes that control winds;
Heat budgets;
Winds on a water-covered, nonrotating Earth;
Coriolis effect and winds on a rotating Earth;
Effects of seasonal and daily changes on winds.

ATMOSPHERIC GASES   Earth's atmosphere is a mix of transparent, odorless gases, predominantly nitrogen (78%) and oxygen (21%). Water vapor, ozone, carbon dioxide, and dust are important but variable constituents (Fig. 4-1).

Water vapor in the atmosphere plays an especially important role. As we see later, it affects the way the atmosphere absorbs heat from the sun. It is also important because it is involved in heat transport by the atmosphere. Evaporation of water from the ocean surface removes heat from the ocean and puts it into the atmosphere. Conversely, condensation of water vapor releases heat to the atmosphere. Let's look at this process in more detail.

When a water surface is heated, water vapor escapes into the overlying atmosphere. At a given temperature, a certain amount of water vapor is in equilibrium with the water surface (Fig. 5-1). When there is less than the equilibrium amount, water evaporates. When there is more than the equilibrium amount of water, it condenses, forming either rain or fog. As the temperature of the system rises, the equilibrium amount of water vapor in the atmosphere increases. An increase of temperature from 0° to 10°C doubles the amount of water vapor in the atmosphere at equilibrium.

Evaporating water removes heat from the water. This is especially important as water surfaces become warmer. Warming water from 20° to 30°C requires 10 calories per gram of water. The amount of water vapor in equilibrium with the water doubles between 20° and

BOX 5-1

The Gaia Hypothesis

Earth's atmosphere is so unlike its neighbors' (Mars, Venus, and the moon) that scientists are hard pressed to account for its uniqueness. Earth's unique moving lithospheric plates can be explained by its size. It is large enough to retain enough heat for continued volcanic activity and mantle convection. The other terrestrial planets are now too cool for volcanism or plate movements.

Processes controlling the composition of atmospheric gases are less obvious. A recent theory explaining this is the **Gaia hypothesis,** named after a Greek Mother Earth. It was published in 1979 by the British biochemist J. E. Lovelock.

Lovelock postulated that physical and chemical conditions on the earth's surface are controlled by biological processes. The Gaia hypothesis states that these processes control the composition of atmospheric gases and sea salts. Furthermore, the theory holds that these processes are self-regulating. That is, surface temperatures have remained nearly constant between 0° and 20°C since life first appeared on Earth, about 3.8 billion years ago. This is difficult to explain otherwise, since the sun is now approximately 30% brighter than when it first formed. Without some form of temperature regulation, one would expect either the newly formed Earth to have been ice-covered or the present Earth to be extremely hot, similar to Venus. Clearly, neither has happened.

Lovelock argues that without life the earth's atmosphere would consist primarily of carbon dioxide, like its neighboring planets'. All the nitrogen would combine with oxygen and dissolve in seawater. He compares the atmospheres on Venus and Mars with that on Earth, with and without life. Without life, Earth's atmosphere would resemble that on Venus.

Sea-salt composition is also affected by these biological processes. Lovelock estimates that the ocean on a lifeless Earth would have a salinity of 130‰. The excess salt would primarily consist of nitrogen compounds, now mostly in the atmosphere.

| | PLANET | | | |
| GAS | VENUS | MARS | EARTH | |
			LIFELESS	WITH LIFE
Carbon dixoide	98%	95%	98%	0.03%
Nitrogen	1.9	2.7	1.9	79
Oxygen	Trace	0.13	Trace	21
Argon	0.1	2	0.1	1
Surface temperature (°C)	477	−53	−20±	18

30°C. Remember that evaporating a gram of water requires about 590 calories per gram. Thus much more energy is required to evaporate water than to warm it. Removal of so much heat from the ocean surface by evaporation is the primary reason that the ocean surface is rarely warmer than 30°C. This is also the source of energy for hurricanes.

Adding water vapor to the atmosphere makes it less dense and causes vertical movements (which we call **convection**). The molecular weight of a water molecule (H_2O) is 18 (oxygen–16, hydrogen–1), while that of nitrogen gas (N_2) is 28 (nitrogen–14). Thus adding water vapor makes the atmosphere less dense.

Ozone—a three-atom molecule of oxygen—is another important but variable constituent of the atmosphere. Near the earth's surface, ozone occurs as a pollutant in smog. It also occurs naturally in the upper levels of the atmosphere, where it absorbs damaging radiation from the sun. (There is more about this in the next section.)

Carbon dioxide is another variable constituent in the atmosphere. It is an important absorber of radiation in the atmosphere. Its increased

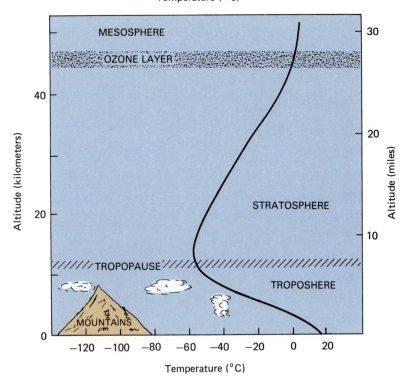

concentration in the atmosphere due to burning of fossil fuels is a major contributor to the apparent warming of the earth's surface—the greenhouse effect (see Box 5-2).

Finally, *dust* occurs in the atmosphere, primarily over land. It is not a major factor in atmospheric circulation; however, transport of dust by winds is a major source of sediment to the ocean far from land. Also, windblown dust is a major source of pollutants, both on land and to the ocean.

ATMOSPHERIC STRUCTURE

The atmosphere is density stratified. In other words, the densest air is nearest the Earth's surface. Since gases are compressible, the higher pressure near the surface compresses the gases, making them denser than those higher in the atmosphere. Atmospheric pressure decreases by half for every 6-kilometer increase in altitude. (Recall that water is only slightly compressible. Thus there is little increase in water density in the depths of the ocean.)

The atmosphere is warmed near the earth's surface and cooled at the top. Thus air temperatures decrease with increased elevation in the **troposphere,** the lower atmosphere (Fig. 5-2). The troposphere is bounded by the **tropopause,** where temperatures begin to rise. In the **stratosphere,** above the tropopause, temperatures rise with elevation, because ozone molecules absorb some ultraviolet radiation from the sun and are excited. Temperature therefore increases in the ozone layer. (Reduction in the ozone layer observed in the Antarctic and Arctic has caused concern, since less ultraviolet is removed from the sunlight there. The effects of this change are not yet obvious.)

There is little vertical movement in the stratosphere, unlike the troposphere, where most of our weather occurs. Thunderstorms are familiar examples of vertical motions in the lower atmosphere. Substances injected into the stratosphere persist there, since there are few mechanisms to remove them.

HEAT BUDGET

One way to study processes is to construct energy *budgets*. Budgets show where energy or materials come from (*sources*) and where they go (*sinks*). We use the same techniques later in studying the ocean.

We start with the *atmosphere*'s *heat budget*. Directly beneath the sun [Fig. 5-3(a)], the top of the atmosphere receives about 2 calories per square centimeter per year (also written cal cm^{-2} min^{-1}). Incoming energy is distributed over the earth's surface as it rotates. On the average, about 0.5 calorie per square centimeter per minute strikes any given point at the top of the atmosphere [Fig. 5-3(b)]. (A small amount of heat comes from the earth's interior. We ignore it here because it is so small compared to solar heating.)

Despite the solar heating, average surface temperatures over the past few centuries (where we have historical records) have been nearly

FIGURE 5-3
Incoming solar radiation varies over the Earth's surface. (a) The effect of latitude on the average solar radiation (compared with the Earth's equator) at the Earth's surface in the Northern Hemisphere at the equinoxes. (b) Relative amount of incoming solar radiation at the Earth's surface during the summer solstice (June 21), when the sun is directly overhead at 23.5°N. Note the changes between the equator and the poles.

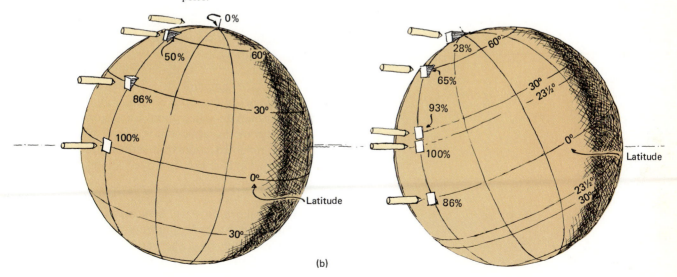

(b)

BOX 5-2

Greenhouse Effect

Earth's atmosphere is nearly transparent to visible light. Thus sunlight easily penetrates to its surface. Atmospheric gases absorb some of the radiation coming from the sun. They also absorb some of the energy radiated back to space from the warm surface of the earth. As we have previously seen, Earth's surface is in equilibrium with the radiation it receives from the sun, radiating back an equivalent amount of energy each year. This accounts for the long-term constancy of the earth's climate.

Gases released by burning fossil fuels, and from agriculture and manufacturing, are altering the composition of the atmosphere and thus changing atmospheric absorption and transmission of energy from the sun. This phenomenon is called the **greenhouse effect** (an analogy with a greenhouse, which is warmer than its surroundings because its thick glass panes keep in heat).

Carbon dioxide is the most important gas. It readily transmits the shorter wavelengths of incoming solar radiation while absorbing much of the longer wavelengths (infrared or heat) emitted by the earth's surface, thus warming the earth's surface.

Burning coal and oil since the 1850s (Industrial Revolution) has increased carbon dioxide concentrations in the atmosphere. In the next century, atmospheric carbon dioxide levels will likely double over preindustrial levels.

About half the carbon dioxide dissolves in the ocean or goes into vegetation; the rest remains in the atmosphere. This has already caused a warming of about 0.6°C of the earth's surface. Further warming—up to 1.5° to 5°C—may occur over the next century.

Other atmospheric gases can contribute to the greenhouse effect. Methane and chlorofluorocarbons (used in refrigerators and in aerosol sprays) are released by agriculture and industrial processes. They apparently are as important as carbon dioxide.

Similar processes may be depleting ozone in the stratosphere, which absorbs ultraviolet radiation from the sun, protecting plants and animals against damage from ultraviolet. A large hole in the ozone layer over Antarctica has grown larger over the past several years. A hole in the stratospheric ozone layer has also been detected over the Arctic. The effects of increased ultraviolet at the surface are not yet known.

Effects of warming the earth's surface are also not known. One concern is that the warming will cause polar ice caps to melt. This would raise sea level 50 meters. Flooding of large areas of low-lying coastal lands, much of Florida, and Egypt, could ensue. It would also likely shift rainfall patterns, seriously affecting agriculture. Grain-growing regions of North America and the Soviet Union might shift northward.

As we have already seen, the cooling of the ocean surface by increased convection probably limits sea surface temperatures to about 30°C, but warm tropical ocean waters could spread farther north and south. This might also increase the number and strength of hurricanes, intensify El Niños, and cause increased year-to-year variability in weather. Effects on droughts are more difficult to predict. Humans are thus conducting a planetary-scale experiment with the earth and its climate, and the outcome cannot yet be predicted with confidence.

---

constant. This indicates that Earth radiates back to space as much energy as it receives from the sun each year.

The sun's surface is extremely hot (approximately 6000°C) and radiates a wide spectrum of wavelengths, primarily in the visible part of the spectrum. Absorbing this energy warms the earth's atmosphere and ocean. Earth radiates heat to space primarily in the infrared (Fig. 5-4) because its surface temperature (18°C) is much cooler than the sun's. Much of the energy from the earth is radiated by cloud tops.

Clouds reflect back to space about one-fourth of the solar radiation striking the top of the atmosphere (Fig. 5-5). The ratio of the amount of energy reflected compared with the amount incoming is called the **albedo.** Clouds typically cover about half of the earth's surface at any time.

On the average, about half of the sun's radiation striking the top of the atmosphere reaches the earth's surface (Fig. 5-6). About 5% of the earth's heat loss is due to radiation from the surface. Clouds absorb most of the back-radiation from the earth's surface. Without clouds and the atmosphere, Earth's surface temperatures might drop as low as

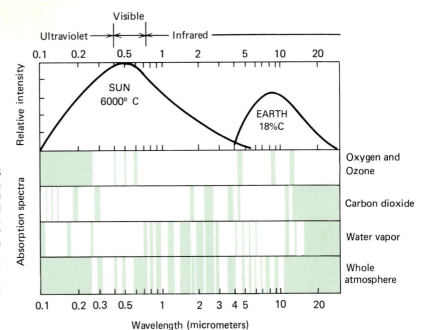

FIGURE 5-4
Emission spectrum of the sun and Earth compared with the absorption spectrum of the atmosphere, showing absorption lines for several important atmospheric constituents. (The dark areas are absorption bands, indicating that energy at that wavelength is absorbed.) Note that the Earth's surface is much cooler than the surface of the sun, so the Earth's radiation occurs in the infrared part of the spectrum, whereas the sun radiates primarily in the visible part of the spectrum. (After P. K. Weyl. 1970).

FIGURE 5-5
Average annual heat budget of the Earth, showing interactions with incoming solar radiation and loss of heat.

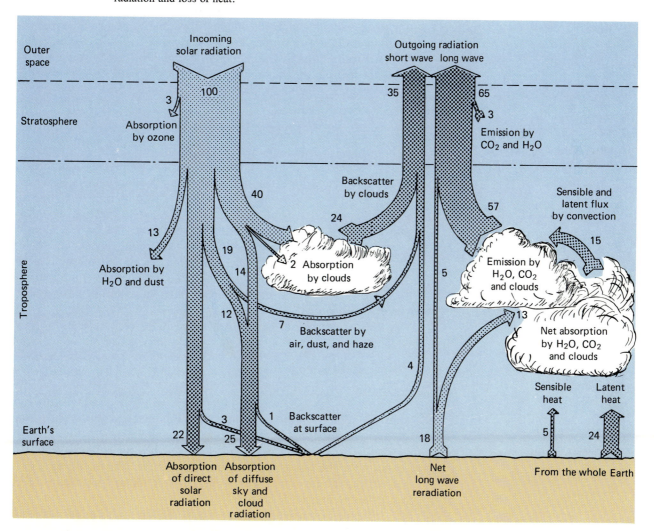

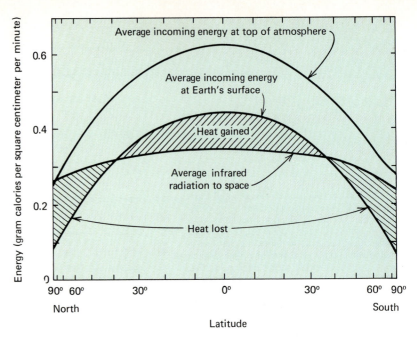

FIGURE 5-6
Primary areas of heat loss and heat gain on the Earth's surface. Note that about four times more solar energy reaches the surface at the equator than at the poles. The radiation of heat back to space is nearly the same over the entire Earth. Net heat loss occurs primarily in polar and subpolar regions, between 50° and 90°.

−20°C, the average temperature of the cloud tops. Instead, ocean surface temperatures average about 17.5°C and the land about 14°C.

The atmosphere is heated in its lower few thousand meters primarily by **latent heat** from condensed water vapor. The heat that originally went into evaporating water is released to the atmosphere when water vapor condenses. Relatively little heat is transferred directly by **sensible heat,** in which a volume of warm air flows into a region and an equal volume of cold air flows out.

WINDS ON A WATER-COVERED, NONROTATING EARTH

The winds are driven by the unequal heating of the earth. On a nonrotating, ocean-covered Earth, wind patterns would be quite simple (Fig. 5-7). The atmosphere would be warmed near the equator and would rise. As air rises, it cools, and water vapor condenses to fall as rain. The air, now drier, would flow toward the poles, where it would cool further. Near the poles, the cold, dry air would sink to flow along the surface to the tropics, where the process would repeat. On such an Earth, there would be a simple two-celled circulation. In the Northern Hemisphere, surface winds would blow from north to south (reversed in the Southern Hemisphere). Such a simple circulation pattern occurs near the equator.

Since the earth rotates, we must consider its effects on atmospheric circulation. We begin by considering the Coriolis effect, which arises from Earth's rotation.

CORIOLIS EFFECT

Earth's rotation profoundly affects winds and currents. We are not conscious of these effects on our activities. If we throw a snowball, we expect it to travel in a straight line and hit our target because we experience the earth as a nonrotating system. Thus a freely moving object should move in a straight line at a constant speed unless acted on by some outside force.

Winds or currents that are freely moving (not bound to the sur-

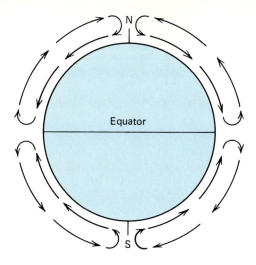

FIGURE 5-7
On a water-covered, nonrotating Earth, heat at the equator causes air to rise. Cooling at the poles causes the denser air to sink. The result is one large convection cell in each hemisphere.

FIGURE 5-8
The Coriolis effect is caused by the Earth's rotation. (a) To an observer standing on a rotating turntable, the path of a freely moving object is apparently deflected to the right as the turntable rotates under the path of the ball. (b) The apparent deflection increases with the speed of the object.

face) are deflected to the right of the motion in the Northern Hemisphere and to the left in the Southern Hemisphere. This is a consequence of our being on a rotating surface. This effect is called the **Coriolis effect,** named after the French mathematician Gaspard Gustave de Coriolis (1792–1843). He explained these deflections as a consequence of a rotating system in which freely moving objects move in circles at constant speed unless acted on by an outside force. The earth's rotation causes the curved paths of rockets and artillery shells as seen by an observer on the ground.

To understand the cause of the Coriolis effect, refer to Fig. 5-8(a). If we stand on a stationary turntable and throw a snowball, it moves in a straight line and hits the target. But if the turntable is moving counterclockwise (viewed from the top, as the earth does when viewed from above the North Pole), the snowball is apparently deflected to the right. Actually the snowball traveled a straight line, but the target moved as the snowball was in flight. From this example it is easy to understand

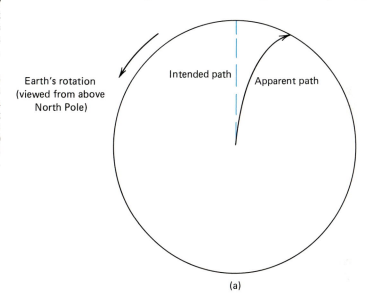

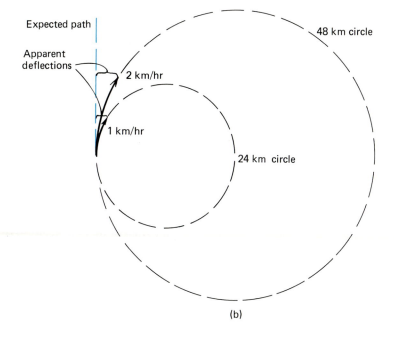

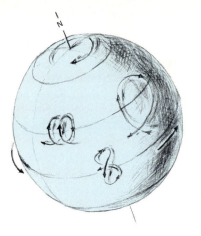

FIGURE 5-9
Particles moving freely over the Earth's surface follow circular paths except when moving east or west along the equator, where they are unaffected by the Coriolis effect.

why the apparent deflection is reversed (to the left) in the Southern Hemisphere. Viewed from above the South Pole, the earth appears to rotate clockwise.

The time for a freely moving object to make one complete revolution depends on its latitude. At a pole, it takes 12 hours, and the time increases as one goes toward the equator. At 30° latitude, it takes 24 hours to complete a circle. At the equator, there is no Coriolis effect for an object moving east or west along the equator because there is no effect of the earth's rotation there. In fact, little effect of rotation is felt within 5° on either side of the equator.

Apparent deflection increases with the speed of the object. An object traveling twice as fast as another will go around a circle with twice the circumference [Fig. 5-8(b)]. After two hours, if we look at the apparent deflections, we find that an object moving 2 kilometers per hour has apparently been deflected more than an object moving 1 kilometer per hour. Both will have gone around one-twelfth of the circumference of its circle.

If a ball is moving on a sloping surface, it also experiences the pull of gravity to make it roll down slope. It is fairly easy to imagine that gravity balances the apparent deflection due to the Coriolis effect as the ball rolls along the sloping surface. In winds and ocean currents, this situation occurs where winds or currents are deflected by the Coriolis effect and by sloping-density surfaces. Particles moving freely over the surface of a rotating Earth, with no outside influences, follow circular paths, as shown in Fig. 5-9. As we shall see in later sections, there are usually other forces acting.

WINDS ON A ROTATING EARTH

The unequal heating of the earth sets the atmosphere in motion, and the Coriolis effect causes a complex six-celled atmospheric circulation (seen on the right side of Fig. 5-10). Let us see how this complex pattern is formed.

When the atmosphere is warmed in equatorial regions, air rises and begins to move toward the poles (Fig. 5-10). The Coriolis effect deflects the rising air to the right in the Northern Hemisphere and to the left in the Southern Hemisphere. Along the equator, there is a persistent band of clouds and high rainfall marking the area of persistently rising air. Winds coming from the west converge at high altitudes in the midlatitudes, around 30° north and south, forming a belt of high pressure. As the air in these zones of convergence sinks, it spreads along the earth's surface, causing the prevailing winds shown in Fig. 5-11. Part of this air moves toward the equator as the **trade winds.** As the air moves along the surface, it is warmed, picking up water vapor. Along the equator, it rises to continue the process.

The band of trade wind convergence and persistently rising air is called the **intertropical convergence zone** (abbreviated ITCZ). This band is usually visible in satellite photos of the equatorial ocean. The low-latitude cell, including the trade winds, is called the **Hadley cell,** named after the meteorologist, George Hadley (1685–1768), who first postulated its existence. The midlatitude cell is called the *Ferrel cell,* after another meteorologist, W. Ferrel.

The rest of the air flows toward the poles, forming the *westerlies,* surface winds blowing from the west. (Note that winds are named for the direction from which they are blowing. Later, when we discuss

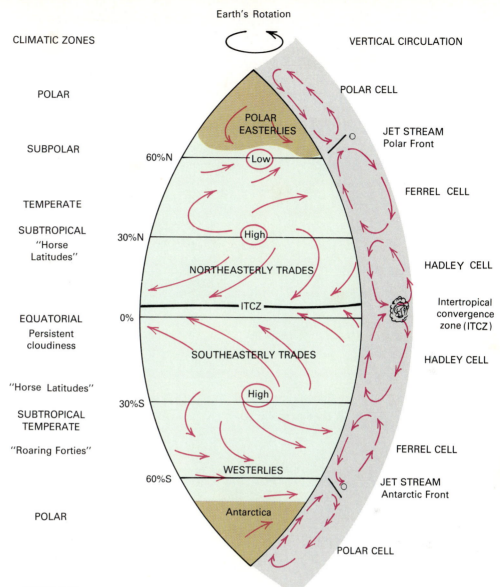

Earth's Rotation

CLIMATIC ZONES

POLAR

SUBPOLAR

TEMPERATE

SUBTROPICAL
"Horse
Latitudes"

EQUATORIAL
Persistent
cloudiness

"Horse Latitudes"

SUBTROPICAL
TEMPERATE

"Roaring Forties"

POLAR

VERTICAL CIRCULATION

POLAR CELL

JET STREAM
Polar Front

FERREL CELL

HADLEY CELL

Intertropical
convergence
zone (ITCZ)

HADLEY CELL

FERREL CELL

JET STREAM
Antarctic Front

POLAR CELL

POLAR
EASTERLIES

60%N — Low

30%N — High

NORTHEASTERLY TRADES

ITCZ

0%

SOUTHEASTERLY TRADES

30%S — High

WESTERLIES

60%S

Antarctica

FIGURE 5-10
Schematic representation of the atmosphere over
an ocean basin on a rotating Earth.

currents, you will find that currents are named for the direction in which they are moving.)

At about 50° north and south, the westerlies meet colder, denser air from the polar regions moving toward the equator. This convergence zone is the *polar front* (called the *Antarctic front* in the Southern Hemisphere), a persistent boundary between the polar and tropical air masses. The latitudinal band affected by the polar front experiences highly variable weather. This accounts for the highly variable weather of much of North America and Europe, which experience a succession of relatively warm, moist subtropical air masses and cold, dry polar air masses.

The polar front has a succession of large waves on it that appear on weather maps as curved warm or cold fronts. A narrow band of strong winds, called the **jet stream,** occurs at elevations of around 10 kilometers.

Climate (weather averaged over a long time, typically 30 years) is determined by the locations of areas of either rising or sinking air

WINDS ON A ROTATING EARTH

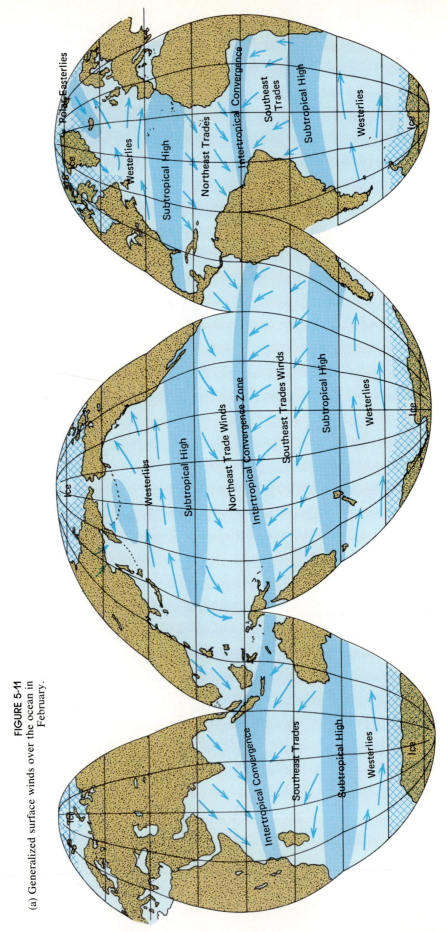

FIGURE 5-11

(a) Generalized surface winds over the ocean in February.

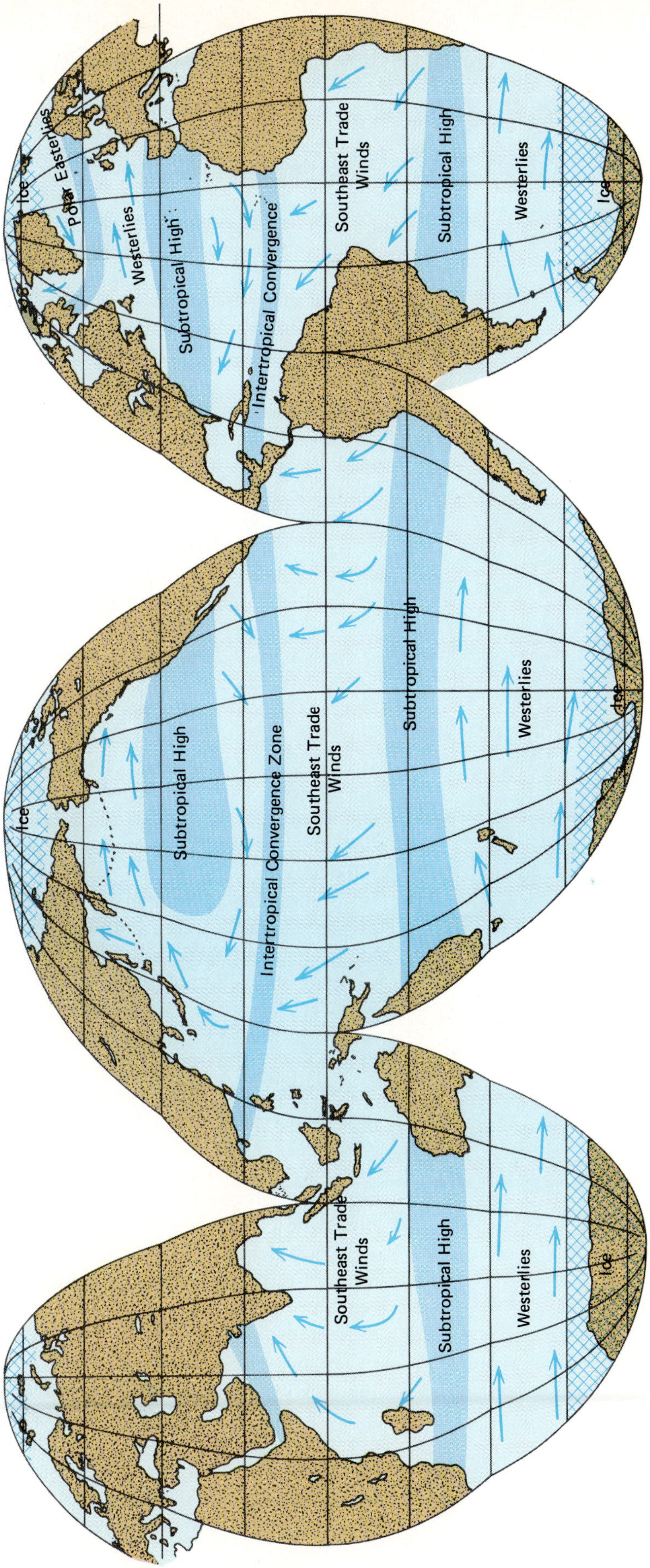

FIGURE 5-11
(b) Generalized surface winds over the ocean in August.

masses. In midlatitudes, where air masses commonly sink over the ocean, the climate is characterized by:

Low rainfall and high evaporation;
Light variable winds, called the *horse latitudes;* and
High atmospheric pressure.

Near the equator, where the air is rising due to warming and high water-vapor contents, the climate is characterized by:

High rainfall and much cloudiness;
Light and variable winds, called the **doldrums;** and
Low atmospheric pressure.

Because of the earth's rotation, the winds in each cell follow spiral paths as the warm air rises, cools, and then sinks. In addition, there is an east-west circulation (not affected by the Coriolis effect) along the equator, called the *Walker cell,* named after George Walker, a meteorologist. This equatorial circulation participates in the large-scale changes (called the **Southern Oscillation**) in atmospheric and oceanic circulation that produce the **El Niños,** which we discuss later.

WEATHER SYSTEMS

So far, we have discussed atmospheric motions on a planetary scale. But most of our experience with weather involves smaller atmospheric motions. Winds move across the earth's surface in a series of turbulent systems. These seemingly chaotic motions involve a spectrum of swirling motions, called **eddies.** We are familiar with small-scale turbulence visible in cigarette smoke in a room. The large-scale atmospheric motions may be as large as enormous hurricanes that fill the entire Gulf of Mexico.

These air movements are caused by differences in atmospheric pressure, temperature, and humidity. Each eddy obeys physical laws that also control movements of ocean waters (which we discuss in Chapter 7). Predicting the movements of these features is the basis of weather forecasts. The unpredictable aspects of atmospheric motions limit weather forecasts to only a week to 10 days. Beyond that, unknowable aspects of these chaotic motions cannot be predicted.

Over tens or hundreds of kilometers there are differences in air pressure. Where upper atmospheric winds diverge, they cause areas of low pressure, called **lows** or low-pressure systems (Fig. 5-12). Where upper level winds converge, they cause areas of high pressure, called **highs** or high-pressure systems. Air moves from high-pressure areas toward low-pressure areas.

Not being connected to the earth's surface, winds are deflected by the Coriolis effect. Eventually the winds are deflected so that they blow more or less parallel to the lines of equal atmospheric pressure; these lines are called *isobars.* When the path of the winds is controlled by the balance of forces between the Coriolis effect and the pressure differences, they are called **geostrophic winds.** By mapping distributions of highs and lows and the atmospheric pressures, meteorologists are able to predict movements of storm systems and to predict local weather. (In Chapter 7, we discuss how oceanographers use the same techniques to map ocean currents.)

Air masses form over land and ocean. Staying in a spot for sev-

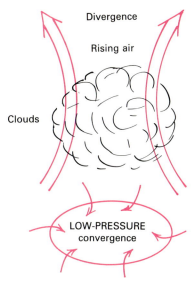

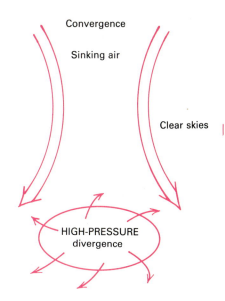

LOW-PRESSURE SYSTEM

Divergence

Rising air

Clouds

LOW-PRESSURE
convergence

HIGH-PRESSURE SYSTEM

Convergence

Sinking air

Clear skies

HIGH-PRESSURE
divergence

FIGURE 5-12

Areas of surface low and high pressure cause surface winds. They are caused by divergence and convergence of winds in the upper atmosphere. Such areas are zones of rising and sinking air, respectively. (These examples are for the Northern Hemisphere.)

eral days or weeks, they acquire the characteristics of the regions in which they form. In winter, cold, dry air masses form over the land in Canada and Siberia. When they move out of these regions, they maintain the temperature and humidities they acquired in their source areas. Likewise, tropical maritime air masses form over equatorial waters. They are distinguishable by their warm air and high humidities. When air masses move into other regions, they are modified, and their temperatures and humidities are altered. Arctic air masses warm somewhat and become more humid. Tropical maritime air masses lose moisture by precipitation and by cooling.

When air masses collide, they are separated by **fronts,** marked changes in temperature and humidity across the boundary. A cold front is an advancing cold-air mass. Since cold air is denser than warm air, it pushes under the warm air. The rising warm air cools and often loses some of the water vapor it was carrying. A warm front is an advancing mass of warm air, displacing colder air. Interactions between such air masses cause much of the weather that we experience in the midlatitudes.

OCEAN INFLUENCE ON WEATHER PATTERNS

Large masses of unusually warm or cold surface ocean waters influence weather on land in two ways. In the first case, the presence of large masses of warm water on the western side of the equatorial ocean causes El Niños and the Southern Oscillation, which we discuss later. The other way is for masses of either warm or cold water to steer weather patterns.

Such water masses occur in the North Pacific, where they have been extensively studied. These water masses are 1000 to 2000 kilometers across and 200 to 300 meters thick. They persist for many years as they move across the basin. The size of these water bodies and the energy they contain permit them to steer weather systems over the continents.

Large cold-water masses in the North Pacific cause shifts in prevailing western winds blowing across the eastern part of North America. Cold, dry air from Canada displaces warm, moist air from the Gulf of Mexico and the Atlantic. As a result, winter temperatures in the southeastern United States are much colder than normal. Shifts in the relative positions of such water masses can bring about the opposite effect.

Some scientists believe that prolonged droughts in Africa are caused by unusually warm surface waters in the Atlantic. If the causes of these relationships could be proven, it would permit predictions of persistent weather patterns such as multiyear droughts.

EXTRATROPICAL CYCLONES

Two types of powerful storms affect the ocean: extratropical storms and hurricanes. **Extratropical cyclones** form at the polar front in the midlatitudes where cold, dense air moving toward the equator meets warm, humid air moving poleward.

Winds blow around a low-pressure area along a stationary front. The cold front circulates around the low-pressure zone faster than the warm front. As a result, it eventually overtakes the warm front and cuts it off, forming an *occluded front*. At this stage the low-pressure cell intensifies greatly as atmospheric pressure drops and winds grow stronger. The intensified low-pressure cell with its associated winds (Fig. 5-13) moves eastward. Such storms have wind speeds greater than 75 miles per hour, comparable to hurricanes. They are common in winter when contrasts in air temperature across the polar front are largest. Extratropical cyclones can be exceedingly damaging in coastal areas. They move more slowly than hurricanes and often cause extensive damage by eroding beaches and causing flooding.

FIGURE 5-13
A large extratropical storm, southeast of New Zealand. (Courtesy NASA.)

Hurricanes (or typhoons) are intense tropical cyclones that occur in local summer and autumn (June through November in the West Indies, peaking in September). They form in all oceans except the South Atlantic and the eastern South Pacific (Fig. 5-14). Hurricanes form when winds converge and concentrate the rotation that comes from their participation in the earth's rotation. A low-pressure disturbance, such as a thunderstorm, in the large-scale winds intensifies. This causes local winds to converge more strongly. As warm, moist air sweeps in, it rises in a helical pattern. As the hurricane develops, wind speeds build up to 300 kilometers per hour. The fastest winds at the sea surface form a band surrounding the hurricane's eye—a relatively cloud-free area of low atmospheric pressure, typically about 25 kilometers across (Fig. 5-15).

Most hurricanes form between latitudes 15° and 20° in the Northern Hemisphere (Fig. 5-14). They rarely form within 5° of the equator, since the Coriolis effect is weak there. Hurricanes form where sea surface temperatures exceed 28°C, usually in the western parts of ocean basins (Fig. 5-14). About 40 hurricanes occur each year. A hurricane's power comes from condensing water vapor in the rising air currents in the storm's center.

Hurricanes move westward in the trade winds at speeds around 10 to 30 kilometers per hour. A typical West Indies hurricane lasts 9 days. Eventually each hurricane moves poleward, carried by the general atmospheric circulation, usually steered by high-level winds. At this stage, its speed and direction often become erratic. Some hurricanes follow looping paths; others become stationary; still others move unusually rapidly with speeds around 1000 kilometers per day.

When a hurricane moves over land or relatively cold surface water, it rapidly loses strength. Winds blowing across the land experience more friction than over the sea surface. So storms lose energy, and their water vapor supply is cut off. Cold waters also deprive storms of their energy source. As a result, the storm dissipates most rapidly over land.

Hurricanes release enormous amounts of energy, typically 300 to 400 billion kilowatt-hours of energy each day. This is 200 times the total electrical power generation in the United States. An average hurricane precipitates 10 to 20 billion tons of water every day, often causing disastrous flooding.

Hurricanes cause millions of dollars in damages each year through the effects of winds, associated waves, and elevated sea levels. Hurricane Camille in 1969 killed 256 people and caused $1.4 billion in damage, the most damaging storm in U.S. history.

Many aspects of hurricanes still remain unexplained, such as their origins or distributions, but there has been some success. For instance, it is now possible to predict paths of hurricanes, which has decreased hurricane death tolls substantially. For instance, Hurricane Gilbert in September 1988 was the strongest hurricane in this century to strike the Caribbean and Mexico. Yet observations by satellites and aircraft permitted detailed predictions that saved the lives of many people who evacuated low-lying areas before the hurricane struck. Hurricanes also spawn tornados, which also cause much damage. Due to increasing populations and buildup of coastal areas, property damage from hurricanes continues to increase.

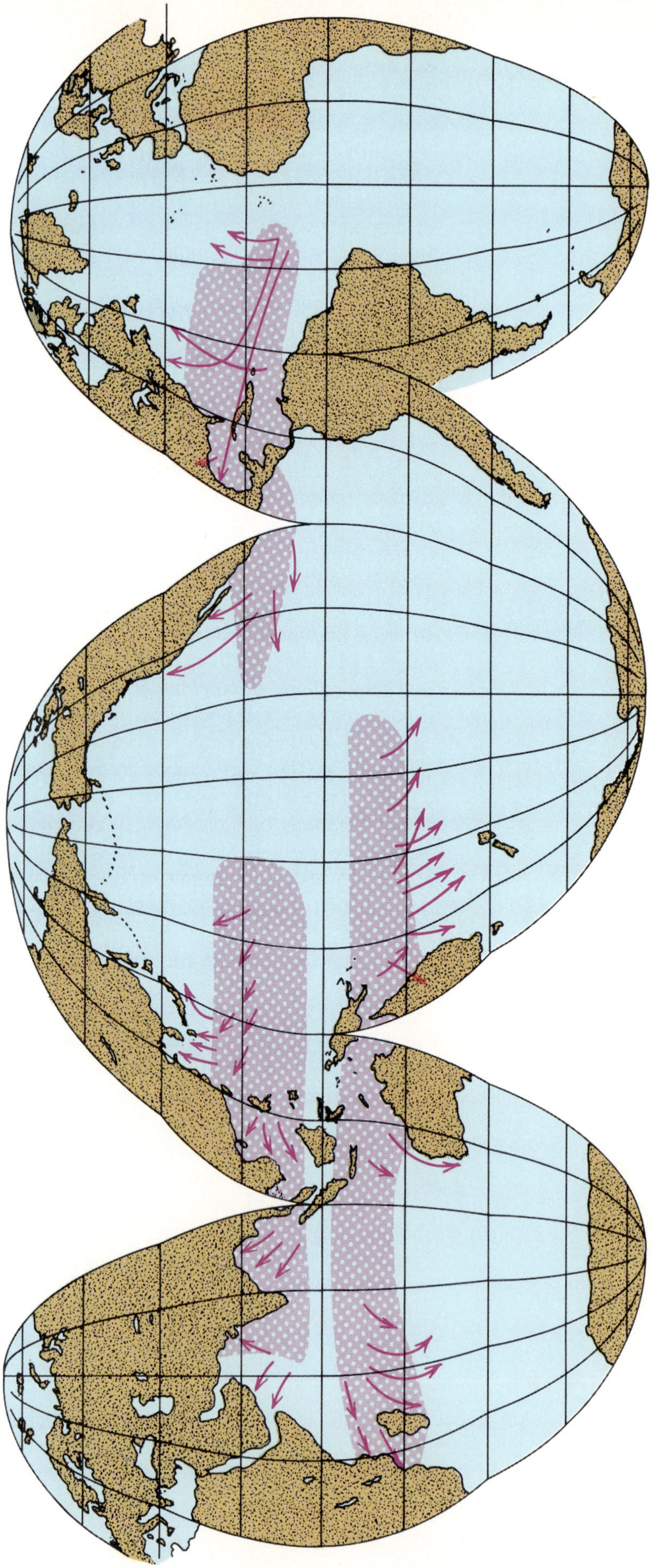

FIGURE 5-14
Areas of hurricane formation and paths of some major storms.

FIGURE 5-15
Upper-level winds in a hurricane near Hawaii observed by SEASAT, whose track is shown by the straight dashed line. Note the eye of the hurricane and the circular winds around it. (Courtesy NASA.)

SEASONAL WINDS

Earth's equator is tilted 23.5° to the plane of its orbit around the sun. Thus the point directly beneath the sun (where the incoming solar radiation is greatest) changes seasonally. At the summer solstice (June 21), the sun is farthest north, directly overhead at 23.5° N (Fig. 5-3). The Northern Hemisphere receives much more solar energy per unit area than does the Southern Hemisphere. At the winter solstice (December 21), the situation is reversed.

The presence of continents alters wind patterns. Seasonal changes in wind patterns are especially pronounced in the Northern Hemisphere because of the large amount of land there. During summer months, land warms faster than the ocean. As the air over the land warms, a low-pressure area forms. Air over the ocean is cooler and forms high-pressure areas. As a result, high-pressure cells form over the ocean areas in midlatitudes. This is quite different from the simple north-south bands predicted by our model of a water-covered Earth.

In winter, the situation reverses. Air over land cools markedly, forming large high-pressure areas. Over the ocean, air warms, forming low-pressure areas. Averaged over a year, air flows from high-pressure areas to low-pressure areas form the general patterns shown in Fig. 5-11.

In the Southern Hemisphere, there is relatively little land to disturb large-scale wind patterns. Therefore, there is little seasonal change in wind patterns there.

MONSOONS

Differences in heating and cooling of land and water produce large-scale seasonal changes in winds (**monsoons**) and local daily reversals of winds (land-sea breezes). These seasonal changes are most obvious in the northern Indian Ocean. In summer in Southeast Asia and India, the warm land causes air to rise over the continent (Fig. 5-16). These vertical air motions draw cooler, moisture-laden winds from the ocean to replace the rising air. As this onshore flow rises over the continent, it produces heavy monsoon rains. Summer monsoon rains are essential to rice crops in India and Southeast Asia.

FIGURE 5-16
Surface winds (a) in summer during the Southwest Monsoon and (b) during winter. Note that the intertropical convergence zone occurs near the equator in February and well north of the equator in August. Thus its average position over the year is about 5°N. This is sometimes called the *meteorological equator*.

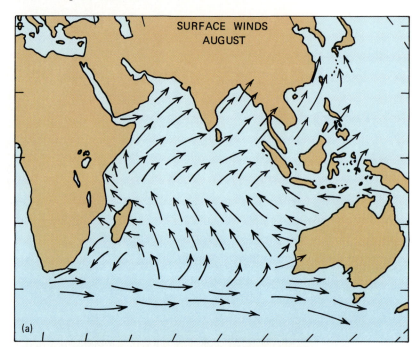

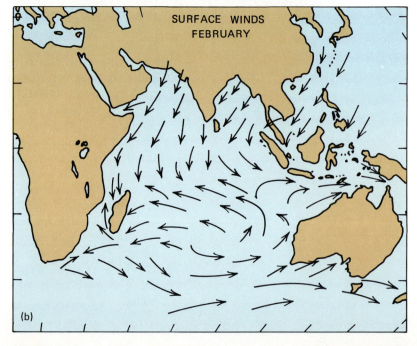

In winter, the winds reverse. Cold air from over the continent is drawn seaward by warm air rising over the ocean. This produces cool, dry weather on land, known as the winter monsoon. For centuries, sailors of the northern Indian Ocean depended on these seasonal wind reversals to carry them back and forth on their trading expeditions between India and Africa. (This was part of the *Silk Route* between Europe, China, and India.)

Winds drive surface ocean currents. Thus these wind reversals also cause major shifts in currents in the northern Indian Ocean, as we shall see in Chapter 7. Monsoonlike wind patterns are also common in other parts of the world. For instance, much of the summer rainfall over the U.S. Gulf Coast comes from monsoonlike atmospheric circulation.

LAND AND SEA BREEZES

Differential heating of land and ocean surface affects most coasts on a daily basis. During the day, the land warms more quickly than the ocean. Thus during the afternoon, air over the land rises, pulling in cooler air from the ocean. This is known as the **sea breeze** (Fig. 5-17).

At night, the land cools more rapidly than the sea. Now the wind blows from the land toward the ocean. This is the **land breeze,** which is strongest in the late night and early morning hours. The Florida peninsula in summer provides an example of this. Frequently there is a north-south line of vigorous thunderstorms in the middle of the state. Around Southeast Asia, fishing boats make use of these wind reversals to take them to sea in the morning for fishing and to bring them home in the afternoon.

FIGURE 5-17
(a) Sea breezes occur when air warmed by the land rises and is replaced by cool air from the ocean. (b) Land breezes occur after the land is cooled below the ocean surface temperatures. Warm air then rises over the ocean and is replaced by cool air from the land.

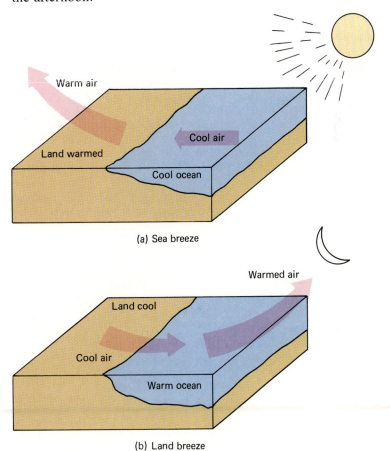

Warm air

Cool air

Land warmed

Cool ocean

(a) Sea breeze

Warmed air

Land cool

Cool air

Warm ocean

(b) Land breeze

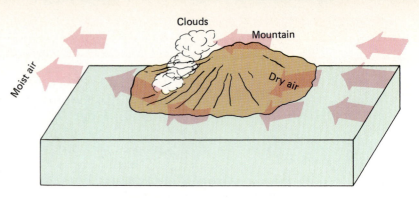

FIGURE 5-18
Moisture-laden air from the ocean rises over a mountainous island, causing clouds and rain on the windward side. As dry air flows down the mountain, it warms, causing the leeward side to be sunny, warm, and dry.

ISLAND EFFECT

Mountains affect near-surface winds. As winds blow across mountains, they are forced upward (Fig. 5-18). These rising air currents cool, and the lower temperatures cause water vapor to condense and fall as rain on the windward side of mountains. After the winds blow across a mountain, they blow down the other side. In the process, the air warms and can hold more water vapor. In areas of descending winds, there is little or no rain. Sides opposite the prevailing winds are often quite dry.

This condition occurs on mountainous islands, such as Oahu in the Hawaiian Islands. On the windward side, there is frequent rain and lush vegetation. On the side away from the winds (the leeward side), the weather is sunny, and there is little rain and sparse vegetation. This is called the *island effect*. Because of such effects, islands often have more rain than the surrounding ocean.

BOX 5-3

Antarctic Ozone Hole

Ozone forms a layer high in the stratosphere that absorbs ultraviolet light from the sun. This shields the earth's surface from the damaging radiation. Humans exposed to excessive ultraviolet develop cataracts on their eyes and skin cancers. Organisms can be damaged, including those living in the near-surface ocean waters.

This protective shield has been greatly reduced in the Antarctic during spring. There ozone levels in the stratosphere have been measured at 40% below their normal levels. The so-called "Ozone Hole" may be getting worse.

The cause of the ozone hole is still a matter for research. But the major points are generally agreed on. The culprits are synthetic chemicals called chlorofluorocarbons. (One brand is called Freon.) They are nearly chemically inert and are used in many applications—coolants in refrigerators and air conditioners, propellants in aerosol sprays, foaming agents, and cleaners in the electronics industry. When released they go into the troposphere where they remain for many decades.

Some of the chlorofluorocarbons dissolve in seawater. They are chemically inactive so they serve as tracers for movements of waters that have recently been at the ocean surface. In the ocean and in the troposphere, the chlorofluorocarbons pose no problem.

Some of the chlorofluorocarbons escape, however, into the stratosphere. There they absorb ultra-violet light which causes them to release their chlorine which actually does the damage. The chlorine acts as a catalyst (that is, it is not affected by the chemical reaction) which destroys the ozone by causing it to form ordinary oxygen. A single chlorine atom can destroy 100,000 ozone atoms before it is destroyed or returns to the troposphere.

Polar stratospheric clouds are involved in the process. In the extreme cold of the Antarctic stratosphere, the clouds remove nitrogen compounds from the stratosphere. These compounds normally inactivate chlorine, thereby reducing the ozone depletion. When nitrogen compounds are removed, the chlorine is free to destroy ozone molecules (Fig. B5-3-1).

While the process, as presently understood, may be unique to Antarctica, its effects might be felt elsewhere. There is some evidence that similar conditions exist in the Arctic stratosphere. This might lead to an Arctic ozone hole which would expose more humans but have less effect on the ocean. There is also the possibility that the process might affect the stratosphere in other areas which would increase the risk to the ocean.

FIGURE B5-3-1
The October 1983 minimum in the Antarctic stratospheric ozone layer is shown by the purple-violet color in the upper picture. Development of the ozone hole between 1979 and 1986 is shown by the set of diagrams below. Yellows, browns, and greens represent high ozone levels. (Courtesy NASA.)

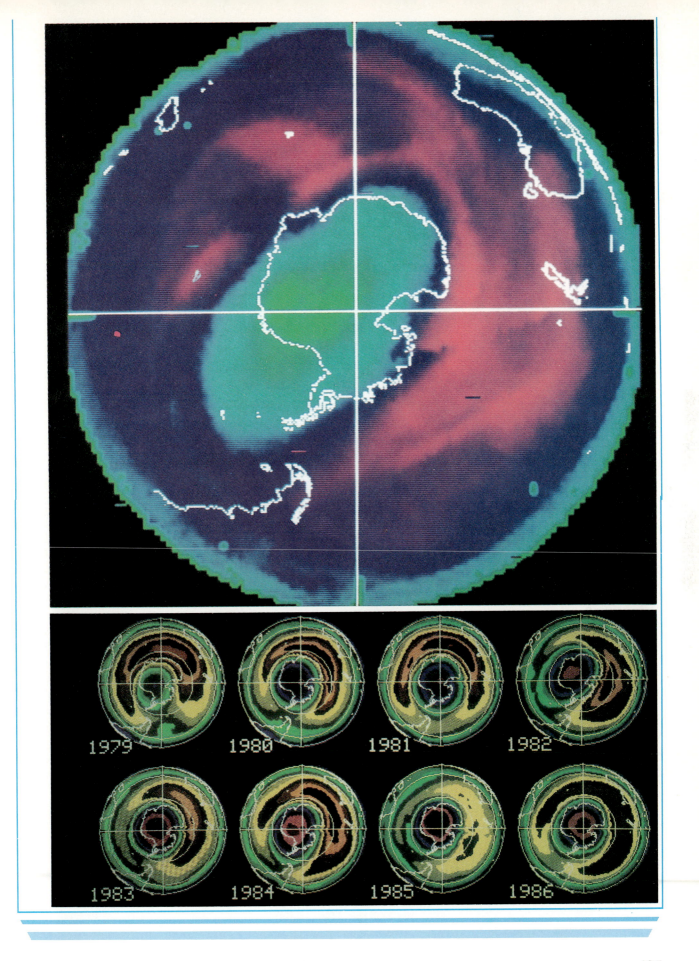

1979 1980 1981 1982
1983 1984 1985 1986

EL NIÑOS AND THE SOUTHERN OSCILLATION

Ocean currents, winds, and weather patterns are closely linked, especially along the equator in the Pacific. Let us see how they interact and how these processes can be observed and predicted.

Waters warmer than 28°C normally cover one-third to one-half of the tropical oceans, primarily in the eastern Indian and the western Pacific oceans. Cooler surface waters normally dominate the Pacific coast of South America and the equatorial ocean. Every 3 to 5 years, this pattern of ocean and atmospheric conditions changes dramatically. Warm waters occur along the equator and the west coast of South America. Wind patterns shift, and the effects are felt over much of the earth (Fig. 5-19).

Before satellite observations were available, the first indication of such changes came when unusually warm surface waters replaced the cold surface waters along the Peruvian coast. Since this usually occurred around Christmas, fishermen called it **El Niño**—Spanish for "the child."

To understand the processes causing El Niños, we begin with the equatorial currents. These move warm surface waters westward, where they accumulate in the western Pacific north of Indonesia. Apparently currents cannot transport and disperse these warm waters effectively throughout the ocean. Thus they accumulate in equatorial regions of the western sides of the basins. These gradually changing ocean conditions affect the overlying atmosphere.

Recall that heat and water vapor for the equatorial atmosphere come primarily from surface ocean waters. Adding heat and water vapor makes atmospheric gases less dense. This causes strong vertical circulation in the atmosphere over very warm waters. This process causes summer afternoon thunderstorms over Florida and the Gulf Coast. The very warm waters along the equator likewise cause Indonesia to have the most thunderstorms of any place on earth.

After a large pool of warm water has accumulated in the western Pacific, this somehow causes the trade winds to reverse. (This process is still not well understood.) Such wind reversals last for a few days to a few weeks and usually occur in November through April. These reversals are sometimes accompanied by twin "super typhoons"—unusually strong typhoons north and south of the equator.

These twin typhoons cause pulses of warm water to move eastward along the equator at speeds of a few tens of kilometers per day (similar to the speed of an eastern boundary current). (These pulses are a special kind of wave that can move only along the equator.) Such pulses moving eastward along the equator release the warm waters accumulated in the western Pacific. The pulses of warm waters moving across the equatorial Pacific encounter the South American coast about 9 months later. There, other similar waves form and move north and south along the coast. These pulses of warm waters cause the El Niño off Peru [Fig. 5-20(a)].

After an El Niño (which lasts 1 to 2 years), the pool of warm water in the western equatorial Pacific is depleted. Normal current patterns return to the equatorial ocean, and the atmosphere resumes its normal wind patterns. Upwelling resumes along the equator and off South America, and organisms begin their recovery.

Warm waters begin to accumulate in the western Pacific. The process begins again, leading to yet another El Niño in 3 to 5 (occasionally 7) years. Using satellite observations and sea-level observations on central Pacific islands, scientists now monitor buildup of warm waters

FIGURE 5-19
The 1982–83 El Niño was the strongest event of the century. It altered sea-surface temperatures from their normal pattern of cold upwelled waters (shown in green) near South America and along the equator. Instead the waters were warm (shown in reds and oranges) throughout the tropics. (Courtesy NASA.)

FIGURE 5-20
(a) In May 1987, during the 1986–87 El Niño, sea-surface temperatures were high along the west coast of South America and along the equator in the eastern Pacific. (Yellow indicates surface-water temperatures 2 to 3 °C above normal.)

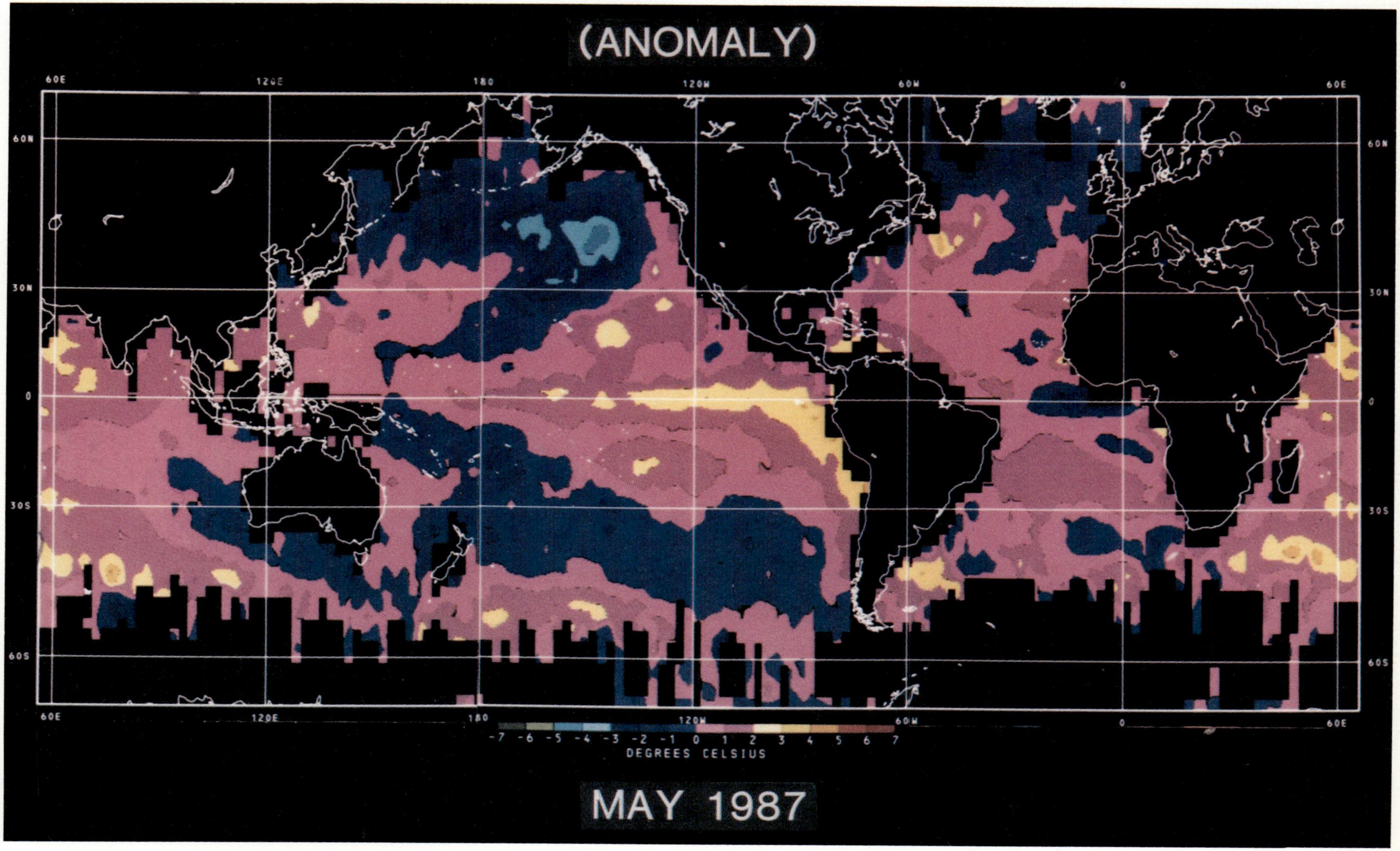
(ANOMALY)
60E
120E
180
120W
60W
0
60N
30N
0
30S
60S
-7 -6 -5 -4 -3 -2 -1 0 1 2 3 4 5 6 7
DEGREES CELSIUS
MAY 1987

FIGURE 5-20

(b) In May 1988, after the end of the El Niño, sea-surface temperatures were unusually cold along the equator. (Light blue colors indicate water temperatures 3 to 4°C below normal.) These cold waters—called La Niña by some to distinguish it from El Niño—caused winds to be displaced further north than usual in the Northern Hemisphere. This displaced the jet stream and may have contributed to the 1988 drought in North America. (Courtesy A. Strong, NASA.)

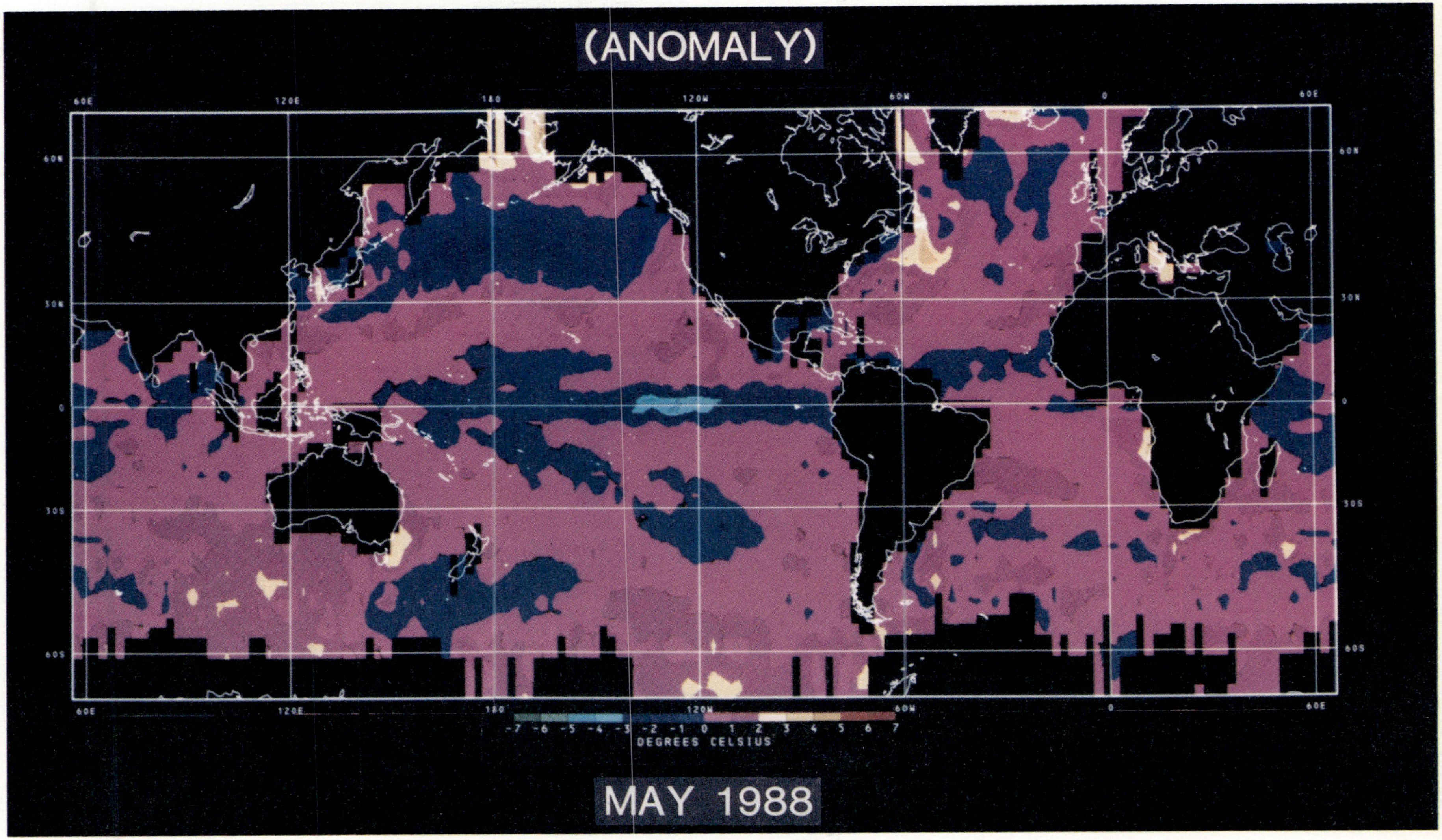

EL NIÑOS AND THE SOUTHERN OSCILLATION

129

and observe the typhoons that initiate El Niños. This permits predictions of El Niño up to 9 months before it affects the coast of South America and the rest of the South Pacific.

Movements of these large pools of warm water along the equator affect the atmosphere over the entire South Pacific. The atmospheric high-pressure zone usually located off South America weakens, as does the low-pressure zone located over Indonesia. In short, the atmospheric circulation is changed across the South Pacific and along the coast of the Americas. These changes in the atmosphere are called the Southern Oscillation.

Changed weather patterns are felt worldwide, especially along the equator and in the Southern Hemisphere. Areas of heavy rainfall shift from the Indonesian area eastward to the normally dry islands of the central equatorial Pacific. The results can be locally catastrophic. Sea level rises because the warmer (less dense) surface layer is thicker. Higher sea level combined with heavy rains floods low-lying islands. There are fewer hurricanes during El Niños; therefore, those areas which depend on hurricanes to supply summer rain experience droughts.

The 1982–1983 El Niño was especially severe. Droughts in Australia caused billions of dollars in damages through massive losses of crops and livestock. Heavy rains and storms caused flooding and destroyed highways and bridges in normally arid coastal regions of South America.

El Niños also disrupt biological processes. Thick layers of warm water disrupt the food production that supports abundant growth of marine organisms, especially fish (discussed in Chapter 11). This reduces fish catches by seabirds and fishermen, with catastrophic effects. For example, a strong El Niño, combined with overfishing, destroyed the Peruvian anchoveta fishery in 1972.

Effects on organisms are equally dramatic on the equatorial islands. Seabird nests on the ground are flooded and destroyed by the higher sea level and by heavy rains. When they cannot find food, adults have to abandon their nests and chicks and flee to avoid starvation. Recovery of populations of organisms takes years, especially after an intense El Niño.

After an El Niño, the currents return to their former condition. But they can "overshoot," resulting in unusually cold waters along the Peruvian coast and in the eastern equatorial Pacific. This happened in 1988, following the mild 1986–87 El Niño [Fig. 5-20(b)].

The unusually cold water along the equator caused the intertropical convergence zone to shift northward. This in turn displaced the jet stream, disrupting weather patterns in the United States and Canada. This set of circumstances contributed to the unusually severe drought of the summer of 1988.

Somewhat similar processes occur in the Atlantic. They affect rainfall and droughts over Africa and South America.

SUMMARY

The atmosphere, like the ocean, came from volcanic action. It consists of a mixture of gases, primarily nitrogen and oxygen, with traces of dust and water vapor. The atmosphere is density stratified—densest near the surface. In the troposphere, the lowest part of the atmosphere, there are strong vertical motions. Most of the weather occurs here. The atmosphere is warmed at the earth's surface and cooled at the top. There is little vertical motion in the stratosphere. Materials injected there remain for a long time.

The atmospheric heat budget is used to describe sources and sinks of energy and matter. The earth's heat comes from the sun, about 0.5 calorie per square centimeter per minute. This amount varies seasonally because of the earth's inclined axis of rotation. Most of the heat is received in the equatorial zone, least near the poles, where heat is lost. Latent heat of evaporation of water dominates heat transport processes.

On a nonrotating, water-covered Earth, the atmosphere would have a simple two-celled circulation pattern. Warm air would rise at the equator and sink at the poles. On a rotating Earth, winds are deflected by the rotation—to the right in the Northern Hemisphere, to the left in the Southern. The effect is greatest near the poles and absent along the equator. Winds and currents tend to move in circles unless acted on by an outside force.

Because of the Coriolis effect, the earth has a complicated six-celled circulation pattern—three in each hemisphere. The simple circulation pattern persists near the equator, with warm air rising there and sinking near 30° north and south. A polar front separates warm and cold air at the polar front. A secondary circulation occurs along the equator.

Patterns of prevailing winds shift seasonally and are deflected by the continents. Winds in the Southern Hemisphere are least affected by the presence of land. The intertropical convergence zone shifts northward in northern summer.

Monsoons are seasonal wind shifts caused by warm air rising over continents in summer. The rising air is replaced by cool, moisture-laden air from the ocean, which produces heavy rains on land. In winter, the circulation is reversed, causing cool, dry weather on land. These seasonal wind shifts cause current patterns to change.

Extratropical cyclones form on the polar front when a low-pressure system is cut off. These storms move generally westward, relatively slowly. Hurricanes are intense, fast-moving cyclones. They form over water warmer than about 28°C in all tropical oceans except the eastern Pacific and the western South Atlantic. Storm tracks are controlled by upper-air winds. The storms are powered by the condensation and release of heat from water vapor. Hurricanes lose power over land or over cold waters.

Near islands, winds are forced to rise as they blow over mountains. This causes local precipitation on the windward side of the island and dry weather on the opposite side.

Ocean waters affect weather over large areas by steering storm tracks. Pools of unusually warm or cool water in the North Pacific affect weather over large areas of North America. El Niños—occurrences of warm surface waters off the Pacific coast of South America—are caused by shifts in winds and movements of warm equatorial waters. They also affect weather patterns over large areas, causing droughts over North America and Africa.

STUDY QUESTIONS

1. Sketch a vertical cross section showing the major features of the atmosphere.
2. Describe how budgets are used to study atmospheric processes.
3. Discuss the atmospheric heat budget.
4. What causes seasonal changes in atmospheric temperature?
5. Describe the greenhouse effect and what causes it.
6. Describe atmospheric circulation on a water-covered, nonrotating Earth.
7. Sketch the pattern of prevailing winds on a rotating Earth.
8. Describe how the Coriolis effect deflects winds.
9. Describe the causes of the monsoon winds; of land-sea breezes.
10. Where do hurricanes form? What controls their paths?
11. How does the ocean influence weather on land?
12. Describe the causes of El Niños.

SELECTED REFERENCES

EAGLEMAN, J. R. 1985. *Meteorology: The Atmosphere in Action*, 2d ed. Belmont, Calif.: Wadsworth Publishing Company. 394 pp. Elementary.

IMBRIE, J., AND K. P. IMBRIE. 1979. *Ice Ages: Solving the Mystery*. Short Hills, N.J.: Enslow Publisher. 224 pp. Lively account of the causes of ice ages.

LOVELOCK, J. E. 1979. *Gaia: A New Look at Life on Earth*. Oxford: Oxford University Press. 157 pp.

MILLER, A., AND R. A. ANTHES. 1985. *Meteorology*, 5th ed. Columbus, Ohio: Charles E. Merrill Publishing Company. Elementary, nontechnical.

RIEHL, H. 1972. *Introduction to the Atmosphere*. New York: McGraw-Hill. 516 pp. Elementary.

SIMPSON, R. H., AND H. RIEHL. 1981. *The Hurricane and its Impact*. Baton Rouge, La.: Louisiana State University Press. 398 pp.

WEISBERG, J. S. 1976. *Meteorology: The Earth and its Weather*. Boston: Houghton-Mifflin. 241 pp.

WELLS, N. 1986. *The Atmosphere and Ocean: A Physical Introduction*. London: Taylor & Francis. 347 pp.

An iceberg with sea ice in the foreground. (Courtesy U.S. Navy/Photo Researchers.)

Ocean and Climate

OBJECTIVES —————————————————————

1. To understand Earth's heat budget and the ocean's role in transporting heat;

2. To understand the relationship between Earth's heat and water budgets;

3. To grasp the vertical structure of the ocean and its importance for oceanic processes;

4. To understand how and where water masses form;

5. To understand how sea ice forms and its importance.

*I*n this chapter, we consider the ocean without currents. In other words, we examine the effects of ocean processes averaged over very long times—essentially a static ocean. And we examine how the ocean affects **climate**—weather averaged over a long time, typically 30 years.

Two different approaches are used to describe the long-term behavior of ocean waters. First is the **budget** (which we also used in studying the atmosphere). Heat and water budgets are constructed to show major sources and sinks. Regarding the water budget, it is important to remember that the amount of water on the earth's surface is fixed. New sources need not be considered, only redistribution of water on the earth (except over hundreds of millions of years). The second approach determines connections between the ocean surface and subsurface waters. This procedure includes studying *distributions of water temperature and salinity*. We use both approaches in this chapter.

Topics discussed in this chapter include:

Interactions of light and the heating of the ocean;
Oceanic budgets of heat, water, and salt;
Characteristics of depth zones in the ocean;
Water masses, how they form, move, and are studied;
Sea ice, and ocean effects on climate.

In Chapter 7, we discuss the complexities introduced by ocean currents.

LIGHT IN SEAWATER

The sun's radiation striking the earth's surface supplies energy to heat the ocean surface and to warm the lower atmosphere, as we have already seen. Part of this incoming solar radiation is in the visible part of the spectrum. After passing into the surface of the ocean, most of this energy is converted into heat, either raising water temperatures or evaporating water.

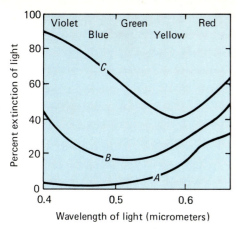

FIGURE 6-1
Spectrum of visible light at 1 meter below the surface in seawater from different sources. (a) Pure open-ocean water; (b) clearest coastal-ocean water; (c) turbid coastal-ocean waters. All wavelengths are transmitted in open-ocean waters. Near coasts, the presence of dissolved pigments strongly inhibits light transmission in the violet and blue ranges. (After G. Dietrich and others. 1980. *General Oceanography: An Introduction*. New York: Wiley-Interscience.)

Energy from the sun is filtered as it passes through the atmosphere and is again filtered in surface ocean waters. Within the first 10 centimeters of even pure water, virtually all the infrared portion of the spectrum is absorbed and changed into heat. Within the first meter of seawater, about 60% of the entering radiation is absorbed, and about 80% is absorbed in the first 10 meters. Only about 1% remains at 140 meters in the clearest subtropical ocean waters.

In coastal waters, abundant marine organisms, suspended sediment particles, and dissolved organic substances absorb light at even shallower depths. Near Cape Cod, Massachusetts, for instance, only 1% of the surface light commonly penetrates to 16 meters. In such waters, the maximum transparency shifts from the bluish region typical of clear oceanic waters to longer wavelengths, as shown in Fig. 6-1. In turbid coastal waters, peak transparency is in the yellow range. In highly polluted waters, absorption of all light takes place within a few centimeters of the water surface.

Far from the coast, ocean water often has a deep luminous blue color quite unlike the greenish or brownish colors common to coastal waters. The deep blue color indicates an absence of particles. In these areas, the color of the water results from **scattering** of light rays within the water. A similar type of scattering is responsible for the blue color of the clean atmosphere.

The amount of light reflected from the ocean is controlled by the state of the sea surface and the angle at which the sun's rays strike the water. When the sun is directly overhead, only about 2% of incoming radiation is reflected; the remainder enters the water. When the sun is near the horizon, nearly all incoming radiation is reflected. Waves on the sea surface generally increase the amount of light reflected by as much as 50%, but when the sun is very near the horizon, the presence of waves decreases the amount of sunlight reflected at the water surface.

SEA SURFACE TEMPERATURES

In an ocean without currents, bands of equal surface temperatures would run east-west. Water temperatures would be highest along the equator, because of the warming of the earth there, and become cooler toward the poles. These relationships can be seen in maps of ocean temperatures (Fig. 6-2, p. 138). There are, however, complications due to currents transporting cold waters toward the equator and warm waters toward the poles. Here we will deal only with general temperature distributions. We discuss currents in the next chapter.

Ocean temperatures change with the seasons, due to variations in the amount of incoming solar radiation (Fig. 6-3, p. 138). In the equatorial ocean, water and air temperatures change little seasonally. In the high latitudes, the small changes in surface-water temperatures are due to the year-round presence of ice.

Differences in ocean surface temperatures between February and August are greatest in the midlatitudes. By contrast, on land the largest temperature differences occur in the high latitudes where ice covering the ocean surface prevents it from moderating air temperatures as it does elsewhere (shown in Fig. 6-3).

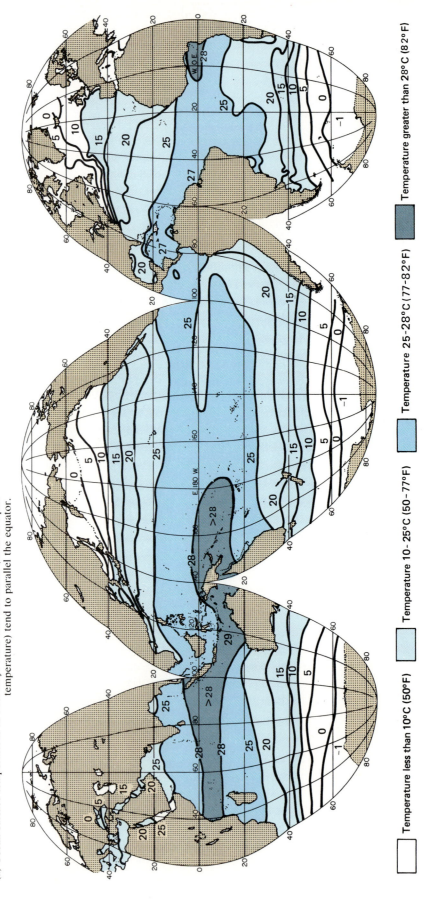

FIGURE 6-2

(a) Ocean surface temperatures in February. Note that isotherms (lines of equal temperature) tend to parallel the equator.

Temperature less than 10°C (50°F)

Temperature 10–25°C (50–77°F)

Temperature 25–28°C (77–82°F)

Temperature greater than 28°C (82°F)

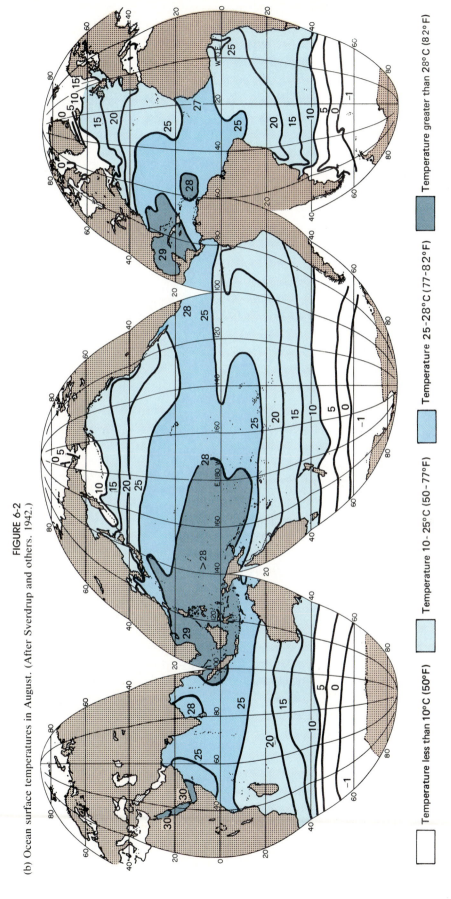

FIGURE 6-2

(b) Ocean surface temperatures in August. (After Sverdrup and others. 1942.)

Temperature less than 10°C (50°F)

Temperature 10-25°C (50-77°F)

Temperature 25-28°C (77-82°F)

Temperature greater than 28°C (82°F)

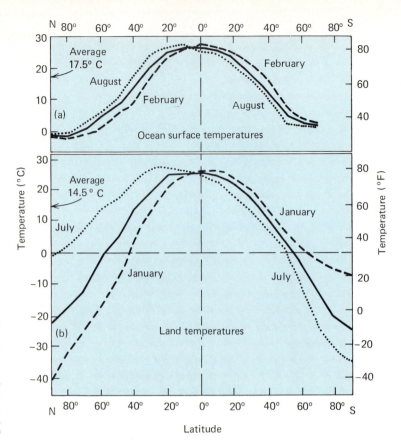

FIGURE 6-3
Average and range of temperatures at the Earth's surface. (a) Average temperature ranges for the ocean surface. Note that the greatest seasonal change in temperature occurs in the midlatitudes, around 40°N and 40°S. (b) Average temperature ranges for land or ice-covered areas. Note that the extreme temperature ranges occur in the polar regions, in contrast to the oceans. (Data from *Smithsonian Physical Tables. 1964.*)

SEA SURFACE SALINITIES

Sea surface salinities are distributed quite differently than water temperatures. Highest salinity waters occur in the centers of the ocean basins (Fig. 6-4). Lowest salinities occur in high latitudes and near continents where the ocean receives fresh water from melting ice or from river discharges.

Evaporation from the sea surface is greatest in subtropical areas. The highest surface-water salinities occur in the centers of ocean basins, where there is no dilution by river discharges. Also, the sides of ocean basins where dry winds blow off the continents are areas of high evaporation. River outflows often dilute surface waters, obscuring effects of high evaporation rates, as in the Gulf of Mexico. As shown in Fig. 6-5, p. 140, the relatively low salinities of the high northern latitudes are due, in part, to low rates of evaporation there. Fresh water from melting sea ice also dilutes surface waters during summer.

Salinity differences among the ocean basins are caused by variations in evaporation and precipitation. Evaporation from the sea surface is greatest in subtropical areas. The highest surface-water salinities occur in the subtropics in the centers of ocean basins, where there is no dilution by river discharges.

The relatively high salinities in the Atlantic result from interactions of prevailing winds and the mountain ranges of North and South America. Winds blowing across the western mountain ranges lose water vapor on the Pacific sides of the mountains. Thus winds blowing toward the Atlantic are relatively dry. In the equatorial region, winds

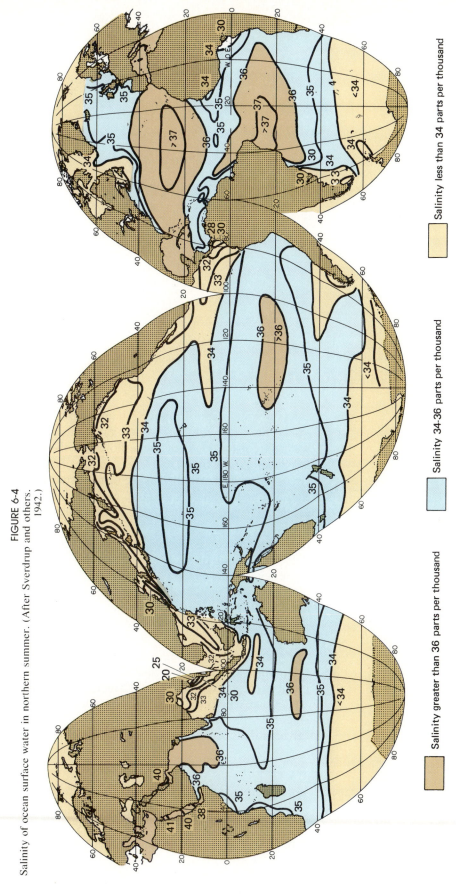

FIGURE 6-4

Salinity of ocean surface water in northern summer. (After Sverdrup and others. 1942.)

Salinity less than 34 parts per thousand

Salinity 34-36 parts per thousand

Salinity greater than 36 parts per thousand

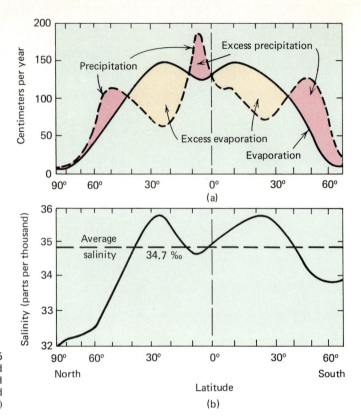

FIGURE 6-5
Relationships between oceanic evaporation and precipitation at various latitudes (a) and surface-water salinities (b). (After Wüst and others. 1954.)

carrying water vapor from the Atlantic blow across the low-lying Central America region to carry water vapor into the Pacific. Finally, saline waters from the Mediterranean flow into the North Atlantic through the Strait of Gibraltar.

The highest salinities in the ocean occur in the nearly landlocked Red Sea and Arabian Gulf (Fig. 6-4). There the loss of water due to evaporation is especially large, and there are no river outflows to compensate.

Precipitation is highest near the equator. This region, called the doldrums (about 10°N), lies between the northern and southern trade wind belts. Heavy rainfall, resulting from the condensation of warm, moist air rising in the equatorial atmospheric circulation, dilutes equatorial surface waters. Low salinities of the marginal seas of Southeast Asia are due to large river discharges there, especially during monsoons.

In summary, highest salinities occur in areas of excess evaporation, either in subtropical central regions of the ocean basins or in the landlocked seas of arid regions. The lowest salinities occur where precipitation exceeds evaporation, primarily coastal or equatorial regions (Fig. 6-5). Fresh water comes from rivers and from melting sea ice in the high latitudes.

WATER AND HEAT BUDGETS

Water and heat budgets are intimately related, as we saw when discussing the atmosphere in Chapter 5. Evaporation of water from the ocean surface removes heat. This heats the atmosphere when water vapor rises, cools, condenses, and falls back to Earth as rain or snow.

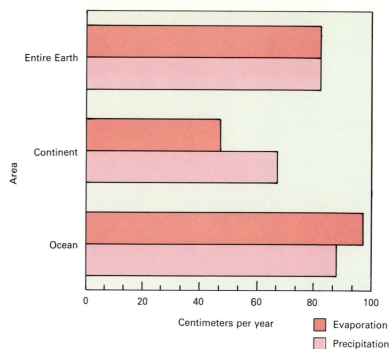

FIGURE 6-6
Water budget of the Earth. (After A. Defant. 1961. Vol. 1, *Pergamon Press: Physical Oceanography, p. 235.*)

(This is the latent heat we discussed earlier.) In this section we examine water and heat budgets and show how they interact. We know that the amount of water on the earth's surface has remained constant for billions of years. So we can make a budget to keep track of where the water goes, how it moves, and where it is stored.

The ocean is the primary reservoir for water on the earth's surface. The small quantities of water in rivers, lakes, and the atmosphere are in transit back to the ocean. Groundwater also finds its way back to the sea, but it takes much longer.

One convenient way to visualize the earth's water budget (shown in Fig. 6-6) is to consider the thickness of the layer of water evaporated each year from the ocean surface. About 97 centimeters of water is evaporated each year. About 88 centimeters, or 91%, returns as rain on the ocean. The remainder falls on land (along with rain from local evaporation) and eventually makes it way through rivers to return to the sea. Still another way is to express the total amount of water in thousands of cubic kilometers, as in Fig. 6-7

A water budget for a region is written as:

Evaporation = precipitation + runoff + current transport

If the region under consideration is large (the North Atlantic, for example), the amount of river runoff compared with the amount of seawater is small. And the currents in most regions simply move water around but do not contribute much fresh water. Thus we can ignore both terms. So the water budget for a large ocean region simplifies to:

Evaporation = precipitation

We can make a heat budget because we know that the Earth loses as much heat as it gains each year. Heat budgets are slightly more

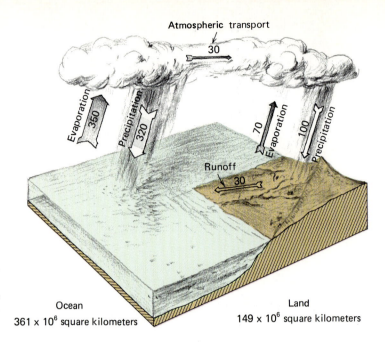

Atmospheric transport
30

Evaporation 350

Precipitation 320

Evaporation 70

Precipitation 100

Runoff 30

Ocean
361 x 10⁶ square kilometers

Land
149 x 10⁶ square kilometers

FIGURE 6-7
Schematic representation of the hydrologic cycle, showing the total amounts of liquid water involved in each, in thousands of cubic kilometers.

complicated because of interactions between the ocean and atmosphere. In its simplest form, the heat budget is:

$$\text{Heat gained} = \text{heat lost}$$

But there are more terms which must be considered in the full budget, which is written as follows:

$$\begin{array}{l}\text{Heat from} \\ \text{solar radiation}\end{array} = \begin{cases}\text{evaporation} + \text{radiation} \\ + \text{heating atmosphere} + \text{current transport}\end{cases}$$

If the ocean region is large (say an entire ocean basin), the amount of heat gained or lost by current transport (Fig. 6-8) can be ignored. Thus the heat budget says that the ocean loses heat through evaporation, radiation back to space, and by directly heating the atmosphere.

Areas of high surface salinity supply water vapor to the atmosphere, which is transported by winds to equatorial or subpolar regions. There water vapor condenses, forming either rain or snow and releasing heat to the atmosphere. Thus the poleward transport of heat and water is reflected in low surface-water salinities in high latitudes. Regional salinity differences of surface ocean waters are manifestations of the planetary transport of heat and water in the ocean and atmosphere.

In summary, heat is transported by the ocean and atmosphere. In the low latitudes, the ocean dominates heat transport. The atmosphere dominates heat transport in the mid- and high latitudes of the Northern Hemisphere.

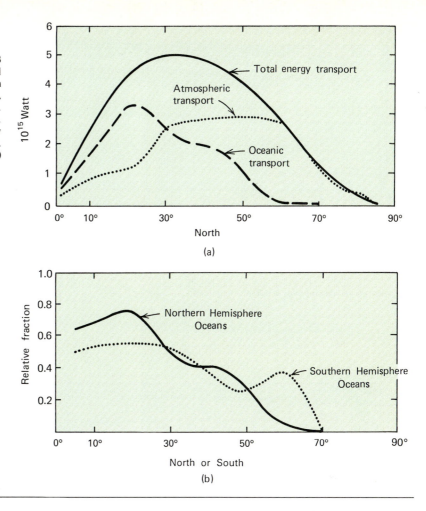

FIGURE 6-8
Ocean currents carry more than half the total energy transported near the equator (between about 30°N and 30°S). At higher latitudes, atmospheric transport exceeds oceanic transport. (After T. H. Von der Haar and A. H. Oort. 1973. New estimate of annual poleward energy transport by Northern Hemispheric oceans. *J. Phys. Oceanogr.* 3:169–72.)

DEPTH ZONES

Now we consider the vertical structure of the ocean. There are three principal depth zones: surface, pycnocline, and deep zones, shown schematically in Fig. 6-9, p. 146.

The **surface zone** is in contact with the atmosphere. It changes seasonally because of variations in precipitation, evaporation, cooling, and heating. The surface zone contains the warmest (> 10°C) and least dense waters in the ocean (Fig. 6-10, p. 146). (Average surface-water temperature is 17.5°C.)

The surface zone is 100 to 500 meters thick and contains about 2% of the ocean volume. Near-surface waters usually have near-neutral stability and are therefore well mixed by winds, waves, and cooling or heating of the surface. For this reason, the surface zone is also called the **mixed layer.**

The **pycnocline zone** is where water density changes markedly with depth (Fig. 6-9). The top of the pycnocline (base of the surface zone) corresponds approximately to the 10°C contour in Fig. 6-11, p. 145, and its bottom can be taken as the 4°C contour. Thus the pycnocline is typically about a kilometer thick.

Changes in either temperature or salinity (or in both) can cause the marked change in density of the pycnocline. Where changes in temperature dominate, it is called a **thermocline** (Fig. 6-9). Where salinity dominates, it is known as a **halocline.** Thermoclines are important in the open ocean, where salinity changes little. Haloclines are more important in coastal ocean areas, where salinity changes dominate.

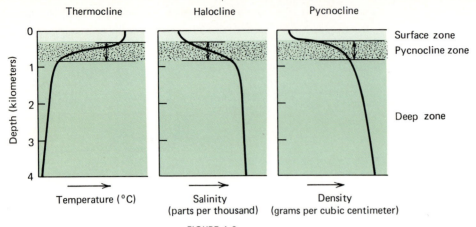

FIGURE 6-9
Variation in temperature with depth creates the thermocline, while comparable changes in salinity create the halocline. The pycnocline is a result of either one occurring independently or of both occurring together.

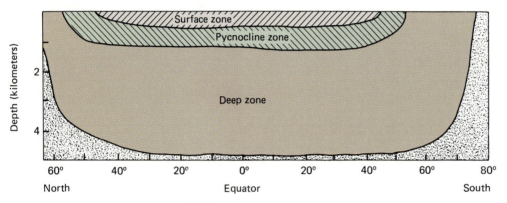

FIGURE 6-10
Schematic representation of the density structure of the ocean. Note that the pycnocline zone isolates the deep ocean from contact with the atmosphere in the equatorial and midlatitude areas. Only near the North and South Poles does the deep ocean have any contact with the atmosphere.

The pycnocline is a zone of marked stability. Thus the vertical movements of the surface zone and seasonal changes in temperature or salinity do not penetrate the pycnocline. Waters in and below the pycnocline move primarily along density surfaces. Except in the high latitudes, there is little or no vertical movement.

Below the pycnocline is the **deep zone,** which contains about 80% of the ocean's volume. Except in the high latitudes (Fig. 6-10), the deep zone is separated from the atmosphere by the pycnocline. This accounts for the relatively low average temperature for ocean waters—3.5°C. Also, temperature and salinity of deep-ocean waters (Fig. 6-11) are unaffected by surface processes, so temperature and salinity act like conservative properties.

Distributions of temperature (Fig. 6-12) and salinity (Fig. 6-13) in all three ocean basins are rather similar. They show the deep zone exposed in the high latitudes, except in the North Pacific, where the surface zone extends up the northern continental margin. The sections also show marked differences from our schematic ocean (Fig. 6-10), especially in the North Atlantic. These differences are caused by deep-ocean currents, which we discuss in the next chapter.

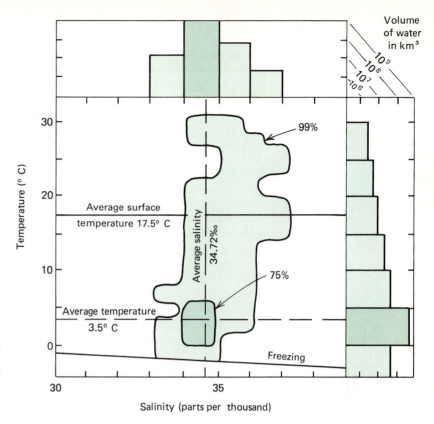

FIGURE 6-11
Temperature and salinity of 99% of the ocean water are represented by points within the stippled area enclosed by the 99% contour. The shaded area represents the range of temperature and salinity of 75% of the water in the ocean.

Sound in the Ocean

Seawater is essentially transparent to sound. If seawater were as transparent to light as it is to sound, we could see the bottom of the ocean. Sound is used in *sonar* (sound navigation and ranging), the underwater equivalent of radar to detect submarines. Scientists use sound to determine ocean depths, to locate organisms, and to communicate. Animals use sound even more—to communicate, to locate food, and to detect predators. Sound in the ocean is as important as light is in the atmosphere.

Sound is a form of mechanical energy. It consists of the regular alternation of pressure or stress in an elastic medium, in this case, water. The behavior of sound waves in the ocean is controlled by temperature, pressure, and salinity. The speed of sound in the ocean is about 1480 meters per second. Changes in temperature have a marked effect on sound speed.

Increasing temperature increases the speed of sound. Over most of the ocean, changes in temperature control the behavior of sound. Sound speed also increases with depth due to increased pressures. Salinity has little effect. (Sound is used in **acoustic tomography** to monitor temperature changes in subsurface waters over very large areas.)

Sound speeds are thus relatively high in the surface zone because of warm waters and high at depth because of higher pressures. Consequently, there is a minimum in sound speed around 1 kilometer in most ocean basins. Sound waves are bent toward regions of lower sound speed. Thus this zone of reduced sound speed functions as a wave guide. This channel is extremely efficient in transmitting sound. It was once proposed as an emergency signalling channel.

In brief, the open ocean has a three-layer structure: surface, pycnocline, and deep zones. The surface zone responds quickly to changes in the overlying atmosphere. The pycnocline inhibits exchanges between the atmosphere and the deep. The deep zone is exposed to the atmosphere only in the high latitudes, which causes its waters to be cold. Vertical distributions of temperature and salinity are similar in all three ocean basins. In the North Atlantic, the surface layer and pycnocline are thicker than average.

FIGURE 6-12
Vertical distribution of temperature in the three ocean basins. Vertical exaggeration is approximately one thousand times. (After Dietrich and others. 1980.)

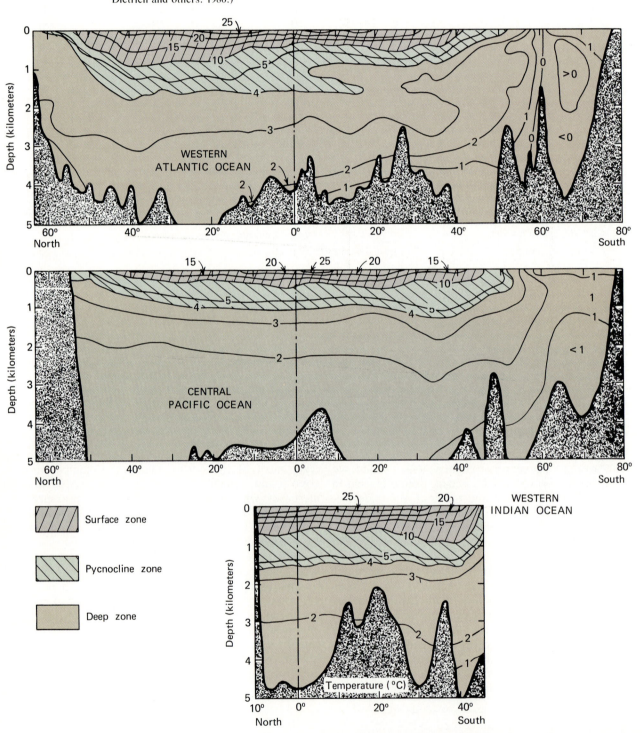

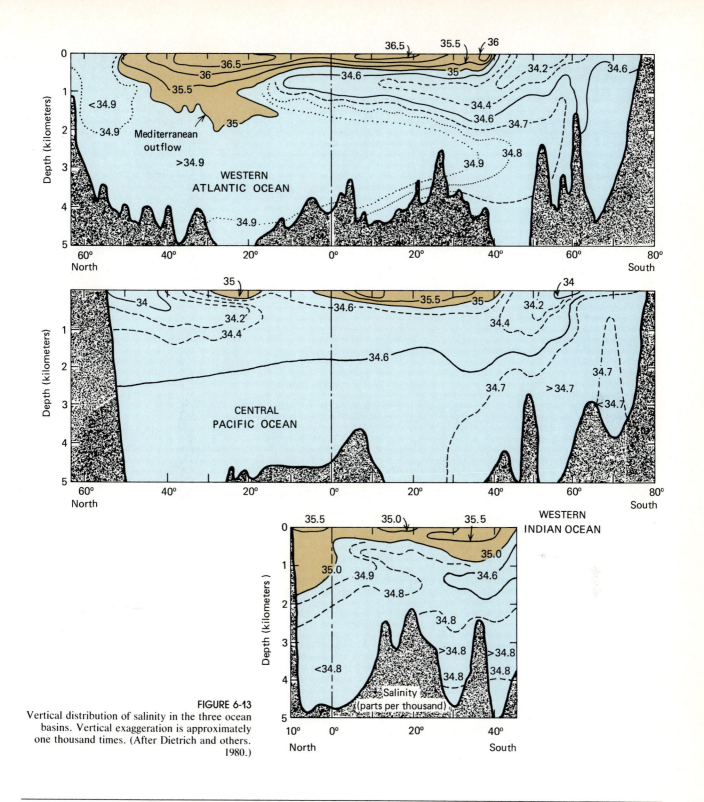

FIGURE 6-13
Vertical distribution of salinity in the three ocean basins. Vertical exaggeration is approximately one thousand times. (After Dietrich and others. 1980.)

TEMPERATURE-SALINITY RELATIONSHIPS

We have seen that temperature and salinity vary markedly at the ocean surface and with depth. Yet temperature and salinity fall within limits determined by various physical processes. For example, we have already discussed the effect of ice on the temperature of seawater. Indeed, we find that the lowest water temperatures are determined by the freezing point of seawater (Fig. 6-11). The highest temperatures are controlled by the evaporation of water, which removes so much heat

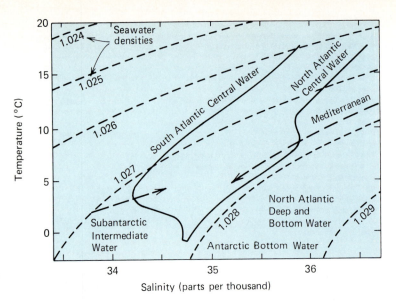

FIGURE 6-14
Temperature-salinity curves for major water masses in the Atlantic Ocean. The South and North Atlantic central water masses form in the subtropical gyres (indicated by the relatively high temperatures and salinities). The Mediterranean waters are recognizable by their high salinities. The other water masses form in the high latitudes, as shown by their relatively low temperatures.

from the surface that it rarely exceeds 30°C (as we discussed in Chapter 5) and then in very restricted areas.

Salinity also shows a relatively limited range, between 33 and 37‰. Waters with higher or lower salinities usually occur in small quantities and in isolated basins. Thus they have little effect on the ocean as a whole.

Below the surface zone, temperatures and salinities tend to vary together in characteristic ways. These reflect the processes occurring in the region where the waters in a particular layer formed. Since both temperature and salinity behave like conservative properties in the deep ocean, these characteristic relationships can be used to identify water masses (having fixed temperatures and salinities). The curve of the temperatures and salinities, called a *T-S curve* (shown in Fig. 6-14), can be used to identify that water mass.

The relatively high salinities of waters from the Mediterranean can be used to identify them as they flow through the Strait of Gibraltar (Fig. 6-15) and throughout the North Atlantic (Fig. 6-16). (The *T-S* curve for Mediterranean waters is shown in Fig. 6-14 in the middle of the right-hand side.)

T-S curves can also be used to determine how water masses mix. To see how this works, consider two water masses [Fig. 6-17(a)]. As

FIGURE 6-15
Waters from the Mediterranean Sea flow out through the Strait of Gibraltar into the North Atlantic. There it forms a thin layer about 1000 meters below the sea surface. (After Judson, Kauffman, and Leet, 1987.)

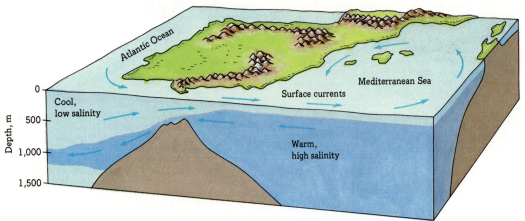

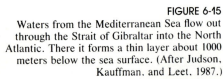

FIGURE 6-16
Waters flowing out from the Mediterranean spread out as a layer a few hundred meters thick that can be recognized (salinity greater than 35‰) all the way across the North Atlantic at depths around 1000 meters. (After L. V. Worthington and W. R. Wright. North Atlantic Atlas. *Woods Hole Oceanographic Institution Atlas Series, vol. 4.*)

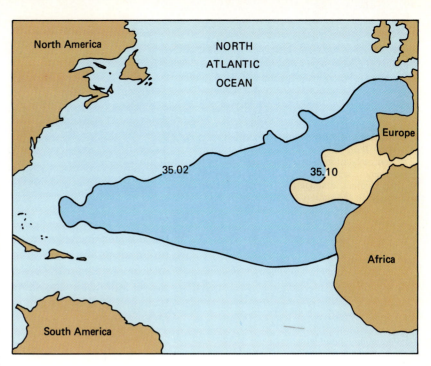

FIGURE 6-17
Temperature-salinity relationships resulting from mixing of water masses. (a) Water mass 1 (depth 0 to 300 meters; $T = 10°C$, $S = 34.8‰$) mixes with water mass 2 (depth 300 to 600 meters, $T = 2°C$, $S = 34.2‰$). Initially, the boundary between the two water masses is sharp and represented by marked changes in temperature and salinity with depth. After mixing, the boundary is more diffuse, and the changes in temperature and salinity occur over a depth range of nearly 600 meters. On the $T–S$ diagram, mixing of these two water masses is shown by the straight line connecting the $T–S$ points characteristic of the two original water masses. (b) Mixing of three water masses can be shown in a similar manner. Water mass 3 (depth 600 to 1500 meters, $T = 2°C$, $S = 34.8‰$) mixes with water mass 2 but not with water mass 1. The sharp boundaries between water masses are gradually obliterated. On the $T–S$ diagram, the mixing is shown by two straight lines, indicating that water masses 1 and 2 mix and that water masses 2 and 3 mix, but not 1 and 3.

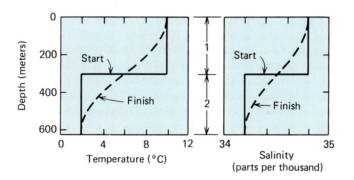
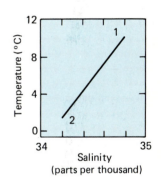

(a) Mixing of two homogeneous water bodies

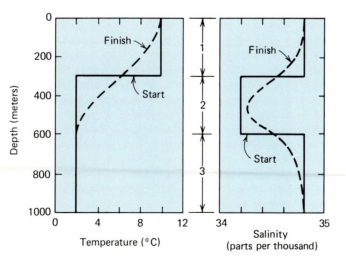
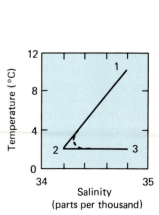

(b) Mixing of three homogeneous water bodies

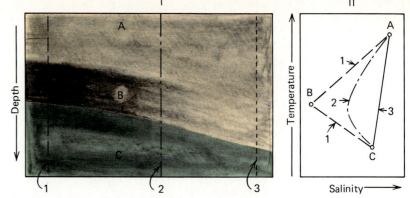

Use of a *T–S* diagram to study subsurface water masses. Water mass B (intermediate density) flows beneath surface layer A and above the denser water mass C. At the three locations shown, the *T–S* curves show the effects of mixing. At station 1, closest to the source of B, its characteristic temperature and salinity plot at point *B* on the *T–S* diagram. Farther from the source (station 2), the water mass mixes with those above and below, so that the *T–S* diagram changes although the original shape is still discernible. At station 3, water mass B does not occur, and water masses A and C mix directly.

these water masses mix, the newly formed mass will have a temperature and salinity intermediate between those of the two original water masses. Adding a third water mass (Figs. 6-17 and 6-18) complicates the situation, but the same general principles hold. We will use this technique in the next chapter to study movements of subsurface water masses.

SEA ICE

Sea ice forms where waters cool below their initial freezing point. Sea ice is permanent in the central Arctic Ocean and around Antarctica. Each winter, sea ice covers the entire Arctic Ocean. It also forms in bays and along the coast of Alaska, Canada, and the Atlantic coast of North America, as far south as Virginia. In spring, much of the ice melts, but large areas in the polar oceans remain ice-covered throughout the year. The annual expansion of ice-covered ocean areas and its retreat in local summer affect climate worldwide. In this section, we discuss how and why sea ice forms.

When seawater chills to its temperature of initial freezing, clouds of tiny needlelike ice particles form. This makes the water surface slightly turbid. The ocean surface dulls and no longer reflects the sky (Fig. 6-19). As ice particles grow, they form hexagonal **spicules,** 1 to 2

FIGURE 6-19
Newly formed ice makes the sea surface look greasy.

centimeters long. The needles and platelets of newly formed ice are called *frazil ice*. When the surface is stirred by winds and waves, the ice forms a soupy layer, known as *grease ice*.

As sea ice continues to form, ice crystals eventually form a blanket. When the surface is calm, an elastic layer of ice forms, called *nilas*. These ice sheets, a few centimeters thick, are easily moved by winds. When the sheets are pushed together they form interlocking fingers.

Since salt is excluded from ice, the remaining water becomes more saline, and its freezing point is lowered. Some brine pockets remain trapped in the ice. Salinities of newly formed ice are typically 7 to 14‰, but this depends on temperature. The slower the ice forms, the easier it is for brines to escape, resulting in lower ice salinities. Conversely, at very low temperatures, ice forms rapidly, and brines cannot easily escape. This results in higher ice salinities. Salinity of sea ice is always lower than that of the surrounding waters. As sea ice ages, the brine cells are expelled. Thus a multiyear ice sheet will typically have salinities around 0‰ at the top and around 4‰ at the bottom.

As freezing continues, rejected salts mix with underlying waters, making them more saline and therefore denser. Very cold water from the ice also chills the underlying waters. Eventually, convection occurs underneath the ice as the dense waters sink. Less dense water from below rises in finger-shaped parcels to replace the sinking brines. Such processes occur in both the Greenland and Labrador seas as well as near Antarctica, especially in the Weddell Sea. (As we see later, these processes form the coldest and densest water masses in the ocean.)

Newly formed ice is broken by waves. As the pieces smash into one another, they form *pancake ice,* so called because individual pieces resemble pancakes (shown in Fig. 6-20). As pancakes coalesce, their outlines can still be seen in the young sea ice.

Waves, and especially winds, break the ice sheets into large pieces, called **floes** (Fig. 6-21, p. 156). Floes constantly move and shift, freeze together and then later break loose, buckle up, or flatten out as the ice moves.

Snow accumulates on top and freezes onto the ice surface. Thus sea ice grows from the top and bottom. The newly formed *first-year ice*

FIGURE 6-20
Pancake ice near Antarctica.

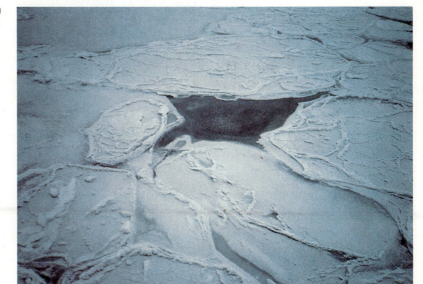

BOX 6-2

Polynyas

Each winter (June through September) in the Southern Hemisphere, the ocean surrounding Antarctica is covered with sea ice. In all, an area twice the size of the continental United States freezes over. With ships, it was impossible to study the ice pack in any detail—only the margins of the ice were known and then only slightly. With remote-sensing satellites, it is now possible to map the buildup and later melting of the Antarctic ice pack.

As we have already seen, ice packs are broken by leads, narrow bands of open water that can stretch for many kilometers. Ships try to find leads and use them to move through the ice. Satellites have shown much larger areas of open water, called **polynyas,** within the ice pack. Polynyas can be as large as the state of Colorado and persist over several years.

There are two different kinds of polynyas: coastal and open ocean. *Coastal polynyas* occur along the coast of Antarctica, formed by winds blowing the ice away from the shore. These were known before satellite images were available. Satellite images showed them to be more common and larger (50 to 100 kilometers across) than was previously known. These polynyas are important because of the large amount of sea ice that forms in them during the winter. Most of the ice in the ice pack is thought to form in coastal polynyas.

Ice excludes salts in seawater when it freezes. The rejected salts sink and mix with the waters below to form the dense, cold Antarctic Bottom Water mass. Enough salt is released to supply the Antarctic Bottom Waters at a rate of 10 million cubic meters per second (roughly equal to the flow of all the world's rivers). Bottom waters formed in this way do not come to the surface long enough to exchange gases with the atmosphere. Thus this process is one that chills the waters and adds salt while keeping them away from the atmosphere. Coastal-ocean polynyas (Fig. B6-2-1) are also called *latent heat polynyas* because they release heat to the atmosphere from the freezing of ice.

Open-ocean polynyas form in the midst of the ice pack (Fig. B6-2-2). They often re-form in roughly the same location over several years. Then in other years, they do not appear at all. The forces controlling their formation are more complicated than for coastal polynyas.

Convection cells are vertical water movements in which warm subsurface waters come to the surface, cool, and sink toward the bottom. The cells may be

FIGURE B6-2-1

Sea ice occurs around Antarctica, expanding greatly in areal coverage during winter. The maximum ice abundance is shown by the dark purple colors. The ice-free area in the upper part of the photo is called a polyna. Such ice-free areas do not occur every year. (Photograph courtesy NASA.)

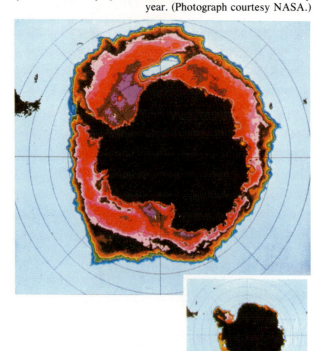

many kilometers across. The heat they bring to the surface melts the sea ice and keeps the surface ice-free. There are probably many individual cells involved in forming and maintaining a single polynya.

These polynyas are called *sensible heat polynyas* because the heat is directly associated with changes in temperature, that is, it can be sensed. Heat from these polynyas is transferred to the atmosphere. Since the waters involved also come to the surface, they can exchange gases. This is especially important for such gases as carbon dioxide.

During the mid 1970s, an exceptionally large (350 by 1000 kilometers) polynya persisted from 1974 through 1976. The effects of this polynyas was detectable for several years in elevated salinities and lower water temperatures.

The location of these open-ocean polynyas is apparently controlled, at least in part, by ocean-bottom topography. The large, persistent polynya was associated with a submarine ridge, called the Maud Rise. Perhaps some of the others are also associated with bottom features.

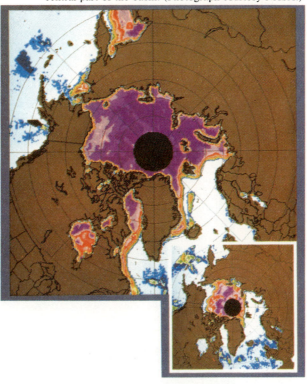

FIGURE B6-2-2
Sea ice covers the central Arctic Ocean throughout the year. During winter, the ice margins extend southward and sea ice forms along the coasts. In summer, the ice is restricted to the central part of the basin. (Photograph courtesy NASA.)

is flat and usually snow-covered. During a single winter, new sea ice can reach a thickness of 2 meters. Where sea ice never completely melts, *multiyear ice* continues to grow in subsequent winters. The older ice has a rough, hummocky surface. In the central Arctic, multi-year ice reaches thicknesses of 3 to 4 meters. Conversely, ice melts during the summer, down to about 2 meters in the central Arctic. The fresh water released by melting ice forms a thin layer of low-salinity water at the surface.

Currents and winds move large pieces of sea ice together, forming hummocks or *pressure ridges* (Fig. 6-21), the most conspicuous features of the ice pack. At these pressure ridges, the ice is deformed and thickened, up to 20 meters thick. Hummocks can extend many tens of meters below the ice.

When floes move apart, they expose open waters in *leads*. Leads range from a few inches to many hundreds of meters wide and can extend for many kilometers. Ships moving through sea ice utilize leads where possible to avoid having to break ice. Mammals stay near the leads and near holes in the ice. This permits them to catch fish and other food in the underlying waters (Fig. 6-22).

In shallow coastal waters, sea ice forms readily during cold winters. Ice attached to the shore which doesn't move is called *fast ice* (Fig. 6-23). **Pack ice** forms at sea and moves with currents and winds. The marginal ice zone occurs seaward of the fast ice. It is a mixture of open water, some first-year ice, and floes of multiyear ice, typically 3

FIGURE 6-21
Pressure ridges in first-year ice near Antarctica.

FIGURE 6-22
Seals on the ice near the holes they use to enter the water.

FIGURE 6-23

(a) This area of Arctic Sea near Point Barrow, Alaska, shows the movable pack ice and the fast ice attached to the shore. Pack ice consists of floes (tens to hundreds of meters across) separated by leads. Open-water leads appear black in this satellite photograph; refrozen leads appear gray. Pack ice is moved by winds and currents. The fast ice is unbroken and is attached to the shore; it does not move. (Photograph courtesy NASA.) (b) Sea ice is moved by winds. The ice on Chesapeake and Delaware bays is pushed onto the eastern shores by strong winds from the west during the exceptionally cold winter of February 1977. (Photograph courtesy NASA.)

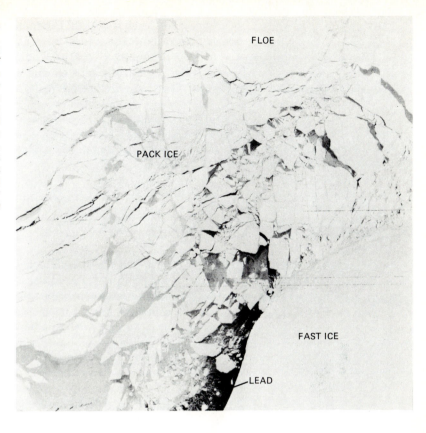

to 5 meters thick. The main polar pack consists of floes, some tens of kilometers across. Pack ice covers the central Arctic and surrounds Antarctica.

Strong currents and winds near Antarctica usually keep the pack ice close to the continents (south of 55°S) except in the coldest winters. During the southern summer (December through February), much of the Antarctic pack ice melts, except in large embayments. Some of the ice around Antarctica includes large tabular *icebergs,* which come from the outflows of large glaciers covering the continent. These floating **ice shelves** border the Ross Sea and the Weddell Sea. They supply flat-topped icebergs, which are most abundant in the Southern Ocean.

To sum up, sea ice is a major feature of the ocean. Its freezing in winter and melting in summer dominates the surface waters in the polar oceans. As we see in later chapters, it is a major factor then in biological processes as well.

Sea ice also influences the deep-ocean circulation. The coldest and densest water masses form in the polar oceans. Freezing sea ice expels salt, which increases the density of water masses, which is especially important near Antarctica. (We discuss this further in the next section.)

OCEANIC CLIMATIC REGIONS

Open-ocean climatic zones run east-west. These regions, indicated in Fig. 6-24, p. 160, have characteristic ranges in surface-water properties. Temperatures and salinities of surface waters in these zones are controlled primarily by incoming solar radiation and by the relative amounts of evaporation and precipitation.

The Arctic Ocean and the band around Antarctica are the *polar ocean* areas. Sea ice occurs most of the year, keeping surface temperatures at or near the freezing point despite long days of strong sunshine

BOX 6-3

Icebergs and the International Ice Patrol

Masses of freshwater ice in the ocean, called **icebergs,** break off the glaciers that cover Greenland and Antarctica. The process of iceberg formation is called *calving.* Icebergs from the glaciers of western Greenland (Fig. B6-3-1) are irregularly shaped, since they come from mountain glaciers along the coastal fjords.

In the Antarctic, the icebergs come from the ice sheets that flow from the continent out over the coastal ocean waters. These floating *ice shelves* are flat-topped, and the flat-topped icebergs they form are called *tabular icebergs* (Fig. B6-3-2). Antarctic icebergs are often much larger than those from Greenland. One formed in 1988 that was as big as the state of Rhode Island.

Icebergs typically last about 4 years before they melt. Thus they can be moved long distances by currents. However, few move out of the polar regions. Around Antarctica, currents keep icebergs relatively close to the continent. Some move northward past

the tip of South America to about 40°S in the Atlantic but rarely reach even 50°S in the Pacific. In 1894 an iceberg reached 26°S, only about 350 kilometers from the Tropic of Capricorn, the edge of the subtropical ocean. Long-lived icebergs are rare.

In the Atlantic, currents move icebergs southward along the North American coast. Large ones can travel as much as 2500 kilometers, reaching the Grand Banks, off Newfoundland. The *Titanic* sank in April 1912 after striking an iceberg near the Grand Banks. The accident took more than 1517 lives.

As a result of the sinking of the *Titanic,* the International Ice Patrol was formed soon afterward. It now tracks icebergs and issues regular bulletins during iceberg season. These predictions have greatly reduced the number of ship collisions with icebergs—only one ship has been lost. This occurred during World War II when the activities of the Ice Patrol were curtailed, and communications with ships were restricted.

FIGURE B6-3-1
An iceberg near the Grand Banks of Newfoundland is tracked by a U.S. Coast Guard airplane for the International Ice Patrol. (Photograph courtesy U.S. Coast Guard.)

FIGURE B6-3-2
A tabular iceberg in Nelville Bay, Greenland. The icebreaker approaching the iceberg is approximately 90 meters long. The iceberg extends 24 meters above the water and is about 1 kilometer long and about 0.75 kilometer wide. The iceberg extends 150 meters below the surface. Such tabular icebergs are most common around Antarctica. (Photograph courtesy U.S. Coast Guard.)

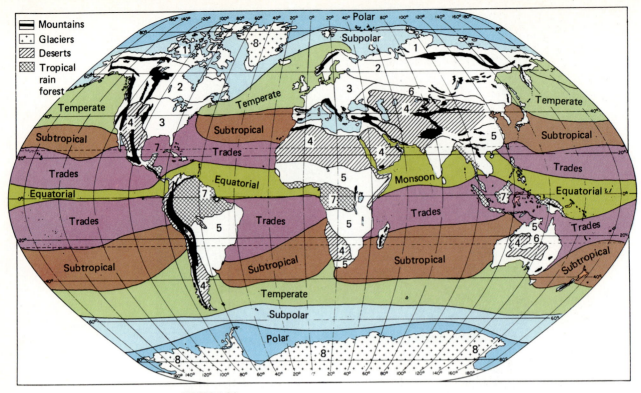

FIGURE 6-24
Climate zones over the ocean and vegetation zones over the land: (1) tundra, (2) boreal forest, (3) temperate, (4) desert, (5) savannah, (6) steppe, (7) tropical rain forest, (8) icecap. (After D.V. Bogdanov. 1963. Map of the natural zones of the ocean. *Deep Sea Res.* 10:520–23.

during local summer. In winter, there is little sunlight. Relatively low surface salinities and ice cover inhibit mixing.

Subpolar regions are affected by seasonal formation of sea ice. In winter, chilling of surface waters and freezing of sea ice lead to the formation of dense water masses that flow into the deep-ocean basins. Sea ice disappears during summer as solar heating melts the ice and raises surface-water temperatures to about 5°C.

The *temperate* region corresponds to the band of strong westerly winds. It has many storms, especially during winter. *Subtropical* regions coincide with the nearly stationary midlatitude high pressure cells. Winds are weak, so surface currents are also. Clear skies, dry air, and abundant sunshine cause extensive evaporation, so surface salinities are generally high. These were called the *horse latitudes* by sailors. Tradition has it that sailing ships carrying horses often had to throw them overboard when they ran out of animal food after their ships were becalmed.

The *tropical* (trade wind) regions have persistent winds blowing from the northeast in the Northern Hemisphere and from the southeast in the Southern Hemisphere. These winds cause the equatorial currents. Tropical ocean waters originate in the subtropical regions. They are therefore more saline than average seawater. Near the equator, precipitation increases, causing lower surface salinities there.

In *equatorial* regions, surface waters remain warm throughout the year. Annual temperature variations are small. Warm, moist air generally rises near the equator, causing heavy precipitation and therefore relatively low surface salinities. In the Atlantic and much of the Pacific oceans, winds tend to be weak. For this reason, sailors named the region the *doldrums* because their ships were often becalmed there.

CHAPTER SIX OCEAN AND CLIMATE

LAND CLIMATES

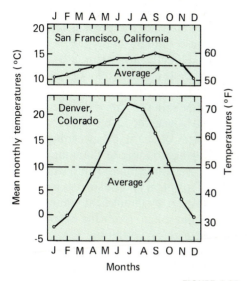

FIGURE 6-25
Variations in temperature at two locations: San Francisco, California, has a maritime climate with cool summers and relatively warm winters. Denver, Colorado, has a continental climate with warm summers and cold winters.

The ocean moderates climate on land. Coastal areas typically have **maritime climates,** where annual temperature ranges are much smaller than in inland areas, as shown in Fig. 6-25. Far from the ocean, **continental climates** prevail, with characteristically large temperature ranges.

The ocean stores large amounts of heat because of water's large heat capacity. This permits substantial heat storage without an accompanying large temperature increase. Furthermore, the ocean (or any other large water body) stores heat in a layer several meters thick, due to mixing of surface waters. This retards the loss of heat at night by keeping the water surface relatively cool. In contrast, rocks have low heat capacities. Thus land surfaces become quite hot during the day, and the heat is not readily transferred to rocks beneath the surface. Most of this heat is lost at night by radiation from the surface. Such a situation can be observed in deserts, where intense daytime heat is followed by chilly nights during clear weather.

Mountain ranges can block winds off the ocean, thereby affecting climate. For instance, the north-south ranges of the Americas partially block winds from the Pacific Ocean. Coastal areas have a maritime climate, but inland, the climate is continental. In contrast to the high rainfall on the ocean side of the mountain ranges, the interior is often desert. (The process is the same as in the island effect.) In Europe, the highest mountains run east-west and thus do not block winds off the Atlantic. There maritime climates extend far inland.

ANCIENT CLIMATES

Over the past 600 million years, the earth has had two distinctly different climatic states. (Remember that climate is weather averaged over a long time, usually 30 years or more. Or as one person has commented: "Climate is what you expect. Weather is what you get.") During most of the past 600 million years (the portion of the earth's history for which we have good fossil records) the earth has had a warm, humid climate (Fig. 6-26) with relatively little temperature difference between the equator and the poles. For example, during Cretaceous time (about 100 million years ago), crocodiles and palm trees were living in Spitzbergen, now only 1000 kilometers south of the North Pole.

FIGURE 6-26
Comparison of generalized sea levels and general climatic conditions over the past 600 million years. Note that times of high sea level have generally corresponded with periods of warmer climate.

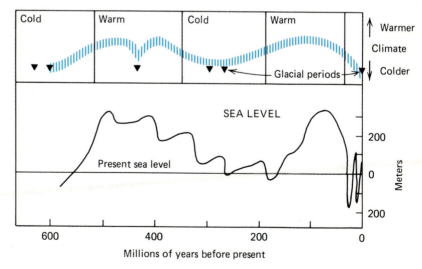

The second climatic state is a cold or glacial climate, much like the climate the earth has experienced over the past 3 to 5 million years, beginning perhaps 20 million years ago. During such glacial times, there are marked differences between the temperatures of the polar and equatorial regions. (We have more to say in the next section about glacial climates during the most recent stage of the present Ice Age.)

It is still not clear what causes these climatic shifts. One hypothesis (by no means proven) is that warm climates are relative to times of high sea level, when low-lying parts of continents are flooded, forming shallow seas. Such seas in the normally arid subtropics would supply large amounts of water vapor to the atmosphere. This theory leaves us with the question of what causes the sea level to be high. Another hypothesis is that greatly increased rates of sea-floor spreading cause the ocean basins to be shallower, causing flooding of the continents.

Large amounts of water vapor in the atmosphere could shift heat-transport mechanisms between the subtropical and polar regions. During times of warmer climate, heat transport by latent heat would be far more important than at present. The heat taken up during evaporation would be released in high latitudes through precipitation. Cooler, low-salinity waters would then return toward the equator in surface currents. Thus the transport of heat through the atmosphere would be more important than the present ocean circulation, where surface currents move warm waters poleward, and subsurface currents return dense, cold water from polar regions toward the equator.

Since our knowledge of ocean processes comes from studies made over the last few decades, during a time of cold climate, it is difficult for us to comprehend how the ocean worked during times of warm climates. Thus we can only speculate about ocean currents, and especially about subsurface currents.

ICE AGE CLIMATES

The most dramatic recent climatic change was the last glacial stage—which ended about 10,000 years ago—of our present Ice Age. The record of that glacial stage has been studied to determine what causes these shifts in the earth's climate and to improve our predictive ability.

Fossils in marine sediments were used to reconstruct distributions of ocean surface salinity and temperature 18,000 years ago, when the most recent glacial advance was at its maximum. Using sediment from carefully dated layers in sediment cores, scientists analyzed the chemical composition, abundance, and distribution of foraminifera, radiolarians, and other planktonic microfossils. Knowing the conditions under which these planktonic organisms now grow, scientists were able to determine the salinity and temperatures of the ocean surface waters in which these organisms grew.

The ocean surface during the last glacial stage was quite different from present conditions (Fig. 6-27). For example, surface ocean waters were cooler by about 2.3°C on the average; the land was about 6.5°C cooler. But the central waters of the major ocean basins were little changed in position or temperature. The Gulf Stream flowed directly eastward from the Carolinas to Spain. Cold polar waters advanced southward; thus temperature changes between subtropical and subpolar waters were much more pronounced than at present. There was much more upwelling along the equator, causing the equatorial ocean to be much colder than it is today. Temperature changes were especially noticeable in the North Atlantic, which was markedly cooler (Fig. 6-28, p. 164).

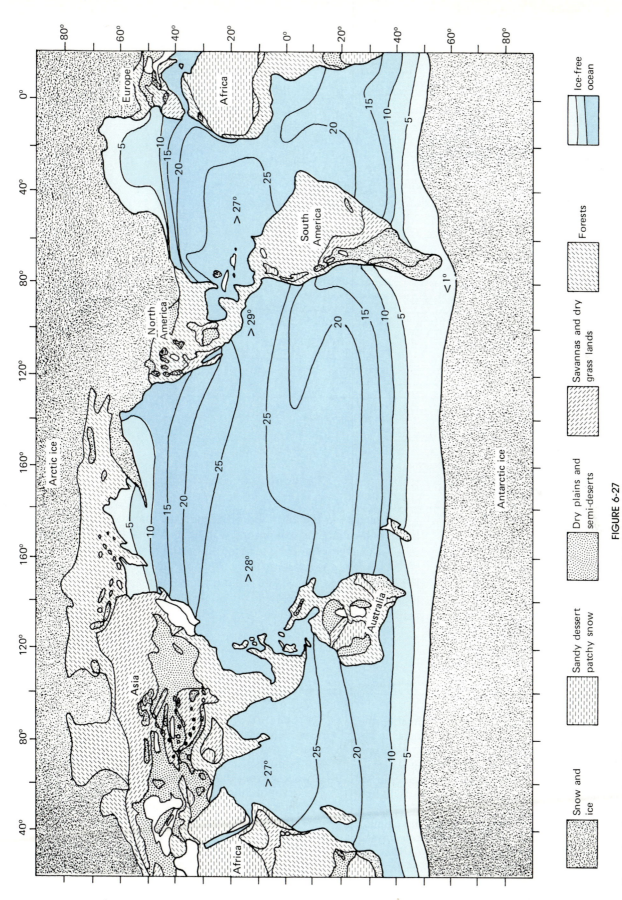

FIGURE 6-27

Sea surface temperatures (°C) and the extent of ice during August 18,000 years ago, when the ice had reached its greatest extent during the last glacial advance. Continental outlines correspond to sea level 85 meters below its present level. (After CLIMAP. 1976. The surface of the ice-age earth. *Science* 191:1131–44.)

Snow and ice

Sandy dessert patchy snow

Dry plains and semi-deserts

Savannas and dry grass lands

Forests

Ice-free ocean

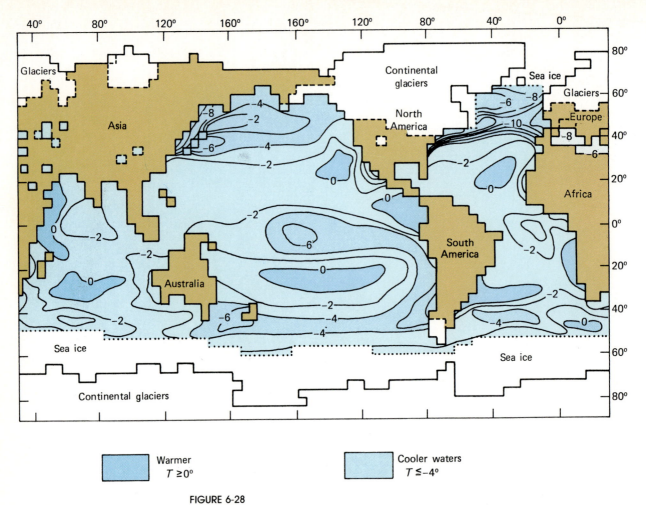

FIGURE 6-28
Differences between sea surface temperatures 18,000 years ago and modern August temperatures. Dotted lines show the extent of sea ice. Dashed lines indicate the limits of glaciers on land. (After CLIMAP. 1976. The surface of the ice-age earth. *Science* 191:1131–44.)

Warmer
$T \geq 0°$

Cooler waters
$T \leq -4°$

Changes in the earth's orbit around the sun seem to be a major factor controlling the waxing and waning of the glacial stages. Because of these changes, the volume of ice on Earth reaches a maximum every 100,000 years. At present, the earth's surface is warmer than it has been for 98% of the past half million years. Increasing levels of carbon dioxide (and other gases) in the atmosphere may cause warmer climates and higher sea levels (Fig. 6-25).

SUMMARY

Solar radiation is the source of energy for heating the ocean and atmosphere. Infrared radiation (heat) is absorbed in near-surface waters. Turbidity increases light absorption and scattering. Pure ocean water is most transparent to blue-green light. The blue color of water is caused by scattering of light in particle-free water.

Sea surface waters are warmest near the equator and coldest near the poles. Bands of equal sea surface temperatures are oriented east-west. Irregularities are caused by currents. Seasonal changes in water temperatures are greatest in the midlatitudes.

Sea surface salinities are highest in the centers of the ocean basins. Lowest salinities occur near the coasts, near river mouths, and in tropical areas with high rainfalls. Salinities are controlled by the balance between evaporation and precipitation.

Evaporation is highest in subtropical regions. Precipitation is highest in equatorial regions and in high latitudes. Poleward heat transport is reflected in low surface-water salinities in high latitudes.

Seawater density is controlled primarily by temperature and salinity. Pressure effects are detectable only at great depths in the ocean. Waters are usually in stable density configurations—densest at the bottom, least dense at the surface. This stability inhibits vertical water motions and mixing between layers.

The ocean is divided into three zones: surface, pycnocline, and deep zones. The thermocline is a marked change in temperature with depth. The halocline is a marked change in salinity with depth. And the pycnocline is a marked change in density, caused by changes in either temperature or salinity, or both.

Temperature-salinity relationships are used to identify water masses. They are also useful in studying mixing between different water masses. Finally, temperature and salinity relationships can be used to study how waters move below the surface zone. Mediterranean waters can be detected thousands of kilometers away from Gibraltar.

Sea ice forms in chilled surface waters. The first stage is freezing of individual crystals, which later form a plastic layer. Eventually pancakes form and then freeze together, forming floes, which are moved by winds and currents. Icebergs are freshwater ice broken off glaciers. They are moved primarily by currents.

Climatic regions on Earth are oriented east-west. Ocean surface waters generally correspond in temperature and salinity to the climatic zones in which they occur. Nearness to the ocean moderates climates on land. Ocean surface waters affect land climates over large areas.

Earth has experienced two different climatic states over the past 600 million years of its history. One is a glacial climate similar to the present. The other is a warm climate with little variation over the earth.

During the last glacial advance, ice on both land and sea extended much farther south than at present. There was little change in temperatures at the equator. The greatest temperature changes occurred in the mid- and high latitudes. Times of warm climate generally correspond to times of high sea level; glacial periods correspond to low stands of sea level.

STUDY QUESTIONS

1. Describe how the color of light changes with depth below the ocean surface. What causes these changes?

2. What factors control seawater density?

3. Why are seasonal temperature differences much greater over the land than over the ocean?

4. Describe a stable density distribution; an unstable density distribution.

5. Draw a diagram showing the layered structure of the ocean.

6. Draw a sketch showing the general distributions of sea surface water temperatures.

7. Draw a simple diagram showing the generalized distribution of surface salinities.

8. Explain why most of the bottom water forms in the North Atlantic and near Antarctica.

9. What techniques are used to trace movements of subsurface water masses?

10. Describe how sea ice forms.

11. Describe how temperature-salinity diagrams are used to identify water masses and to trace their movements through the ocean.

12. Where do icebergs come from? Where are they most abundant?

SELECTED REFERENCES

PERRY, A. H., AND J. M. WALKER. 1977. *The Ocean-Atmosphere System.* London: Longman. 160 pp. Discusses air-sea interactions.

PICKARD, G. L. 1979. *Descriptive Physical Oceanography,* 3d ed. Elmsford, N.Y.: Pergamon. 233 pp. General discussion of ocean features.

SVERDRUP, H. U., M. W. JOHNSON, AND R. H. FLEMING. 1942. *The Oceans.* Englewood Cliffs, N.J.: Prentice-Hall. 1087 pp. Classic text in oceanography; discusses ocean structure.

The Gulf Stream is shown by the surface water temperatures (red colors are warmest waters, blues are the coldest). The warm waters (1) off Florida move northeastward in a narrow swift current (2). After leaving Cape Hatteras the current meanders. Some meanders pinch off, forming either rings of warm water (3) north of the Gulf Stream or rings of cooler waters (4) south of the current. (Courtesy NASA.)

Currents

OBJECTIVES _____

1. To learn the basic features of ocean surface and sub-surface circulation;

2. To understand the interactions between winds and sur-face currents;

3. To understand the processes driving subsurface cur-rents;

4. To grasp the smaller-scale features of ocean circula-tion.

*C*urrents—large-scale water movements—occur everywhere in the ocean. Surface currents are driven primarily by winds. Subsurface currents, on the other hand, are driven by the sinking of chilled waters from the subpolar oceans. These spread out to flow through all the oceans and eventually return to the surface to be warmed. Both surface and subsurface currents are caused by the unequal heating of the earth's surface. Ocean currents transport heat from the tropics toward the poles, thereby partially equalizing surface temperatures over the earth. Ocean currents play an important but little-understood role in controlling climate (long-term weather conditions).

Our knowledge of currents is increasing rapidly because of ocean observations made by new instruments on satellites. Maps of surface currents up until now have come primarily from compilations of the effects of currents on ships, primarily nineteenth-century sailing ships. With such data we can detect only large currents. Aside from Indian Ocean monsoon currents, we know little about seasonal changes in currents.

In this chapter, we discuss:

The major ocean currents and what causes them;
Interactions between winds and currents;
Vertical water movements;
Boundary currents and the rings they form;
Density-driven currents in the deep ocean;
Ancient ocean currents; and
Uses of current predictions.

OPEN-OCEAN SURFACE CURRENTS

Current patterns are similar in all three major ocean basins (Fig. 7-1). They closely resemble surface winds (see Fig. 5-11 and compare with Fig. 7-1). The equatorial ocean is dominated by westerly flowing waters in the *North* and *South Equatorial Currents,* driven primarily by the trade winds. Between them is the narrow, eastwardly setting *Equatorial Countercurrent,* which occurs in a region of light and variable winds called the **doldrums.** Associated with these equatorial currents is

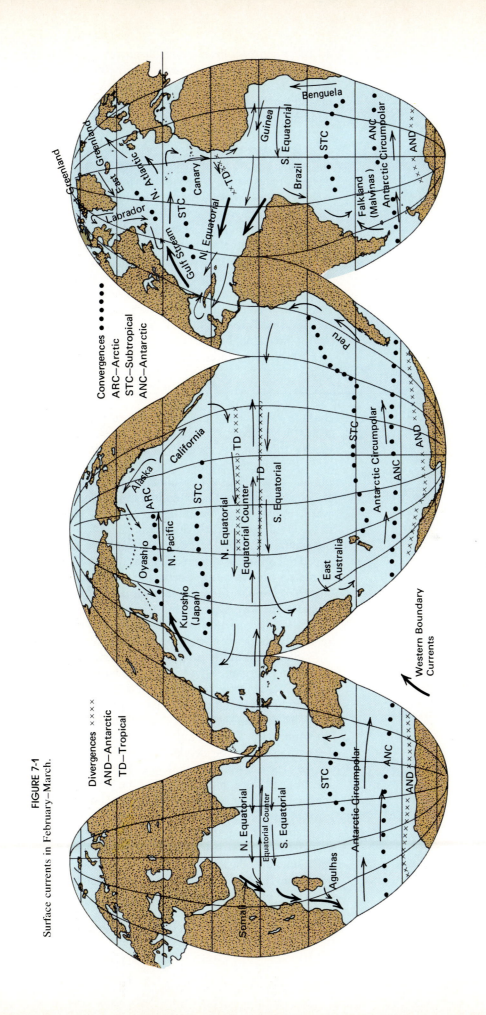

FIGURE 7-1
Surface currents in February–March.

Convergences ●●●●●●
ARC–Arctic
STC–Subtropical
ANC–Antarctic

Divergences ××××
AND–Antarctic
TD–Tropical

Greenland

East Greenland

N. Atlantic

Labrador

Gulf Stream

STC

N. Equatorial

Canary

TD

Guinea

S. Equatorial

Benguela

STC

Brazil

Falkland (Malvinas)

ANC

Antarctic Circumpolar

AND

Alaska

ARC

N. Pacific

Oyashio

Kuroshio (Japan)

California

STC

N. Equatorial

Equatorial Counter

TD

S. Equatorial

TD

East Australia

Antarctic Circumpolar

STC

ANC

AND

Peru

N. Equatorial

Equatorial Counter

S. Equatorial

STC

Antarctic Circumpolar

ANC

AND

Agulhas

Somali

Western Boundary Currents

167

a current **gyre**—a nearly circular current system. The east-west-elongated gyres are centered in the subtropical regions north and south of the equator. In addition to an equatorial current, each current gyre includes a major east-west current at higher latitudes, flowing in the opposite direction to the equatorial currents. These east-west currents are relatively slow.

Boundary currents flow parallel to the continental margins, usually north-south. Boundary currents on the western sides of ocean basins are especially strong. Eastern boundary currents are much weaker. (We discuss the reasons why in a later section.)

In the subpolar and polar regions of all ocean basins there are smaller current gyres. These high-latitude gyres circulate in the opposite sense from the subtropical gyres. Because of the positions of the continents, subpolar gyres are well developed in the Northern Hemisphere. There currents flow in a counterclockwise sense (Fig. 7-1).

Subpolar gyres occur in the Southern Hemisphere as well, primarily near Antarctica. Since there are no land barriers, currents flow uninterrupted around Antarctica in the *Antarctic Circumpolar Current*. As we have already discussed in Chapter 2, this is the primary connection between the three ocean basins.

EKMAN SPIRAL

Surface winds and ocean currents are intimately related. But how winds drive ocean currents is not so obvious. As we shall see, the earth's rotation plays a complicating role.

The process begins when winds blow across the water and drag on the surface. This sets a thin layer in motion which, in turn, drags on the one beneath, setting it too in motion. This process continues downward, transferring momentum (product of mass and velocity) to successively deeper layers. Such transfers of momentum between layers are inefficient, and energy is lost in the process. As a result, current speed decreases with increasing depth below the surface. In an infinite ocean on a nonrotating Earth, the waters would always move in the same direction as the wind that set it in motion. (This is seen in **storm surges,** large waves caused by strong winds that cause flooding in hurricanes. We discuss this in a later section. Here distances and times are relatively short, so Earth's rotation has little effect.)

Since the earth rotates, movements of surface waters are deflected to the right of the wind in the Northern Hemisphere, as shown in Fig. 7-2. This was noted by Fridtjof Nansen while studying the drift of Arctic ice. He found that the ice moved 20° to 40° to the right of the wind. To explain such effects, we assume a simple, uniform ocean with no boundaries. In such an ocean, each layer sets in motion the layer beneath, so that the deeper layer moves more slowly than the one above. The motion of each deeper layer is deflected to the right of the one above. These movements can be represented by arrows (*vectors*) whose orientation shows current direction and whose length indicates current speed. The change in current direction and speed with increasing depth forms a spiral when viewed from above (Fig. 7-2). This is called the **Ekman spiral** after the physicist who first explained the phenomenon. Figure 7-2 shows an Ekman spiral for the Northern Hemisphere. A spiral for the Southern Hemisphere exhibits the opposite sense of deflection, but current speeds still decrease with increasing depth.

Water-column stability controls the depths to which wind effects

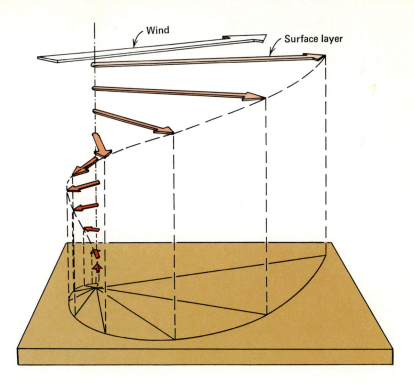

FIGURE 7-2
An Ekman spiral formed by a wind-driven current in deep water in the Northern Hemisphere. Note that current speeds decrease with increased depth, and the water movements in each layer move more to the right with depth. Surface currents move at speeds about 2% of the wind speed that set them in motion.

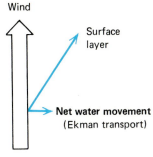

FIGURE 7-3
Schematic representation of the relationships of wind, surface current, and net water transport (also called Ekman transport) in the Northern Hemisphere. Again, current speeds are approximately 2% of the wind speed.

penetrate below the sea surface. A strong pycnocline inhibits transfer of momentum from the surface to waters below the pycnocline. The limit for wind effects is usually taken to be the depth at which the subsurface current is exactly opposite to the surface current. Under strong winds, wind-drift currents may be as deep as 100 meters below the surface. Surface currents move at about 2% of the speed of the wind that caused them. For instance, a wind blowing at 10 meters per second causes a surface current of about 20 centimeters per second.

In shallow waters, wind-generated currents are deflected less than predicted by Ekman's model. The ocean is obviously not uniform; we have already mentioned the effects of the pycnocline. Furthermore, winds do not always blow long enough to produce a fully developed Ekman spiral. (It apparently requires a day or two of steady winds to generate a fully developed Ekman spiral.) Under these conditions, the deflection is less than the 45° predicted by the simple case shown in Fig. 7-3.

GEOSTROPHIC CURRENTS

Prevailing winds move surface waters toward subtropical regions. (This is an example of Ekman transport.) As Fig. 7-1 shows, these midlatitude areas are zones of convergence for surface waters. Two things happen in a convergence (Fig. 7-4): water piles up, forming a hill, and the surface layer thickens. (This process of winds moving waters toward or away from a gyre center is sometimes called *Ekman pumping*.)

In the opposite case, a **divergence,** winds blow surface waters away from an area, as illustrated in Fig. 7-5. In a divergence, subsurface waters move upward to replace waters moving away from the region, thereby thinning the surface layer. A prominent divergence occurs around Antarctica, and divergence occurs in the subpolar gyres.

As a result of convergences and divergences, the ocean surface

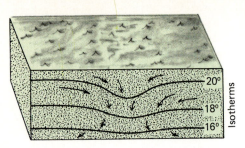

FIGURE 7-4
Schematic representation of a convergence zone in the open ocean. A single hill of water forms and the surface layer thickens because of the accumulation of surface waters.

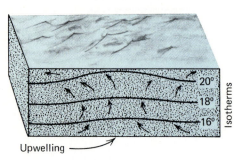

FIGURE 7-5
Schematic representation of a divergence zone in the open ocean. Note that the surface layer is thinned by the divergence.

has a subtle topography—hills in areas of convergences, and valleys at divergences. The difference in height between the midlatitude convergences on the western side of the ocean basins and the divergence surrounding Antarctica is about 2 meters. This occurs over a distance of about half the earth's circumference. These are very gentle slopes compared with those on land.

Ocean topography (controlled by density distributions) responds slowly to wind shifts, so ocean currents are equally slow to change. In effect, ocean topography is the flywheel that stores pulses of energy from winds and storms.

Water responds to the oceanic topography as it would on land—by running downhill. Consider the case of a hill in the Northern Hemisphere. We will follow a water parcel (see Fig. 7-6) to see how the earth's rotation changes its path.

The easiest way to visualize this process is to imagine a water parcel moving initially down slope. The Coriolis effect then deflects it to the right. This process continues until the water follows a path that allows it to flow downhill just enough to keep moving. Most of its motion parallels the side of the hill. (There is no evidence that this actually happens in the ocean. In fact, we do not know how currents are started by winds.)

If our hill were contoured to show lines of equal sea surface elevation, the path of the water parcel would nearly parallel these contour lines. On a frictionless ocean, water movements would exactly parallel the side of the hill. This balance—between the down-slope component of the gravitational force and the Coriolis effect that deflects it—gives rise to **geostrophic currents.** (These are analogous to the geostrophic winds discussed in Chapter 5.) Such currents are strongest on the steepest slopes (where the lines of sea surface elevation are closest together) and weakest on gentle slopes. (Remember that sea surface slopes are extremely slight, so the forces acting along the sea surface are essentially horizontal.)

Where a western boundary current passes near land, it is possible to measure directly such sea surface slopes. Thus we know that the *Florida Current* (part of the *Gulf Stream* system between Cuba and the Bahamas Banks) has slopes of about 20 centimeters over about 200 kilometers. At this location, current speeds are 150 centimeters per second (about 3 miles per hour). This rapid current has a much steeper slope than other currents where waters move only a few kilometers per day.

It will soon be possible to survey the ocean surface by satellites; then oceanographers will be able to map ocean currents everywhere in the ocean. But until direct measurement techniques are available, oceanographers will still use indirect techniques, based on observations of temperature and salinity.

First, a depth is chosen—called the **depth of no motion**—where

FIGURE 7-6
Schematic representation of water flow and balance of forces in a geostrophic current in the Northern Hemisphere.

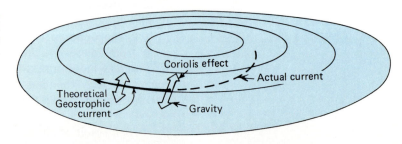

CHAPTER SEVEN   CURRENTS

no currents are presumed to exist. Second, it is assumed that the weights of all water columns overlying this chosen surface are constant. Then it is possible to calculate the height of the sea surface above the reference depth. This calculated **dynamic topography** is the surface topography to which a water parcel responds.

A brief example will show how dynamic topography is calculated. Temperature and salinity are accurately measured at two stations (we will call them A and B). Assume that the average salinity is the same at both stations but that the average temperature at station A is greater than at B. Thus the average water density at B is greater than at A.

Since the mass of each water column is equal, a column of less dense water occupies more space (stands higher) than a corresponding column of denser water. From this information you could determine that the water surface at station A is higher above the reference level than at station B. Using this technique for many pairs of stations, oceanographers map ocean surface topography. From such maps, current patterns and speeds are determined, using the relative slopes. Based on such maps, movements of floating objects, such as icebergs, can be predicted. Comparable techniques are used to determine flows of subsurface waters.

BOUNDARY CURRENTS

Western boundary currents are the strongest currents in the ocean. They are especially well developed in the Northern Hemisphere—for instance, the *Gulf Stream* in the Atlantic and the *Kuroshio* in the Pacific. The Gulf Stream has been studied most, so we will focus on it.

The Gulf Stream separates open-ocean waters from coastal waters (Fig. 7-7). Coastal water masses exhibit seasonably variable salinity and temperature. In contrast, Gulf Stream waters have fairly warm temperatures (20°C or higher) and relatively high salinities (around 36‰). There is some movement (convergence) of surface waters toward the center of the North Atlantic subtropical gyre. This area, also known as the *Sargasso Sea,* is bounded by currents. High salinities in the Sargasso Sea (36.6‰) are caused by evaporation exceeding precipitation.

Western boundary currents extend to depths of around 1 kilometer, too deep to come up on the continental shelf. The Florida Current is a relatively fast current, with speeds of 100 to 300 centimeters per second. Speeds are highest at the surface in a relatively narrow band, 50 to 75 kilometers wide.

Separating the Gulf Stream from adjacent slower-moving waters are **oceanic fronts,** which are marked by changes in water color, temperature, and salinity (Fig. 7-7). At the North Wall of the Gulf Stream, water colors change from the greenish colors of coastal waters to the intense cobalt blue of the Gulf Stream. Such fronts mark **convergences**—areas toward which surface waters flow from different directions and sink.

Eastern boundary currents are much weaker than their western counterparts. There the waters move much more slowly, and the boundaries between the boundary current and coastal waters are much more diffuse. Eastern boundary currents are relatively shallow and therefore can readily flow over continental margins.

Boundary currents are more changeable than the major currents shown in Fig. 7-1. Irregularities of continental coastlines cause eddies in coastal ocean currents (compare Fig. 7-8 with Fig. 7-7).

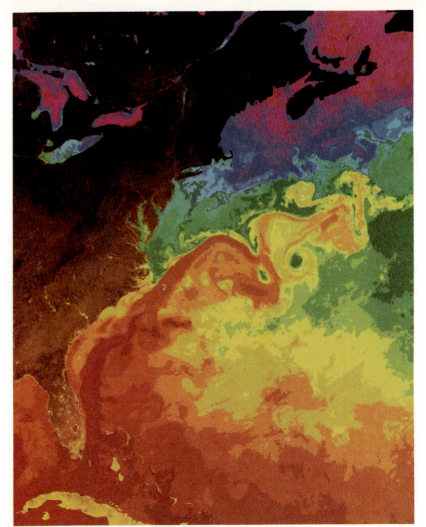

FIGURE 7-7
(a) Sea surface temperatures in June 1984 show the warm waters of the Gulf Stream (shown in false colors as orange and red) and cooler waters (shown in blue and purple colors) flowing southward along the coast. Warm core rings, shown by the yellow colors northwest of the Gulf Stream, are surrounded by cooler waters (greens and blues). Cold core rings occur southeast of the Gulf Stream, shown by green cores surrounded by yellows and reds. The image was obtained by processing data from a NOAA satellite (AVHRR) at the Rosensteil School of Marine and Atmospheric Science at the University of Miami and printed at NASA's Goddard Space Flight Center. (b) Sea surface temperatures on June 14, 1979, show a sharp front (called the North Wall) separating warm Gulf Stream waters (3) from cooler waters near the continent (4). A warm core ring (5) is moving southwestward. (Data courtesy NASA.)

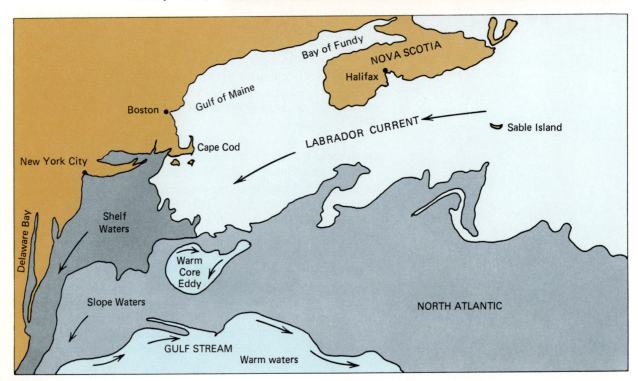

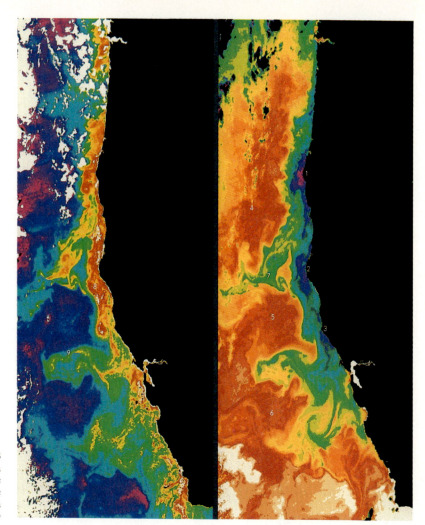

FIGURE 7-8
Surface-water temperatures in the California Current (an eastern boundary current) are dominated by upwelling at the capes along the shore. Note the absence of rings here. (Data courtesy NASA.)

TABLE 7-1
Boundary Currents in the Northern Hemisphere

TYPE OF CURRENT (example)	GENERAL FEATURES	SPEED	TRANSPORT (millions of cubic meters per second)	SPECIAL FEATURES
Eastern boundary currents California Current Canaries Current	**Broad,** ≈ 1000 km **Shallow,** ≤ 500 m	**Slow,** tens of kilometers per day	**Small,** typically 10–15	Diffuse boundaries separating from coastal currents Coastal upwelling common Waters derived from midlatitudes
Western boundary currents Gulf Stream Kuroshio	**Narrow,** ≤ 100 km **Deep**—substantial transport to depths of 2 km	**Swift,** hundreds of kilometers per day	**Large,** usually 50 or greater	Sharp boundary with coastal circulation system Little or no coastal upwelling; waters tend to be depleted in nutrients, unproductive Waters derived from trade wind belts

WESTERN INTENSIFICATION OF CURRENTS

Strong western boundary currents and weaker eastern boundary currents are major features of the ocean's surface current patterns. Four processes acting together intensify western boundary currents.

First is the earth's rotation, which displaces gyres toward the west. This compresses them against the continents. Consequently, sea surface slopes are steeper on the western than on the eastern side of the basin (Fig. 7-9).

Trade winds are the second factor. They blow generally westward along the equator (Fig. 6-12). This piles up the surface waters on the western sides of basins and makes the surface layer thicker on the western than on the eastern side of ocean basins. Put another way, the pycnocline is deeper on the west than on the east side of gyres.

The third factor is also due to the winds—the westerlies. These strong winds force surface waters in the midlatitudes to flow toward the equator as they move across the basin. Thus the return flow toward the equator occurs over a broad zone on the eastern side of ocean basins. It is not confined to eastern boundary currents.

The fourth factor is a bit more complicated. It involves apparent changes in the state of rotation of objects transported north or south along the earth's surface. To illustrate this effect (Fig. 7-10) consider a bicycle wheel at the equator, sitting motionless on frictionless bearings. When we carefully transport the wheel (by its base) to the North Pole (being careful not to start it rotating), we find that it was apparently rotating clockwise once every 24 hours. Actually the wheel remains motionless, but the earth rotates beneath it (counterclockwise). At intermediate latitudes the wheel appears to rotate slower than once a day.

The same thing happens to parcels of air or water moving north or south. Their rotational state apparently changes because they move between areas rotating around the earth's axis at different speeds. On the western side of ocean basins, the currents and the apparent spin

FIGURE 7-9
A current gyre showing westward displacement, resulting from the earth's rotation. Note that the strong current is associated with the relatively steep slopes, while the weaker current is associated with the more gentle slopes.

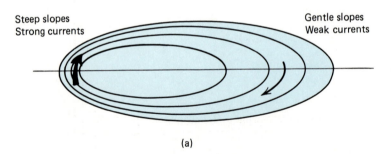

Steep slopes
Strong currents

Gentle slopes
Weak currents

(a)

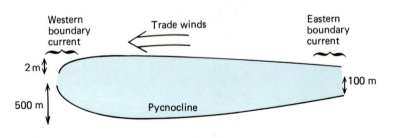

Western boundary current

Trade winds

Eastern boundary current

2 m

500 m

Pycnocline

100 m

(b)

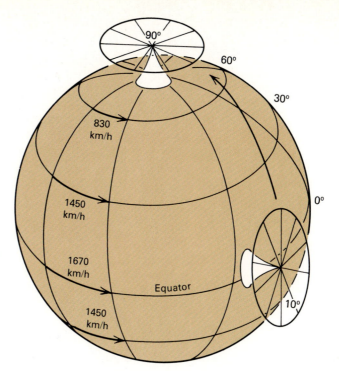

FIGURE 7-10
A motionless bicycle wheel on a frictionless stand at the equator is carefully moved to the North Pole, without disturbing the wheel. At the North Pole, the wheel exhibits an apparent clockwise rotation because the Earth is rotating counterclockwise beneath it. (After K. Stowe. 1983, *Ocean Science* 2nd ed. New York: John Wiley and Sons.)

due to the earth's rotation are in the same direction. Thus they add, giving currents a higher velocity. As a water parcel moves into higher latitudes, it seems to move faster. On the eastern side, the waters are moving closer to the equator, so current speeds decrease.

UPWELLING AND DOWNWELLING

Winds cause vertical water movements in addition to driving the horizontal surface currents. To understand such vertical water movements, we must again refer to the Ekman spiral (Fig. 7-2). Net movement of the surface layer (called the *Ekman transport*) is 90° to the right of the wind in the Northern Hemisphere (Fig. 7-3). Normally, surface waters respond to these forces by moving as a slab.

Near the coast, winds blowing parallel to the shoreline cause a layer of surface waters a few tens of meters thick to move either away from or toward the coast. In the first case, when surface waters move away from the coast, they expose subsurface waters, which then move upward toward the surface. This is called **upwelling** (Fig. 7-11).

The equator is another major upwelling area, which too is caused by the winds and by the change in the sign of the Coriolis effect at the equator [Fig. 7–11 (d)]. Consider the trade winds near the equator. North of the equator, Ekman transport is to the right of the winds (to the northwest) in the northeast trades (see Fig. 6-12 for the location of the winds), and surface waters move away from the equator. South of the equator, the Ekman transport associated with the southeast trades is to the left of the winds (toward the southwest) and again away from the equator. Thus the equator is a divergence zone, where subsurface waters are brought up into the photic zone. Later we shall see how this upwelling causes increased biological productivity along the equator.

Where surface waters move toward the coast, they cause the surface layer to thicken, a process called **downwelling.** In this case the resulting sea surface slopes create currents parallel to the coast.

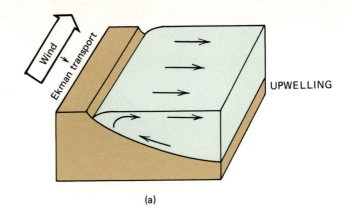

(a)

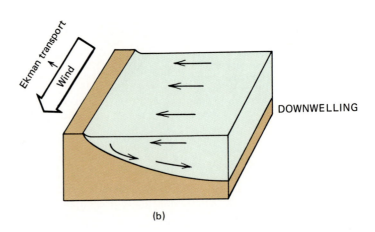

(b)

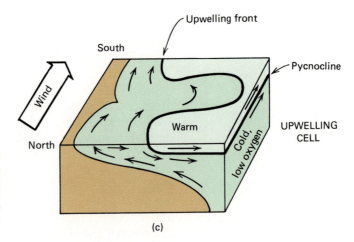

FIGURE 7-11
(a) Upwelling and (b) downwelling caused by winds blowing along the shore. (c) Upwelling cell at a cape. Note that the upwelled waters come from below the pycnocline. Also note that the plume of upwelled waters extends from the cape and is moved by surface currents.

Upwelling is especially conspicuous on the eastern side of ocean basins, where the surface layer is relatively thin. Much upwelling occurs near coasts, usually in areas a few tens of kilometers across, often located near capes or other irregularities in the coastline. The cells of upwelled water often have plumes of cold waters extending offshore for tens of kilometers (Fig. 7-8). The plumes move with coastal currents.

Subsurface waters coming to the surface from depths of 100 to 200 meters are rich in substances, called nutrients, which support abundant plant growth. These upwelling regions support rich growths of marine organisms that provide about half the world's fish catch. (We discuss this further in Chapters 11 and 12.)

MEANDERS AND RINGS

Satellite observations of the ocean surface show events as they happen. One example is the development of current **meanders** and subsequent formation of **rings** in the Gulf Stream and its extension, the North Atlantic Current.

A pronounced temperature change—the so-called *North Wall*—separates the cooler coastal waters from the warmer Gulf Stream waters (see Fig. 7-7). On the other side, the Gulf Stream forms the northern boundary of the Sargasso Sea, where the waters are slightly cooler than in the Gulf Stream.

Gulf Stream meanders, while present most of the time off the U.S. coast, are especially well developed after a storm. The meanders move slowly northeastward with the Gulf Stream at speeds of 8 to 25 centimeters per second (7 to 22 kilometers per day). If a meander becomes too large, it forms a ring and detaches itself from the Gulf Stream to move with the waters flowing slowly southwestward on either side of the Gulf Stream.

Rings are 100 to 300 kilometers across and are bounded by a nearly circular system of swift currents (90 centimeters per second, or 78 kilometers per day) that keeps each ring together and contains the waters in them. Rings move with the waters around them, usually southwestward, at speeds of 5 to 10 kilometers per day. Rings and their associated ring currents extend to depths of 2 kilometers; so normally they do not go up on the shelf, where the waters are only 200 meters deep.

Rings form on both sides of the Gulf Stream. Those on the north side have a core of warm water surrounded by colder slope water [shown in Fig. 7-12(a)]. They are easily detected by aircraft or satellites. Cold rings (also called cyclonic rings) form on the south side of the Gulf Stream and inject cooler water into the Sargasso Sea (Fig. 7-12). As the rings move southwestward, the surface waters warm up. And as their temperatures reach those of the surrounding waters, they become more difficult to detect.

A cold core ring was tracked for 7 months as it moved through the Sargasso Sea from south of Cape Cod to near Cape Hatteras, where it was resorbed into the Gulf Stream. When the ring initially formed, the waters in it were 15°C, nearly 10°C cooler than surrounding waters. The surface waters were greenish, smelling of seaweed—in other words, resembling coastal waters rather than the surrounding Sargasso Sea waters. Satellite infrared photographs showed the formation [illustrated in Fig. 7-12(a)], as the ring remained nearly stationary for a month. Later the ring briefly remerged with the Gulf Stream [Fig. 7-12(b)] and moved northeastward as a meander for nearly a month when it again separated and resumed its southwesterly drift. Finally in September, nearly 7 months later, the ring was again resorbed into the Gulf Stream off Cape Hatteras, North Carolina [Fig. 7-12(c)]. Such rings have been tracked for up to 3 years.

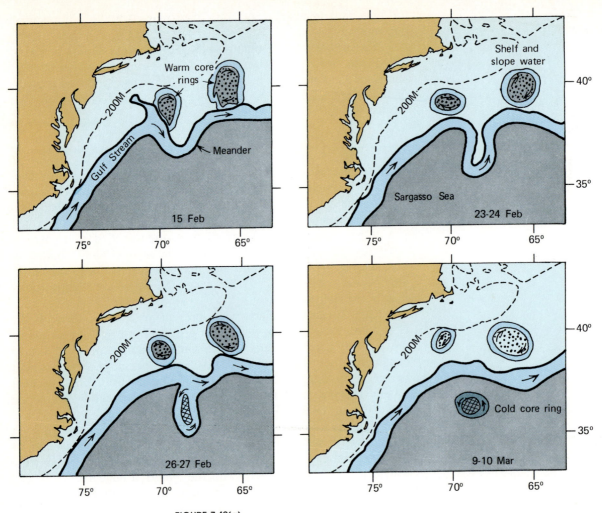

FIGURE 7-12(a)

(a) A cold core ring (called "Bob") formed from a Gulf Stream meander in February 1977 and moved southwestward through Sargasso Sea waters. Note the two warm core rings north of the Gulf Stream. (After P. L. Richardson. 1980. Gulf Stream trajectories. *J. Phys. Oceanogr.* 10:90–104.)

Rings transport heat, momentum, dissolved constituents, and weakly swimming organisms. It is interesting to note that rings apparently do not cause a net transport of water because of their complex history of formation, movement in an opposite direction to the Gulf Stream, and then resorption into the stream. Rings apparently can form, separate, and then be resorbed several times.

Rings are closely associated with the strong western boundary currents, such as the Gulf Stream or the Kuroshio in the Pacific. Other comparable but weaker features called **eddies** are especially abundant in the western portions of the Atlantic (Fig. 7-13) and Pacific oceans. These eddies extend from the sea surface to the ocean bottom, are 200 to 400 kilometers across, and take several months to pass a location. Currents associated with eddies are weaker than those in rings, a few tens of centimeters per second. Still, currents in these eddies are a hundred times more energetic than the average deep-ocean currents. Eddies are abundant near western boundary currents. They are the deep ocean's equivalent of atmospheric storms.

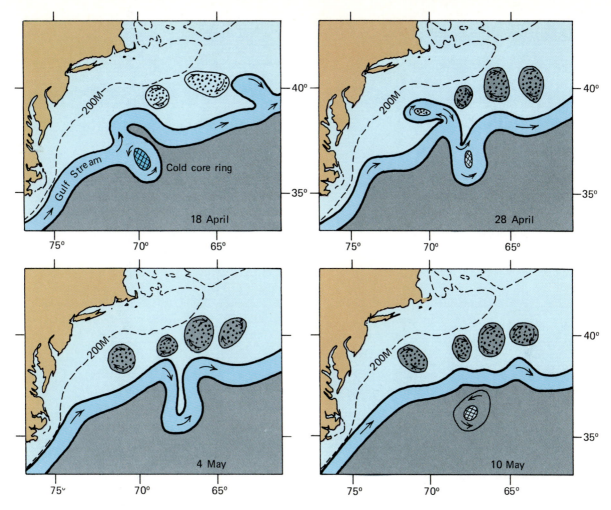

FIGURE 7-12(b)

(b) Cold core ring "Bob" briefly reattached itself to the Gulf Stream, forming a meander on April 18, 1977, that traveled nearly 300 kilometers northeastward before detaching to reform as a ring about 20 days later. Note that two new warm core rings formed during this period.

LANGMUIR CIRCULATION

In addition to the Ekman spiral, winds cause near-surface currents, known as **Langmuir cells** after their discoverer, Irving Langmuir. In Langmuir cells water moves with screwlike motions, in alternately right- and left-handed helical vortices [Fig. 7-14(a)]. The long axes of these cells generally parallel the wind direction. Cells tend to be regularly spaced and are often arranged in staggered parallel rows. Because of the counterrotation of the cells, alternate convergences and divergences form at the surface. Floating debris and foam [Fig. 7-14(b)] collect in the convergences. Between the lines of convergence are lines of divergences where water moves upward and along the surface of each cell toward the next convergence.

Langmuir cells form when wind speeds exceed a few kilometers per hour. The higher the wind speed, the more vigorous the circulation. When evaporation and cooling of surface waters tend to increase water density and favor convective movements, Langmuir cells can form at relatively low wind speeds. Increased stability resulting from surface

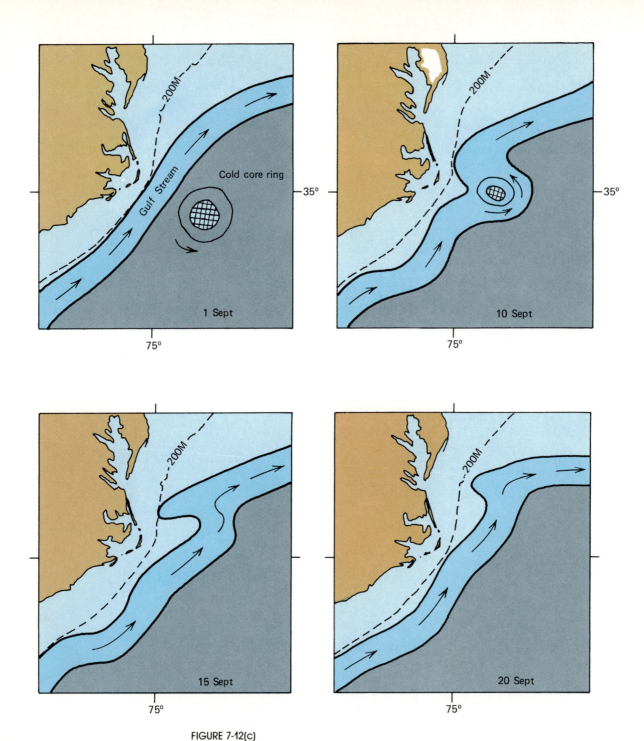

FIGURE 7-12(c)

(c) Cold core ring "Bob" coalesced with the Gulf Stream near Cape Hatteras, North Carolina, about September 10, 1977, forming a meander that moved northwestward with the Gulf Stream.

warming or lowered salinity tends to inhibit cell formation, so stronger winds are required before it can be set up.

Jet streams of water move downward under the convergences. In large lakes, where this type of circulation has been extensively studied, downwelling streams have been observed to extend 7 meters below the surface. Their speeds have been measured at about 4 centimeters per second. In the adjacent divergences, upwelling waters moved at speeds of about 1.5 centimeters per second.

The vertical dimension of these cells is dependent on wind speed and the vertical density structure of the surface layers. If the mixed layer is deep, there is relatively little hindrance to the vertical extent of

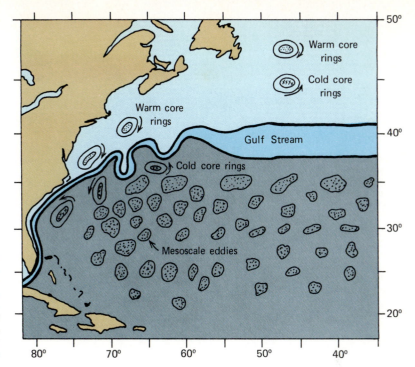

FIGURE 7-13
Relationships of warm core and cold core rings to the Gulf Stream and to eddies. Rings have stronger currents and are closely associated with the Gulf Stream. Eddies have weaker currents and are not obviously associated with the Gulf Stream; they cover 15 to 30% of the Sargasso Sea, southeast of the Gulf Stream. (Courtesy IDOE, National Science Foundation.)

the cells. If the mixed layer is shallow, the cells may not be able to penetrate the pycnocline. Such vertical water movements may partially control the depth of the pycnocline. This is also an important mechanism for transporting heat, momentum, and substances from the surface to layers a few meters deep.

Langmuir cells are one example of the complex short-term responses (occurring over minutes to hours) of the ocean surface to winds blowing across it. Wind waves—considered in the next chapter—are another short-term response. In contrast, Ekman transport, discussed in this chapter, is an example of the ocean's response to winds occurring over hours to a few days. How all these processes relate to each other is still unknown.

THERMOHALINE CIRCULATION

So far we have discussed surface currents, which involve only the surface layer of the ocean. Below the pycnocline, currents are slow and poorly known. Only general patterns, directions, and rates are known. Patterns of subsurface water movements are mapped from density distributions of the deep waters. (The same indirect technique is used for subsurface currents as for surface currents.)

Over most of the ocean, subsurface flows differ markedly from those at the surface. When a water mass reaches its appropriate density level, it spreads out, forming a thin layer. It takes less energy to move a parcel of water along a surface of constant density than across it.

Remember also that the ocean is layered horizontally (Fig. 7-15). Thus surfaces of constant density are essentially horizontal. Because of this, subsurface waters move horizontally along surfaces of constant density.

In the next section, we discuss vertical water movements, which occur primarily in high latitudes where dense water masses form. The

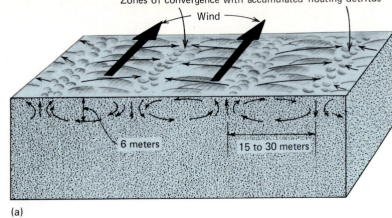

FIGURE 7-14
(a) Oil surface films and debris form windows along convergence lines in the Langmuir circulation set up by a strong, steady wind. (b) Foam lines caused by Langmuir currents near an offshore drilling platform in the North Sea. (Photograph courtesy Mobil Oil Corp.)

Zones of convergence with accumulated floating detritus

Wind

6 meters

15 to 30 meters

(a)

(b)

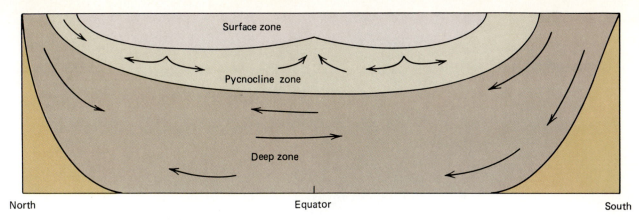

North   Equator   South

FIGURE 7-15
Schematic representation of the thermohaline circulation in an ocean basin. (Compare with Fig. 6-10 showing the density structure in an ocean basin.)

resulting vertical water movements control temperatures and salinities of deep waters throughout the ocean. They also drive currents along the bottom and in the middepths of the ocean. These density-driven currents are the **thermohaline circulation,** so called because temperature and salinity control seawater density.

These subsurface currents transport cold waters from polar regions. The deep waters slowly return to the surface (after hundreds of years). Some move up through the pycnocline throughout the ocean. Others come to the surface through upwelling along the equator and in coastal regions.

Strong, deep currents occur along the western sides of the Atlantic. These currents are slower than western boundary currents at the surface, but they are far faster than most of the slow thermohaline water movements.

BOTTOM-WATER FORMATION

The densest water masses form in only a few locations in the ocean. These polar areas are partially isolated and subjected to cooling of already high salinity waters or the freezing of sea ice. When conditions are right, large volumes of bottom waters apparently are formed.

Large quantities of *Antarctic Bottom Water* form in the Weddell Sea, a partially isolated embayment in Antarctica (Fig. 7-16). There surface waters are chilled to temperatures of −1.9°C. At this temperature and a salinity of 34.62‰, the water sinks to the bottom of the adjacent deep-ocean basin, where it forms the densest water mass in the open ocean. In the process of sinking, it mixes with other waters and is warmed to −0.9°C. After circulating around Antarctica and mixing with other water masses, cold, dense Antarctic Bottom Waters move northward into the deeper parts of all three major ocean basins. Using temperature as a tracer for water masses, we can follow Antarctic Bottom Waters as far north as the edges of the Grand Banks (45°N) in the North Atlantic. In the Pacific, mixtures of these waters reach the Aleutian Islands (50°N).

Because of the outflow of warm saline waters from the Mediterranean, the waters of the North Atlantic are the saltiest of the major ocean basins. This salty water is carried into high latitudes by the Gulf Stream. Near Greenland, it is intensively cooled to less than 0°C. When surface waters reach a critical density, they sink and flow as a mass along the bottom of the North Atlantic basin, especially along the western side of the basin. Such water masses apparently form intermittently in very cold winters.

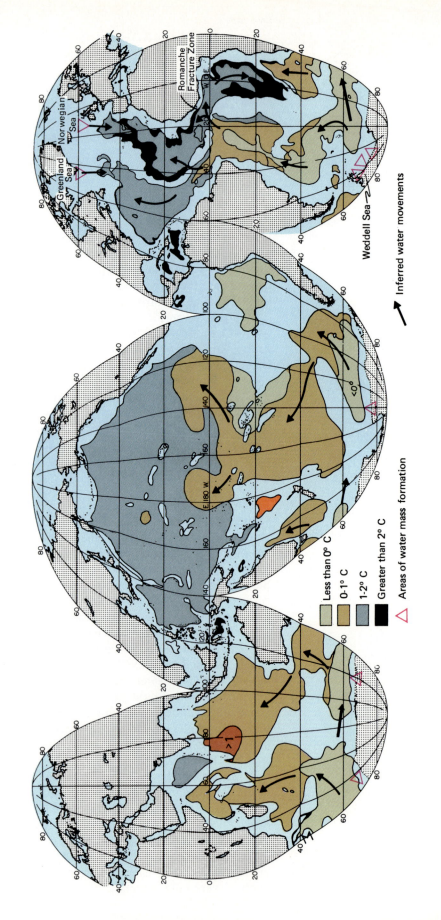

FIGURE 7-16

Variations in bottom-water temperatures at depths greater than 4 kilometers show the movements of waters along the bottoms of the ocean basins. (After G. Wust. 1935. *Die Stratosphare.* Duetsche Atlantische Exped. *Meteor,* 1925–1927. *Wiss.-Erg.* 6(1):288.)

Less than 0° C
0–1° C
1–2° C
Greater than 2° C
Areas of water mass formation

→ Inferred water movements

Greenland Sea
Norwegian Sea
Romanche Fracture Zone
Weddell Sea

Submarine ridges between Greenland and Scotland, the region forming the entrance to the Arctic basin, prevent any bottom waters formed in the Arctic Sea from entering the main part of the Atlantic. The Bering Sill effectively isolates the Arctic from the Pacific Ocean; so deep, cold water masses enter the Pacific only from Antarctica (Fig. 7-16).

Deep-water movement through ocean basins is controlled by ocean-bottom topography. The Romanche Fracture Zone, for example, provides a path for Antarctic waters to flow into the deep basins of the eastern portion of the South Atlantic; direct entry of this water mass into the eastern side of the South Atlantic is blocked, however, by the Mid-Atlantic and Walvis ridges on the sea floor (see Fig. 2.5).

Near-bottom currents commonly move much more slowly than surface currents. Speeds of 1 to 2 centimeters per second are typical—except along the western basin margins, where speeds of 10 centimeters per second have been calculated. This is another manifestation of the strong boundary currents along the western side of ocean basins. These slow water movements are difficult to measure directly, using current meters, because tidal and other currents are much stronger. Therefore, other techniques, primarily tracer techniques, are commonly employed. These are discussed in the next section.

TRACER TECHNIQUES

In addition to temperature and salinity, various tracers are used to map the movements of deep waters. Water temperatures (Fig. 7-16) show the general directions of movements but do not indicate the amount of time involved. Other tracers, such as dissolved oxygen (discussed in Chapter 11) give only a general sense of the rates of water movements. Radioactive tracers from atmospheric testing of nuclear weapons are now used to map water movements and to determine current speeds.

Tritium, a radioactive hydrogen isotope with a half-life of 12.3 years, is one such tracer. Produced during hydrogen bomb tests, tritium immediately reacts with oxygen, forming radioactive water. This labeled water falls as rain or snow and eventually moves into the ocean like normal water. A small amount of tritium occurs naturally in the ocean because of cosmic ray bombardment of the upper atmosphere. Bomb-produced tritium is now far more abundant in ocean surface waters.

Distributions of tritium in the deep ocean illustrate how oceanic processes control substances injected into the ocean surface. Tritium from the bomb tests is incorporated in river waters. When these waters flow into the ocean, initially they remain in surface waters, above the pycnocline (Fig. 7-17). In the high latitudes newly formed water masses incorporate some of this radioactive water when they sink below the surface. Thus they carry a distinctive "signature" of high-tritium contents. In the western Atlantic such water masses form, sink to the bottom, and flow southward to form the North Atlantic deep waters. The high tritium values around 50°N (Fig. 7-17) come from the outflow of Arctic Ocean surface waters, which received large amounts of tritium during the 1960s from hydrogen bomb testing in the northern USSR.

Studies of such tracers also provide information about the ocean's role in taking up carbon dioxide released by the burning of fossil fuels (coal, oil). About half the carbon dioxide released to the atmosphere since the 1850s has apparently gone into the ocean. The remainder has stayed in the atmosphere.

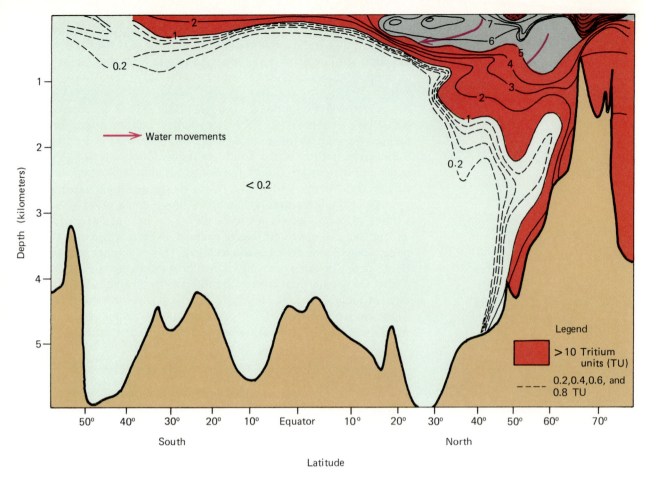

Water movements →

< 0.2

Legend

▮ >10 Tritium units (TU)

- - - 0.2, 0.4, 0.6, and 0.8 TU

Depth (kilometers)

Latitude

South   North

FIGURE 7-17
Tritium distributions in the western Atlantic 1972–1973. Note the high concentrations in surface waters around 60°N and the penetration of tritium into the intermediate and bottom waters of the North Atlantic. No tritium penetrated into the depths of the South Atlantic. (A tritium unit is one atom of tritium in 10^{18} atoms of hydrogen.) (Courtesy G. Ostlund. University of Miami.)

Carbon dioxide injected into deep waters does not return to the atmosphere for hundreds of years, providing long-term storage and thereby reducing the effects of carbon dioxide in the atmosphere. But any carbon dioxide that remains in the surface waters is able to exchange freely with the atmosphere and provides no long-term storage. Thus it is important to know where carbon dioxide is stored in the ocean.

SALT LENSES

We usually think of subsurface water movements as thin water layers moving between adjacent layers—like a card inserted into a stack of cards. We now know that subsurface waters can move in distinct lenses. Differences in temperature, salinity, and nutrient concentrations show that water masses retain the characteristic properties of the region where they formed. Strong currents (up to 25 centimeters per second) surround the lenses and inhibit their mixing with adjacent waters. Thus subsurface waters move in current-bounded masses, much like the rings and eddies found in surface waters.

Lenses of relatively warm saline waters in the North Atlantic have formed from the waters flowing out of the Mediterranean. They had traveled for several years to reach the locations where they were detected. The limited data available suggest that our view of the movements of subsurface waters must be revised just as our pictures of surface currents were altered by the discovery of rings and eddies.

Instruments on satellites can map only the ocean surface. Thus we must use other techniques to detect lenses. One technique is *acoustic tomography* (discussed in Chapter 1), in which sound pulses traveling through subsurface waters are accurately timed. Thus waters of different temperature and salinity can be detected by the changes in velocity of sound rays passing through them.

ANCIENT SURFACE CURRENTS

Surface currents are driven by prevailing winds, and their patterns are modified by each basin's shape. As we have seen, prevailing planetary winds are primarily controlled by the heating of the earth in the low latitudes and cooling it near the poles (see Chapter 5 for discussion of this point). Thus prevailing wind patterns change little over time, despite the changing positions of continents and ocean basins. This allows us to reconstruct ancient surface currents with some confidence, using our knowledge of the shapes and positions of ocean basins and continents (Fig. 7-18).

In narrow basins, winds and wind-driven currents tend to parallel the sides. In the newly formed Atlantic, high mountains bordering the narrow basin restricted prevailing winds to blow from north to south. In such a basin, surface currents would also parallel the basin sides [Fig. 7-18(a)].

As the basin widened, the winds blowing along the axis of the basin weakened, and more complicated current systems developed. Eventually, the current systems resembled those now found in the Atlantic.

Long after the Americas separated from Eurasia and Africa, the connection between North and South America remained submerged [Fig. 7-18(b)]. With the Tethys seaway open between Africa and Eurasia, there was a globe-circling equatorial current system. This current pattern continued until the ends of the Mediterranean Sea closed when Africa and Eurasia came together, about 30 million years ago. Only a few million years ago, the land connection between North and South America emerged and completely disrupted these equatorial currents.

FIGURE 7-18(a)

(a) Inferred surface currents about 160 million years ago, just after the North Atlantic opened, forming a long, narrow basin. Note the dark arrows indicating probable sources of warm saline deep waters. The asterisks indicate upwelling areas.

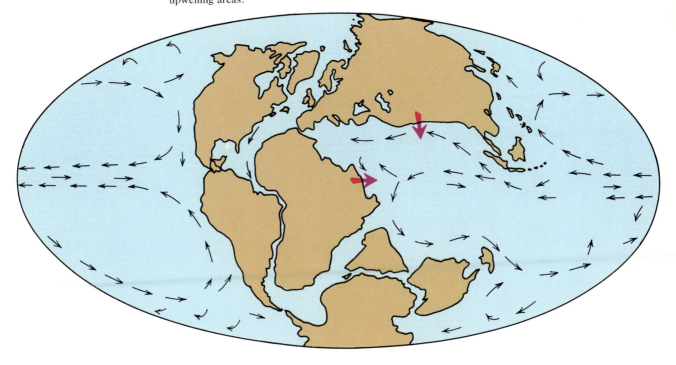

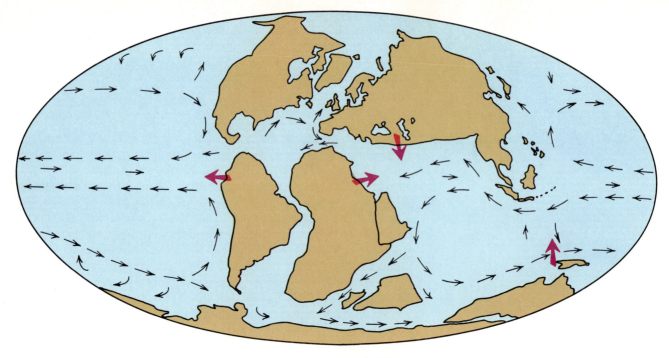

FIGURE 7-18(b)

(b) Inferred surface currents 160 million years ago after the opening of the South Atlantic. Note that the opening at the present Strait of Gibraltar and the submergence of Central America permit surface currents along the equator. Warm saline bottom waters still form in the midlatitudes.

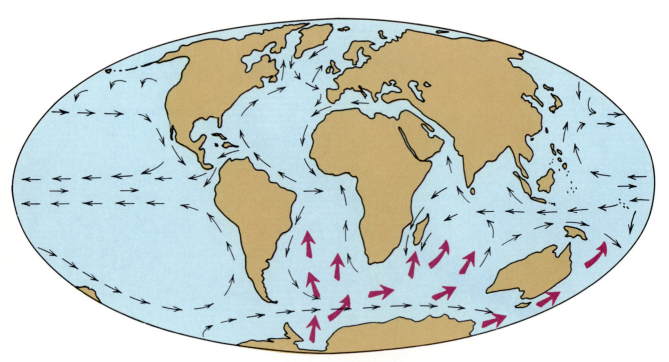

FIGURE 7-18(c)

(c) Inferred surface currents about 30 million years ago. Opening of the Drake Passage (at the tip of South America) and the Tasman Ridge permits the Antarctic Circumpolar Current to flow around Antarctica. Note that cold bottom waters (shown by the heavy arrows) form around Antarctica and flow northward into all three ocean basins. (Modified after B. U. Haq. 1984. Paleoceanography: a synoptic overview of 200 million years of ocean history, pp. 201–31. In *Marine Geology and Oceanography of Arabian Sea and Coastal Pakistan*, B. U. Haq and J. D. Milliman (ed.). New York: Van Nostrand Reinhold Co.)

A major development in the evolution of the present pattern of surface currents was the formation of the *Circum-Antarctic Current* [Fig. 7-18(c)]. The location of this current system was determined by spreading of the midocean ridge surrounding Antarctica. About 40 million years ago, Tasmania separated from Antarctica, forming a connection between the Pacific and Indian oceans. Then about 30 million years ago, the Drake Passage between South America and Antarctica was opened, permitting the Circum-Antarctic Current to flow around Antarctica. This current system isolated Antarctica, probably initiating the formation of the present Antarctic ice cap, marking the onset of the present glacial climate.

ANCIENT SUBSURFACE CURRENTS

Patterns of bottom-water formation and ancient subsurface currents are more difficult to reconstruct. Locations of dense bottom-water formation are controlled by details of ocean basin shape and climate much more than the surface circulation. So we have much more difficulty reconstructing ancient subsurface currents.

When the present spreading cycle began (200 million years ago), the earth had a warmer, more equable climate than the present. While the tropics were similar in temperature, the poles were warmer, and little or no sea ice formed. Thus there was no formation of dense water masses, because there was no chilling of surface waters nor increase in their salinity by salts excluded from newly formed sea ice.

The dense bottom waters formed during these times probably were warm, highly saline waters formed in evaporating basins in the arid midlatitudes (Fig. 7-19). The present Mediterranean and Red seas are probably similar to the areas of bottom-water formation 200 million years ago. Bottom waters then came from the sides of the midlatitude basins. These waters were warm and contained much less dissolved oxygen than bottom waters now do.

During the past 50 million years, the earth's climate has been

FIGURE 7-19
Sources of bottom waters at present (left side), during glacial times, and during times of warm climate when warm saline dense waters formed in midlatitude basins (right side).

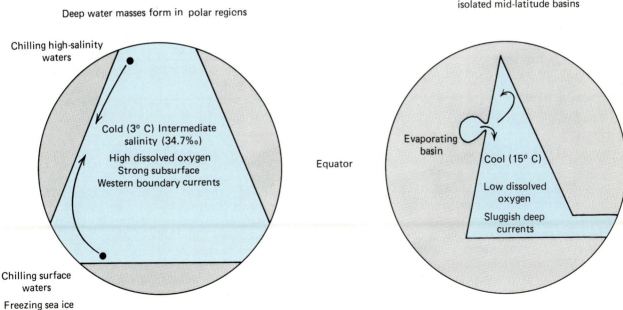

PRESENT

Deep water masses form in polar regions

Chilling high-salinity waters

Cold (3° C) Intermediate salinity (34.7‰)

High dissolved oxygen
Strong subsurface
Western boundary currents

Chilling surface waters

Freezing sea ice

CRETACEOUS (100 million years ago)

Deep water masses form in isolated mid-latitude basins

Equator

Evaporating basin

Cool (15° C)

Low dissolved oxygen

Sluggish deep currents

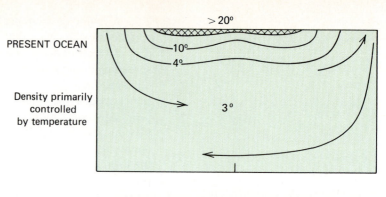

PRESENT OCEAN

Density primarily
controlled
by temperature

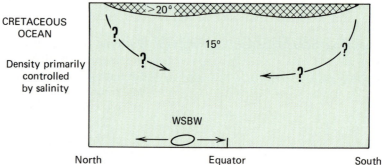

CRETACEOUS
OCEAN

Density primarily
controlled
by salinity

North   Equator   South

WSBW — Warm Saline Bottom Waters

FIGURE 7-20
Present subsurface current patterns (top)
compared with those involving warm saline dense
water masses (bottom).

cooling. At the beginning of the cooling trend, ocean bottom waters (probably the coldest waters in the open ocean) were around 13°C and have cooled to their present temperatures around 0°C. We have already discussed the history of the Circum-Antarctic Current. After this current isolated the shallow seas around Antarctica, and the climate cooled enough to permit freezing sea ice, Antarctic Bottom waters could form in the shallow, partially isolated embayments around Antarctica. From there they could flow northward into all the deep-ocean basins [Fig. 7-18(c)].

Submergence of the Iceland-Faeroe Ridge between Greenland and Scotland permitted subsurface Arctic waters to flow into the North Atlantic. Thus for the past 40 million years, any dense waters formed during cold winters in the Labrador Sea could flow southward in the Atlantic as the North Atlantic Deep Waters. This marked the development of the three-layered ocean (discussed in Chapter 6) with much colder bottom waters (Fig. 7-20).

SUMMARY

Currents—large-scale horizontal water movements—occur everywhere in the ocean. They are driven primarily by winds. Currents (like the winds) result from heating the earth near the equator and cooling it near the poles.

Open-ocean surface-current patterns are similar in all oceans, forming gyres, a nearly closed set of currents. Continents deflect east-west currents. Variable currents occur where prevailing winds shift seasonally, as in the northern Indian Ocean.

Winds blowing across a water surface set the up-

per water layer in motion. When times and distances are short, waters move in the same direction as the wind—for example, in storm surges. When water movements continue over longer times and distances, currents are affected by the Coriolis effect. These water movements show a systematic decrease in current speed and change in direction with depth, called an Ekman spiral. The reduction in speed results from loss in momentum in transfers from layer to layer.

Persistent currents result from the effects of the winds on the oceans. Convergences of surface waters resulting from prevailing winds cause surface waters to accumulate in certain ocean areas, forming subtle hills on the ocean surface. Water responds by flowing downhill, but these movements are deflected by the Coriolis effect. Eventually the water flows around the hills, so the deflection by the Coriolis effect is balanced by the pull of gravity. Since these hills are so low, forces acting on currents are essentially horizontal.

Western boundary currents are the strongest currents in the ocean. They separate coastal ocean waters from the open ocean. Most western boundary currents are so deep that they must flow along the continental margins. Eastern boundary currents are much weaker and shallower.

Intensification of western boundary currents is caused by the earth's rotation. This displaces the gyres toward the west, causing the sea surface slopes to be steeper than on the east. Current strength is controlled by the steepness of the slopes. The trade winds cause western boundary currents to flow northward in a narrow band. The westerlies also cause eastern boundary currents to be weaker. They deflect surface waters toward the equator as currents flow across the ocean basin. Finally, another effect of the earth's rotation is to increase speeds of northward-flowing currents and to decrease speeds of southward-flowing currents.

Upwelling and downwelling are vertical water movements caused by winds. The net water movement in the surface layers from the Ekman spiral is 90° to the right of the wind in the Northern Hemisphere and 90° to the left in the Southern Hemisphere. When surface waters are blown away from a coastline, subsurface waters flowing upward replace them. This process is called upwelling. When winds blow surface waters toward the coast, they accumulate, forming a sloping water surface that rises toward the shore. This results in a current paralleling the coast.

Meanders and rings occur in western boundary currents. These cut off bodies of water, which are enclosed by strong currents. These rings occur commonly in the western portions of ocean basins. Rings move with the currents. Some rings are resorbed back into the boundary currents. Other current-bounded ringlike structures occur widely throughout the western half of ocean basins.

Langmuir circulation—organized sets of horizontal corkscrewlike water motions in the surface layer—is caused by winds blowing across the ocean surface. At the convergences, materials floating on the ocean surface are brought together and attract organisms that come there to feed. Downwelling transports heat and dissolved materials downward.

Thermohaline circulation is caused by differences in water density, which drive vertical circulation. Bottom waters form when surface waters are chilled and freeze near Antarctica and in the North Atlantic near Greenland. Salt released during sea-ice formation increases water density. The dense waters sink to the bottom and flow toward the equator. The return flow occurs in upwelling zones and through the pycnocline all over the ocean.

Tracers for bottom-water movements include changes in dissolved oxygen concentrations, salinities, and water temperatures. Some subsurface waters move in current-bounded lenses.

Ancient surface currents were affected by the changing positions of the continents. The equatorial currents were completely blocked by Central America and the closing of the Mediterranean. Ancient subsurface currents involved warm saline bottom waters during times when Earth's climate was warmer.

STUDY QUESTIONS

1. Draw a diagram showing the principal open-ocean surface currents.
2. What causes the principal open-ocean surface currents?
3. Draw a diagram of the Ekman spiral. Why do the currents change direction and go slower with increasing depth below the surface?
4. Contrast eastern and western boundary currents.
5. Explain Ekman transport and its role in upwelling.
6. How are warm core rings formed by the Gulf Stream?
7. Describe geostrophic currents.
8. What causes western intensification of currents?
9. Describe and contrast upwelling and downwelling.
10. Diagram the formation, movement, and fate of warm core rings associated with the Gulf Stream.
11. Describe thermohaline circulation.
12. Describe Langmuir circulation. What causes it?
13. Describe salt lenses.

SELECTED REFERENCES

DIETRICH, G., K. KALLE, W. KRAUSE, AND G. SIEDLER. 1980. *General Oceanography*, 2d ed. London: Interscience. 1626 pp. Technical.

McLEISH, W. H. 1989. *Painting a Portrait of the Gulf Stream*. Smithsonian 19(12):42–55. Popular account of the Gulf Stream, its causes and effects.

SVERDRUP, H. U., M. W. JOHNSON, AND R. H. FLEMING. 1942. *The Oceans*. Englewood Cliffs, N.J.: Prentice-Hall. 1087 pp. Classic text in oceanography.

A wave strikes the shore at the southernmost point in the United States,
South Point, Hawaii. (© Ray Fairbanks/Photo Researchers.)

CHAPTER 8

Waves

OBJECTIVES _____

1. To describe the basic features of waves and how they form;

2. To differentiate shallow- and deep-water waves;

3. To describe wave effects in shallow water and on beaches.

Waves—disturbances of the water surface—can be seen at any beach. And seafarers have observed waves for thousands of years. Yet despite an abundance of observations, an understanding of sea waves developed slowly. The ancients knew that waves were somehow generated by wind, but not until the nineteenth century were the first mathematical descriptions of waves developed. In this chapter we study the features of waves, how they are formed, and some of the ways that they affect the ocean.

The ocean surface displays a complex and continually changing pattern—a pattern that never exactly repeats itself. Ocean waves come in many sizes and shapes, ranging from tiny ripples formed by light breezes through enormous storm waves, tens of meters high, to the tides (which are also waves, as we shall see in Chapter 9).

Because of their complexity, ocean waves usually do not lend themselves to accurate description or complete explanation in simple terms. Nevertheless, we commonly work with simplified explanations and descriptions that help us understand wave phenomena; moreover, most advances in the study of waves have come through the use of appropriate simplifications.

This chapter discusses:

Features of ideal waves;
Deep- and shallow-water waves in the ocean;
Processes causing waves;
Waves in shallow water and on beaches; and
Energy from waves.

SIMPLE WAVES

Let us consider simple waves and their parts as they pass a fixed point—say, a piling. We can make such waves by steadily bobbing the end of a pencil in a basin of water or a still pond surface. The waves move away from the disturbance, so we call them **progressive waves.** Each wave consists of a *crest*—the highest point of the wave—and a *trough*—the lowest part of the wave. The vertical distance between any crest and the succeeding trough is the **wave height, *H*.** The hori-

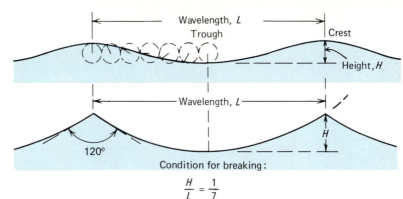

FIGURE 8-1
Two simple waves, showing their parts.
Relatively small waves (and large ocean swell)
can be described as simple sine waves. Larger
waves tend to be more sharp-crested than a
simple sine wave. There are limits to which a
wave can grow. Waves commonly break when
the angle at the crest is less than 120° or the ratio
of wave height to wavelength (*H*/*L*) is 1/7.

Condition for breaking:

$$\frac{H}{L} = \frac{1}{7}$$

zontal distance between successive crests or successive troughs is the **wavelength, *L*.** These wave constituents are illustrated in Fig. 8-1. The time (usually measured in seconds) that it takes for successive crests or troughs to pass our fixed point is the **wave period, *T*.** We can express the same information by counting the number of waves that pass our fixed point in a given length of time. This is the *frequency,* **(1/*T*),** which is expressed in events per second. For individual progressive waves, the *speed, *C** (in meters per second), can be calculated by dividing the wavelength (*L*) by the wave period (*T*) in seconds or (*C* = *L*/*T*).

Where the wave height is low (relative to its length), crests and troughs tend to be rounded. Such waves may be approximated mathematically by a *sine wave* (a smooth, regular oscillation, shown in Fig. 8-1). As wave height increases, sea waves normally have more sharply pointed crests (Fig. 8-1). (We shall see later how these simple curves can be combined to make complicated wave patterns.)

So far we have considered only movements of the water surface—crests and troughs moving together that make up a wave train. But what happens to the water itself as waves pass? How is the motion of the water related to the motion of the waveforms? We can study this by using tanks with bits of material floating on the surface or dyed bits of water or oil droplets below the surface.

When small waves move through deep water, individual bits of water move in vertical circular orbits that are nearly closed. Water moves forward as crests pass, then down, and finally backward as troughs pass. This series of movements is diagrammed in Figs. 8-1 and 8-2. The orbits are retraced as each subsequent wave passes. After each wave has passed, the water parcel is found nearly in its original position. But there is some slight net movement of the water because the water moves forward slightly faster as the crest passes than it moves backward under the trough. This results in a slight forward displacement of the water in the direction of wave motion and perpendicular to the wave crests, as shown in Fig. 8-2. If water moved forward with the waveforms, ships would never have been successful. No ship could withstand the forces exerted by water movements on such a scale.

FIGURE 8-2
Orbital motions and displacement of a water
particle during the passage of a wave.

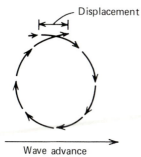

Displacement

Wave advance

You have probably experienced such wave motions yourself. When you float beyond the breakers, there is only a gentle rocking motion as waves pass, because there is little net movement of the water. If, however, you try to stand where waves are breaking, the pounding by breakers demonstrates large-scale water movements. The water in breakers moves with the waveforms. Even large ships are damaged when hit by tons of water from a large breaking wave. Consequently, ships are routed to avoid areas of high waves.

DEEP-WATER WAVES

In deep water, where the water depth is greater than half the wavelength ($L/2$), water parcels move in nearly stationary circular orbits. Such waves, unaffected by the bottom, are called **deep-water waves.** The diameter of these orbits at the surface is approximately equal to the wave height. It decreases to one-half the wave height at a depth of $L/9$ and is nearly zero at a depth of $L/2$ (Fig. 8-3). At depths greater than $L/2$, the water is moved little by wave passage. Thus a submarine is essentially undisturbed by waves when it is submerged to depths greater than $L/2$. When the water is much shallower (less than $L/20$, waves are greatly affected by the presence of the bottom. We discuss these effects when we deal with shallow-water waves later.

So far we have discussed the behavior of simple (or ideal) waves identical with others in a **wave train.** Simple uniform waves are rare at sea, but the concept is useful in analyzing more complicated real waves. Even the most complicated waves may be approximated mathematically by combining simple waves. When added together appropriately, the combinations of simple waves can reproduce the most complex wave patterns.

An observed profile of a sea is shown in Fig. 8-4(a). This group of waves has been analyzed to determine which wave frequencies occur in the wave spectrum. The various components (each a simple sine wave) are shown in Fig. 8-4(b). Combining these waves results in a wave essentially identical to the one observed.

Waves usually occur as wave trains or as a system of waves of many wavelengths, each wave moving at a speed corresponding to its own wavelength. This process, called **dispersion,** separates waves by wavelength.

Suppose that we produce a wave train and observe the results. If we follow a single wave in the resulting wave train, we find that it

FIGURE 8-3
Movements of water particles caused by the passage of waves in deep water (a) and in shallow water (b). Note that the particles tend to move in circular orbits at the surface, with orbit diameters becoming smaller with depth. Near the bottom, the orbits are flattened. Little orbital motion occurs at depths greater than half the wave length ($L/2$).

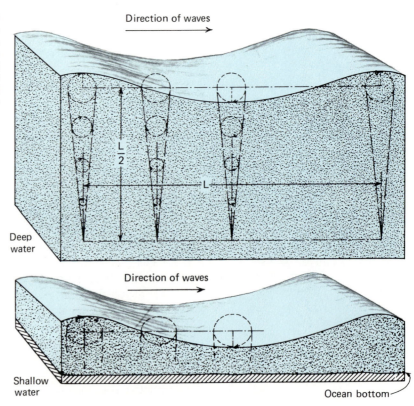

Direction of waves

Deep water

Direction of waves

Shallow water

Ocean bottom

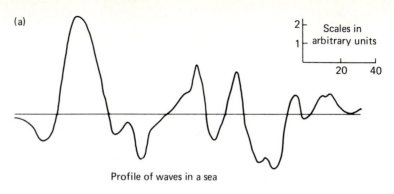

(a)

2 ⌐ Scales in
1 ⌐ arbitrary units

20   40

Profile of waves in a sea

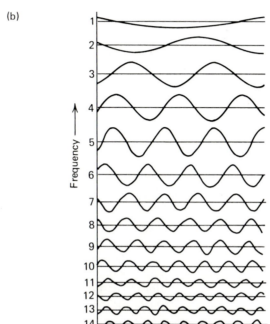

(b)

FIGURE 8-4
An observed wave profile in a sea (a) can be considered to consist of different sine waves (b) superimposed.

advances through the group. Indeed, individual waves move at a speed twice that of the group. As each wave approaches the front, it gradually loses height. Finally, it disappears at the front of the wave train, to be followed by another, later-formed wave that has also moved forward from the rear. New waves continually form at the back of the wave train while others disappear at the front. In deep water, wave energy travels at one-half the speed of individual waves.

FORCES CAUSING WAVES

Wave formation involves two types of forces: those that initially disturb the water and those that act to restore the equilibrium or still-water condition. We are all familiar with the waves formed when we toss a pebble into a pond. If the water surface is initially still, we observe a group of waves changing continuously as they moved away from the disturbance—another example of wave dispersion. Sudden impulses, such as volcanic explosions or large submarine slumps, cause some of the longest waves in the ocean. If the disturbance, such as an explosion, affects only a small area, the waves will move away from that point, much as the waves moved away from our pebble. But if the disturbance affects a large area, as a great earthquake can, the resulting **seismic sea waves** (also called **tsunamis**) behave as if they had been generated along a line.

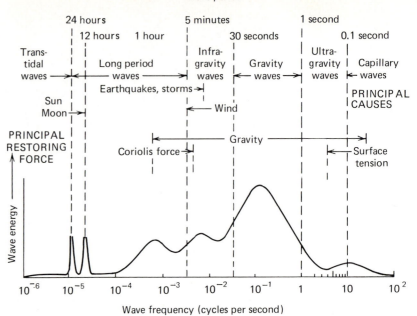

FIGURE 8-5
Schematic representation of the relative amounts of energy in waves of different periods. (After Blair Kinsman. 1965. *Wind Waves: Their Generation and Propagation on the Ocean Surface.* Englewood Cliffs, N.J.: Prentice-Hall.)

Winds are the most common force disturbing the ocean surface. They cause most ocean waves (note the large amount of energy associated with wind waves, as illustrated in Fig. 8-5). Winds are highly variable, so wind waves vary greatly. (More about wind waves in the next section.)

The third disturbing force results from the attraction of the sun and moon on ocean water, which cause the longest waves of all—the **tides.** The tide-producing forces of the sun and the moon act continuously on the ocean water, so the tides are not free to move independently as a seismic sea wave does. Waves where the disturbing force is continuously applied are known as **forced waves**—in contrast to **free waves,** which move independently of the disturbance that caused them. An impact wave is an example of a free wave. Wind waves have characteristics of both free and forced waves.

Depending on the size of the wave, different restoring forces are involved. For the smallest waves (wavelengths less than 1.7 centimeters, period less than 0.1 second), the dominant restoring force is **surface tension.** Recall that the water surface tends to act like a drum head, smoothing out the waves. Such waves, called **capillary waves,** are round-crested with V-shaped troughs. For waves with periods between 1 second and about 5 minutes, gravity is the dominant restoring force. This range includes most of the waves we see. Such waves are known as **gravity waves.**

Waves transmit energy gained from the disturbances that formed them. This energy is in two forms. One-half is *potential energy,* depending on the position of the water above or below the still-water level. The potential energy advances with the group speed of the individual waves. The rest of the wave energy—known as *kinetic energy*—is possessed by the water moving as the wave passes. There is a continual transformation of potential energy to kinetic energy and vice versa.

The total energy in a wave is proportional to the square of the wave height. In other words, doubling the height of a wave increases its energy by a factor of 4. An enormous amount of energy is contained in each wave. A wave 2 meters high, for example, has energy equivalent to 1200 calories per square meter of ocean surface. A 4-meter-high wave has 4800 calories per square meter. Nearly all their energy is dissipated as heat when waves strike a coastline. (The heat is so well mixed into the nearshore waters that the warming is imperceptible.) Sensitive seismographs (to detect earthquakes) record surf hitting distant beaches as faint earth tremors.

BOX 8-1

Energy from Waves

Waves are a potential energy source on many coasts. To give you an idea of the amount of power involved, waves in swell with a period of 10 seconds and a wave height of 2 meters approaching a coastline dissipate approximately 400 kilowatts across each 10 meters of wave front. Thus the energy potentially available on suitably located coasts is enormous. The west coast of Great Britain is a prime candidate for such power generation.

Many designs exist to extract energy from waves. Most involve floating devices that are moved by the waves. Since seawater is so corrosive to metal parts, and the environment along wave-swept coasts is so severe, devices having no moving parts in the water are most desirable. One such scheme, shown in Fig. B8-1-1, uses a partially enclosed chamber in which the rise and fall or the sea surface pumps air in and out of a chamber through a pipe containing a turbine that is connected to an electrical generator. Such devices would be built as part of a breakwater system to protect a harbor or coastline from wave attack. Although cost projections for such systems seem favorable, none have been built because of the high costs involved and the intermittent wave conditions on most coasts.

FIGURE B8-1-1

Simple scheme for generating electrical power, using energy from waves.

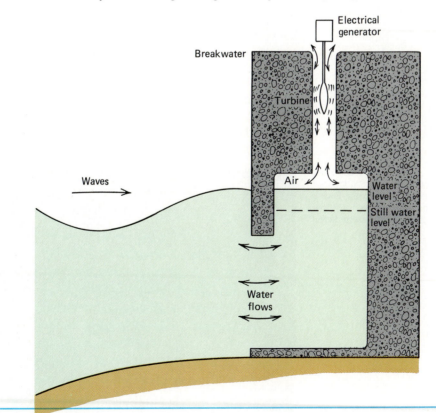

SEA AND SWELL

Waves are manifestations of energy moving across a water surface. Now we shall see how energy in winds acts on the ocean surface, forming wind waves.

Wind-wave formation is easily observed. Even a gentle breeze immediately forms **ripples** or **capillary waves,** more or less regular arcs of long radius, often on top of earlier-formed waves. Ripples play an important role in wind-wave formation by providing the surface roughness necessary for the wind to pull or push the water. In short, they provide the ''grip'' for the wind.

Ripples are short-lived. When the wind dies, they disappear almost immediately. But if the wind continues to blow, ripples grow and are gradually transformed into larger waves, usually short and choppy ones. These latter waves continue to grow as long as they continue to receive more energy than is lost through such processes as wave breaking. Energy is gained through the pushing-and-dragging effect of the wind. The amount of energy gained by the waves depends on such factors as sea roughness, the specific waveform, and the relative speed of the wind and waves. Choppy, newly formed **seas** provide a much better grip for the wind than smooth-crested waves.

The largest wind waves at sea (Fig. 8-6) are formed by storms—often several storms. The size of the waves formed depends on the amount of energy supplied by the wind. The factors operating here are wind speed, the length of time that the wind blows in a constant direction, and the **fetch**—the distance over which the wind blows in a constant direction. The results are shown in Fig. 8-7. Usually some older waves are present on the ocean when new waves begin to form. Either the older waves are destroyed by the storm, or newly formed waves are generated on top of the old ones. There is continuous interaction between waves. Wave crests coincide, forming momentarily new and higher waves. Seconds later the wave crests may no longer coincide but instead cancel each other; then the wave crests disappear.

As the winds continue to blow, waves grow in size (Fig. 8-7), until they reach a maximum size—defined as the point at which the energy supplied by the wind is equal to the energy lost by breaking waves, called **whitecaps.** When this condition is reached, we refer to it as a

FIGURE 8-6
A chaotic sea surface is caused by combined waves of all sizes, shown here in the Pacific Ocean off Sunset Beach, Oahu, Hawaii. (© Van Bucher/Photo Researchers.)

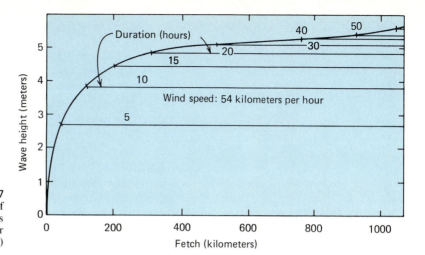

FIGURE 8-7
Growth of wave height under a constant wind of increasing duration acting over different lengths of time and fetches of varying lengths. (After H. U. Sverdrup and others. 1942.)

fully developed sea. In Fig. 8-7 note that initially wave heights increase markedly as the winds continue to blow. After about 10 hours, wave heights do not increase as much. And after about 30 hours, there is little increase in wave height as the winds continue to blow, regardless of the length of the fetch.

Knowing wind speed, duration, and fetch, we can predict the size of waves generated by a storm. Waves of many different sizes and periods are present in a fully developed sea, but waves with a relatively limited range of periods will predominate for a steady wind with a fixed speed (Fig. 8-8). Such predictions are complicated because winds almost never blow at a constant speed; winds are just as likely to be gusty at sea as on land and just as likely to change direction as not.

Initially waves in a **sea** are steep, and sharp-crested, often reaching the theoretical limit of stability ($H/L = 1/7$). Then the waves either break or have their crests blown off by the wind. As waves continue to develop, their speed approaches, then equals, and finally exceeds the wind speed. As this happens, wave steepness decreases as wavelength increases. As waves travel out of the generating area, or if the wind dies, the sharp-crested, mountainous, and unpredictable sea is gradually transformed into smoother, long-crested, longer-period waves—called **swell.** These waves can travel far because they lose little energy due to the viscosity of water.

FIGURE 8-8
In a fully developed sea, most of the wave energy occurs in a relatively restricted range of wave periods (or wave frequencies). Note that changes in wind speeds cause marked changes in wave energy and wave period. (After G. Neumann and W. J. Pierson. 1966. *Principles of Physical Oceanography.* Englewood Cliffs, N.J.: Prentice-Hall.)

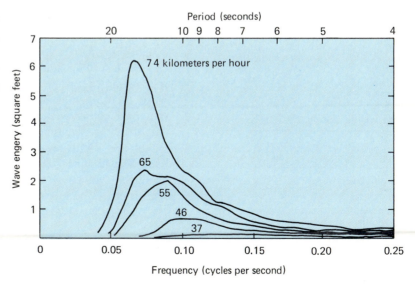

Let us examine some of the processes that cause waves to change from sea to swell as they move through calmer ocean areas. One process is the spreading of waves due to variations in the direction of the winds that formed them. Unless destroyed or influenced in some way by ocean boundaries, waves continue to travel for long distances in the direction that the wind was blowing when they formed. Because winds are rarely constant for long in terms of either speed or direction, waves formed in a storm will move away from the area and "fan out" as they move (*angular dispersion*). In this case, the wave energy is spread over a larger area, causing a reduction in wave height.

At the same time that the waves are fanning out, they are also separating by wavelength, through **dispersion.** Longer waves travel faster than shorter waves. As a result, complex waves of varying wavelengths formed in the generating area are sorted through time as they move away from the storm area, the long waves preceding the shorter waves. Consequently, the first waves to reach a coast from a distant large storm will be those having the longest periods. Island-dwellers, sensitive to the normal wave period on their coasts, may be warned of approaching hurricanes by the arrival of such abnormally long waves, which travel faster than the storm.

Swell travels great distances, crossing entire oceans before encountering a coastline. Storms in the North Atlantic, for example, form waves that end up as surf on the coast of Morocco, 3000 kilometers away. On extremely calm summer days, very long period swell strikes the southern coast of England after traveling about 10,000 kilometers (roughly one-quarter of the way around the earth) from South Atlantic storms. Similarly, waves from Antarctic storms have been detected on the Alaskan coast, more than 10,000 kilometers away.

WAVE HEIGHT

Despite an abundance of data, observations of wave height leave much to be desired. An observer on a moving ship with no fixed reference points for use in making estimates does not provide the most reliable information, but these are still the most extensive data available. More than 40,000 observations made from sailing ships indicate that about one-half of the waves in the ocean are 2 meters or less in height. Only about 10 to 15% of the ocean waves exceed 6 meters in height, even in such notoriously stormy areas as the North Atlantic or in the strong winds of the Roaring Forties in the southern oceans.

How do we report wave height? When we look out over the open ocean, there are waves of many different heights. However, the scene can be described statistically. Ocean waves show a nearly constant relationship between waves of various heights. One useful index is based on the **significant height** of the waves—the average of the highest one-third of the waves present—as shown in Table 8-1. Setting the height of the significant waves at 1, we find that the most frequent waves are about one-half as high, and the average waves are about 0.61. The highest 10% will average about 1.29 times higher than the significant waves. Thus given the wave height for part of the wave spectrum, we can predict other parts of the spectrum.

The largest waves are formed by strong, steady winds blowing for long times in the same direction over large bodies of water. Such waves occur most frequently at stormy latitudes, where storms tend to come in groups traveling in the same direction, with only short periods sepa-

TABLE 8-1
Wave-Height Characteristics*

WAVES	RELATIVE HEIGHT
Most frequent waves	0.50
Average waves	0.61
Significant (average of highest one-third)	1.00
Highest 10%	1.29

* U.S. Naval Oceanographic Office, 1958, p. 730.

FIGURE 8-9
Hurricane winds driving waves against the North Bayshore retaining wall at Bascayne Bay, Miami, Florida. (Photograph courtesy NOAA.)

rating them. Thus the waves of one storm often have no chance to decay or travel out of the area before the next storm arrives to add still more energy to the waves, causing them to grow still larger. Many typhoons and hurricanes (as the one in Fig. 8-9) do not form exceptionally large waves because their winds, although very strong, move in circles and do not blow long enough from one direction because of the rapid movements of the storms.

There are reliable reports of waves up to 15 meters high in the North and South Atlantic and the southern Indian Ocean. It appears that, for several reasons, these ocean regions rarely produce waves much higher. Winds rarely blow from one direction long enough to produce waves that are significantly higher. When winds change directions, waves produced under previous wind systems are destroyed or greatly modified. The stormy areas of all oceans experience equally severe storms at one time or another. Thus the major difference between ocean areas is the maximum fetch over which the wind can act. In the North Atlantic, for example, the maximum effective fetch is about 1000 kilometers. With a 1000-kilometer fetch, a wind blowing about 70 kilometers per hour can produce waves about 11 meters high. With an unlimited fetch, the same wind could produce waves about 15 meters high. A longer fetch and higher wind speeds are needed to produce giant waves.

As might be expected, the Pacific holds the records for giant waves. The largest deep-water wave that has been reliably measured was in the North Pacific on February 7, 1933. The U.S. Navy tanker U.S.S. *Ramapo* encountered a prolonged storm that had winds with an unobstructed fetch of many thousand kilometers. The ship, steaming in the direction of wave travel, was relatively stable. The ship's officers measured one wave (Fig. 8-10) at least 34 meters high. The wave period was clocked at 14.8 seconds and the wave speed at 102 kilometers per hour, somewhat faster than the theoretically predicted wave speed.

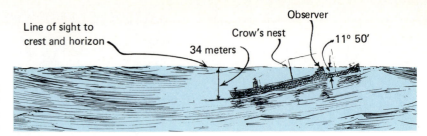

Line of sight to
crest and horizon

Observer

Crow's nest

34 meters

11° 50'

FIGURE 8-10
A wave 34 meters high was measured by the
U.S.S. *Ramapo* in the Pacific, February 7,
1933–the largest wave ever measured reliably.

WAVES IN SHALLOW WATER

Where water depths are less than $L/20$, motions of water parcels are strongly affected by the presence of the bottom (Fig. 8-3). Such waves are called **shallow-water waves.** Orbits of water parcels at the surface may be only slightly deformed, usually forming an ellipse (a flattened circle with its long axis parallel to the bottom). Near the bottom, wave action is felt as the water particles move back and forth. Vertical water movements are prevented by the nearness of the bottom.

Sometimes we observe movements of water parcels in shallow waters as waves pass over them—for example, where attached plants move with the water. The speed of shallow-water waves, C, in meters per second, can be calculated by multiplying the square root of the depth, in meters, by 3.1 or ($C = 3.1\sqrt{D}$). In shallow water, waves are slowed down until the individual wave speed equals the **group speed.**

Wave steepness (H/L), the ratio of wave height to wavelength, is a measure of wave stability. When wave steepness exceed 1/7, waves become unstable and begin to break (see Fig. 8-1) by raveling of the oversteepened crests, forming spilling breakers. The angle at the crest must be 120° or greater for the wave to remain stable. We examine this factor in detail when we discuss breakers and surf.

The largest waves—tides and seismic sea waves—are essentially shallow-water waves and involve substantial water movements. In addition to gravity effects, the Coriolis effect is important for waves with periods longer than 5 minutes.

Waves in shallow waters are most familiar to us. Of these the breaking waves, called **breakers,** are found on all beaches. An impressive amount of energy is dissipated by breaking waves in the surf. A single wave 1.2 meters high, with a 10-second period, striking the entire West Coast of the United States is estimated to release 50 million horsepower. Most of this energy is released as heat, but the heat is not detectable because water has a high heat capacity. Also there is extensive mixing in the surf zone, where waves break, mixing the heat throughout the water.

Some waves are destroyed by opposing winds. Others interact, and some cancel each other, but most end up as breakers when they encounter shallow water. Except for the longest waves, such as seismic sea waves or the tides, most waves move across the surface of the deep ocean unaffected by the bottom. As waves approach the coast, they are increasingly affected by the bottom, changing from deep-water to shallow-water waves. Wavelength and speed decrease, while the wave period remains constant. At the same time, the wave height first decreases slightly and then as the kinetic energy of motion is converted into potential energy, increases rapidly as water depths decrease to one-tenth the wavelength and the wave crests crowd closed together. (This series of events is illustrated in graphic form in Fig. 8-11.)

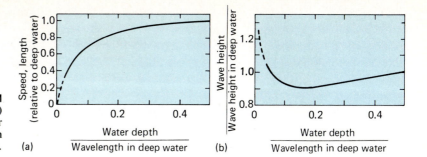

FIGURE 8-11

FIGURE 8-11

Waves change as they enter shallow water. (a) Speed and wavelength decrease as the water shallows; (b) wave height decreases and then increases.

Furthermore, the direction of approach changes as waves enter shallow water. We commonly see breakers nearly parallel to the coastline when they reach the beach, even though they may have approached the coast from many different directions. This process, known as *wave refraction,* occurs because the part of the wave still in deeper water moves faster than the part that has entered the shallower water. The result is that the crest line rotates so that it is more parallel to the bottom depth contours in the shallow water, as shown in Fig. 8-12.

In the simple case just discussed, the ocean bottom was sloping uniformly away from the beach. Obviously such is not always the case. Bottom irregularities cause pronounced wave refraction. Submarine ridges and canyons, for example, cause wave refraction such that the wave energy is concentrated on the headlands and spread out over the bays, as shown in Fig. 8-13. Consequently, headlands are eroded more rapidly than bays. The eroded material is usually deposited in adjacent bays, eventually smoothing the coastline. Note that the wave crest lines and associated *orthogonals* (sometimes called *rays*) bend toward shallower water. Eventually the crest lines become parallel with the depth contours. Complicated bottom topography causes complicated wave patterns on the shore (Fig. 8-14).

As waves encounter shallow water, wave heights increase and wavelength decreases. Consequently, wave steepness (H/L) increases. The wave becomes unstable when the wave height is about eight-tenths of the water depth, and it forms a breaker. The belt of nearly continuous breaking waves along the shore or over a submerged bank or bar is

FIGURE 8-12

Refraction of a uniform wave train advancing at an angle toward a straight coastline over a gently sloping, uniform bottom. Note the bend in the wave crests as they approach the beach. Such waves would cause a longshore current moving to the left near the beach. (After Judson, Kauffman, and Leet, 1987.)

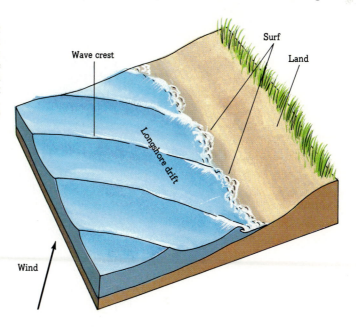

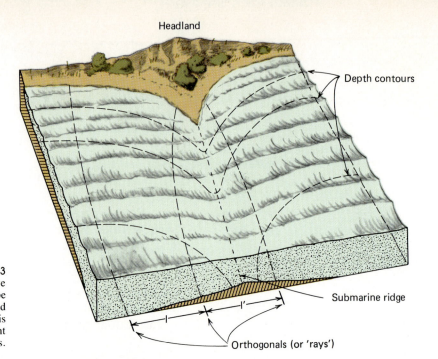

FIGURE 8-13
Wave refraction causes equal amounts of wave energy between orthogonals at 1 and 1 to be concentrated on the headland and to be spread out over the shore or the adjacent bay. This erodes the headland and causes sediment deposition on the bay shores.

FIGURE 8-14
Submarine topography causes complicated wave patterns near the entrance to New York Harbor. Hudson Canyon (a) focuses uniform wave trains on the harbor entrance (b). (After W. J. Pierson, G. Neumann and R. W. James. 1955. *Practical Methods for Observing and Forecasting Ocean Waves by Means of Wave Spectra and Statistics.* U.S. Navel Oceanographic Office Publ. 603., Washington, D.C.)

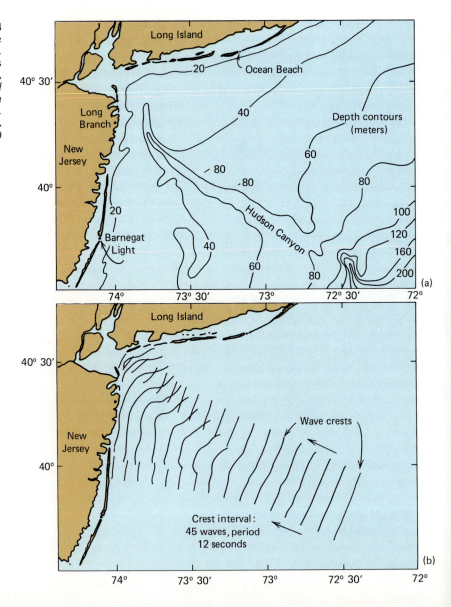

known as **surf.** These breaking waves are distinctly different from the breaking of oversteepened waves in deeper water, where the tops are blown off by the wind or the waves break to dissipate excess energy added by the wind.

SURF

Surf is a mix of breakers. It forms as different types of waves approach a shore and interact with the shallow bottom. Different types of breakers are shown in Fig. 8-15 and classified in Table 8-2. Spilling and plunging breakers behave differently and form under different circumstances. *Spilling breakers* (Fig. 8-16) are easily visualized as oversteepened waves where the unstable top spills over the front of waves as they move toward a beach. In spilling breakers, waveforms advance, but wave heights (i.e., wave energy) are diminished.

Plunging breakers are most spectacular (Fig. 8-17). The wave crest typically curls over, forming a large air pocket. When a wave breaks, there is a large splash of water, and foam is usually thrown into the air. These waves are excellent for surfing. Plunging breakers tend to form from long gentle swells ($H/L = 0.005$) over a gently sloping but irregular bottom. On even steeper bottoms they may break over the

FIGURE 8-15
Characteristics of different types of breakers.

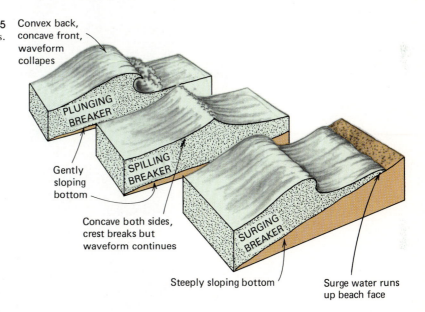

Convex back, concave front, waveform collapes

Gently sloping bottom

Concave both sides, crest breaks but waveform continues

Steeply sloping bottom

Surge water runs up beach face

TABLE 8-2
Types of Breakers and Beach Characteristics Associated with Each*

BREAKER TYPE	DESCRIPTION	RELATIVE BEACH SLOPE	RATIO OF WATER DEPTH TO WAVE HEIGHT
Spilling	Turbulent water and bubbles spill down front of wave; most common type	Flat	1.2
Plunging	Crest curls over large air pocket; smooth splashup usually follows	Moderately steep	0.9
Collapsing	Breaking occurs over lower half of wave; minimal air pocket and usually no splashup; bubbles and foam present	Steep	0.8
Surging	Wave slides up and down beach with little or no bubble production	Steep	Near 0

* After C. J. Galvin. 1968. Breaker type classification of three laboratory beaches. *J. Geophys. Res.* 73(12):3655.

FIGURE 8-16
Spilling breakers in winter surf on Clatsop Spit, near Astoria Oregon. (© David Weintraub/Photo Researchers.)

FIGURE 8-17
A plunging breaker on the beach at Cancun Mexico. (© Tom Hollyman/Photo Researchers.)

lower half of the wave with little upward splash. This is known as a *collapsing breaker* (see Fig. 8-18).

When a wave strikes a barrier, such as a vertical wall, it may be reflected, its energy being transferred to another wave, traveling in a different direction. *Wave reflection* may be seen when small waves in a bathtub are reflected from its sides. In other cases, a wave striking a steeply sloping beach may form a surging breaker or a turbulent wall of water, which moves up the barrier as the wave advances and back down when it retreats.

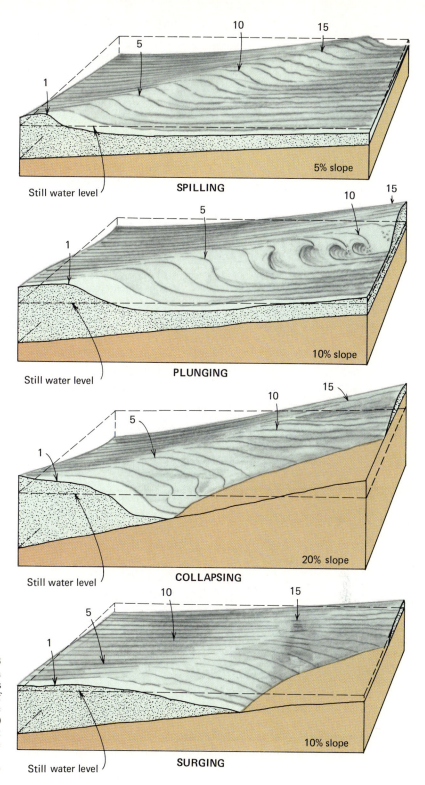

FIGURE 8-18
Changes in the water surface through time as a breaker advances toward a beach. The numbers above the profiles indicate the number of 0.06-second intervals elapsed since the profile shown on the front of the block. Thus 10 indicates a profile 0.6 second after the first profile. (After C. J. Galvin. 1966. Breaker-type classification on three laboratory beaches. *J. Geophys. Res.* 73:3651–59.)

Surf height depends on the height and steepness of the waves offshore and, to a certain extent, on the offshore bottom topography. Breakers and surf may be only a few centimeters high on a lake or a protected ocean beach or many meters high on an open beach. Lighthouse keepers report spectacular breakers at their exposed locations. Minot's lighthouse, 30 meters tall, on a ledge on the south side of Massachusetts Bay, for instance, is often engulfed by spray from breakers. The glass in the lighthouse of Tillamook Rock, Oregon, 49 meters above the sea, has often been struck by waves. We have no observations of the waves that cause such surf.

Breakers 14 meters high have twice damaged a breakwater at Wick Bay, Scotland, moving blocks weighing as much as 2600 tons. Breakers about 20 meters high have been reported at the entrances to San Francisco Bay and the Columbia River (Oregon, Washington) when onshore gales were blowing. Waves and breakers at a river or harbor are likely to be especially high when the incoming waves encounter a current setting in the opposite direction. Under these conditions, ships must wait, often for days, before they can safely enter (or leave) the harbor.

SEISMIC SEA WAVES

Tsunamis, or seismic sea waves, are caused by sudden movement of the ocean bottom resulting from earthquakes or volcanic eruptions (Fig. 8-19). They have very long periods and behave like shallow-water waves even when passing through the deepest parts of the ocean. An earthquake in the Aleutian Islands on April 1, 1946, for example, caused a tsunami with a 15-minute period and a wavelength of 150 kilometers. Being in the Pacific Ocean, where the average depth is about 4300 meters, the wave speed was controlled by the bottom (L/D = 150/4.3); yet it still traveled about 800 kilometers per hour. In deep water such wave crests were estimated to be about a half meter high, which would be virtually undetectable to ships at sea.

Tsunamis eventually encounter a coast (Fig. 8-20), often with catastrophic results. For instance, as waves from the 1946 Aleutian earthquake hit the Hawaiian Islands, they were driven ashore as a rapidly moving wall of water, in a few places up to 6 meters high. Where the water was funneled into a valley, these waves also formed enormous breakers. More than 150 people were killed in Hawaii, and property damage was extensive.

Often tsunamis reach their greatest heights near their source. In the 1946 tsunami, the wave was highest in the Aleutian Islands, the location of the earthquake. A concrete lighthouse and radio tower 33 meters above sea level were destroyed at Scotch Cap, Alaska.

BOX 8-2

Rogue Waves

Waves interact with currents, winds, and other waves. Usually waves interact and cancel each other after leaving the area where they were generated.

When waves interact with currents, they are sometimes intensified, forming enormous waves that can sink even the largest ships. Off the coast of South Africa, for instance, large waves generated in the Southern Ocean arrive as swell and locally combine with waves from other storms. The Agulhas Current moving from the north and west meets these waves, causing them to steepen and to become shorter. Some waves become so steep that their forward face is close to breaking, leaving a deep hole in front of the wave. Ships have been known to sail into such holes, break up, and sink. In the midst of many large waves, such dangerous waves can often be seen only as the ship encounters them. Sometimes, such large waves can be seen at a great distance; then a ship can take action to avoid it. Many ships disappear each year, some probably because of meeting rogue waves at sea. Indeed, such waves may be the reason why some ships are lost each year in the Bermuda Triangle, near the Gulf Stream.

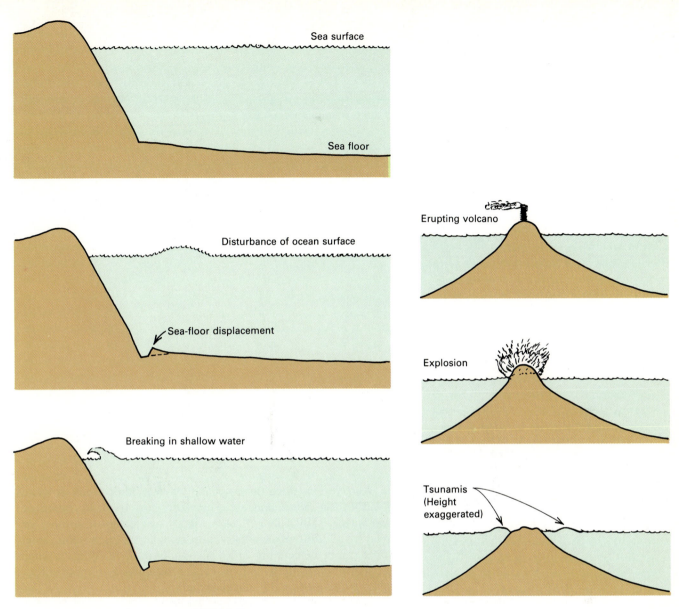

FIGURE 8-19
Tsunamis are caused by (a) displacements in the sea floor due to earthquakes or (b) volcanic explosions.

Tsunamis can be extremely destructive. The Hoei Tokaido–Nankaido tsunami of 1707 killed 30,000 persons and destroyed 8000 homes. Japan has recorded about 150 tsunamis. Today, Japan operates warning systems to predict tsunamis generated by nearby earthquakes. Some of the most vulnerable coastlines have extensive barriers to protect especially vulnerable locations.

Because of the frequent occurrence of tsunamis around the Pacific (Fig. 8-20), an international system, headquartered in Hawaii, issues warnings when earthquakes occur that might cause destructive tsunamis. As a result of these warnings, the number of deaths has been greatly reduced. For example, the 1957 tsunami killed no one in Hawaii even though water levels were higher than in 1946. Still, hundreds of people are killed each year by tsunamis.

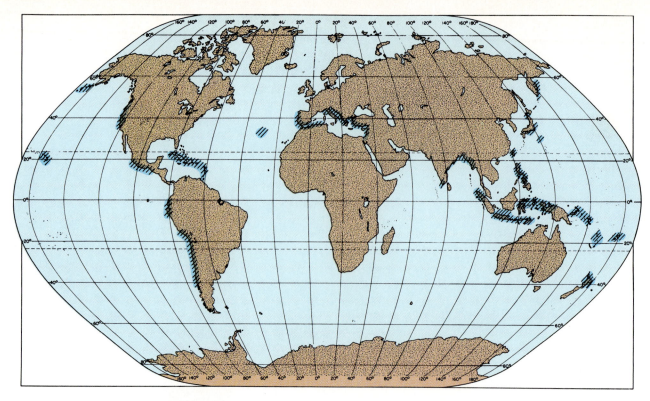

FIGURE 8-20
Areas affected by seismic sea waves generated by sudden movements of the sea floor, such as earthquakes or volcanic explosions.

INTERNAL WAVES

So far we have examined only surface waves. Other types of waves occur within the ocean and are, therefore, not as easily observed. **Internal waves,** shown in Fig. 8-21, occur on density interfaces between layers of different densities—for example, the pycnocline. Internal waves can be detected by studying temperature or salinity changes at a location. Some internal waves can be detected by remote sensing through their effects on the ocean surface.

Internal waves act like shallow-water waves. Surfaces where internal waves form involve only a small density difference between two water layers rather than the large density difference between air and water. As a result, internal waves usually have greater amplitudes than surface waves. They generally move much slower and have less energy than surface waves.

Internal waves can also break, just like surface waves. In the process they can cause mixing between the waters above and below the interface.

STANDING WAVES

Standing (or stationary) **waves** are yet another type of wave. They are also known as **seiches** (pronounced "saysh"). Such waves are important in lakes and in tidal phenomena. A simple standing wave can be made by tilting a round-bottomed dish of water and then setting it flat on a table. The surface will first tilt toward one side and then toward the other. This type of wave motion is distinctly different from progressive waves which move across water surfaces.

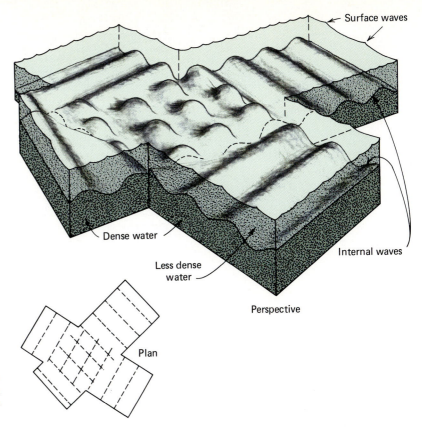

Surface waves

Internal waves

Dense water

Less dense water

Perspective

Plan

FIGURE 8-21
Simple internal waves at interfaces between water layers of different densities interact to form more complex wave patterns.

The water surface at the edge of the dish moves vertically. But along a line, usually in the middle of the dish, the surface does not move vertically. Instead, it acts as a hinge about which the rest of the water surface tilts. This stationary line (or sometimes a point) is called a **node.** The parts of the surface showing the greatest changes in elevation are called **antinodes.** More complicated stationary waves have more than one node and several antinodes.

In a seiche (shown in Fig. 8-22), maximum horizontal water movements occur at the nodes when the water surface is horizontal. When the water surface is tilted most, there is no water motion. In contrast to the continued orbital motion of the water in progressive waves, in a standing wave the water flows for a distinct period, stops, and then reverses its direction of flow. Also, the waveform alternately appears and disappears. The water movements are mostly horizontal rather than the circular or nearly circular and continuous orbits associated with progressive waves. Standing waves are characteristic of steep-sided basins and are well known in many lakes and in the tidal phenomena of nearly closed basins, such as the Red Sea.

Like progressive waves, standing waves are modified by their surroundings. They are reflected by vertical boundaries and absorbed by gently sloping bottoms. Standing waves are reflected by moving into depths substantially less than one-half their wavelength. Standing waves in large basins, such as the North American Great Lakes, are influenced by the Coriolis effect. The resulting wave, instead of simply sloshing back and forth, has a rotary motion around the edges of the basin.

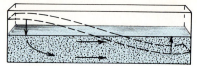

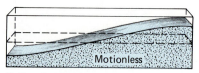

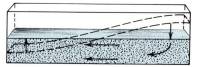

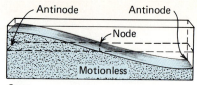

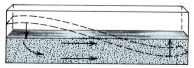

FIGURE 8-22
Simple standing wave with one node, shown at quarter-period intervals.

STORM SURGES

Storm surges are elevations of sea level caused by storm winds, usually extratropical cyclones and sometimes hurricanes. A storm surge is a large wave that moves with the storm or hurricane that caused it.

A storm surge can cause flooding of low-lying coastal areas, especially if the storm that caused it is slow moving. First comes a gradual change in water level, the **forerunner,** a few hours ahead of the storm's arrival. It is apparently caused by the regional wind system and can cause sea level to fall slightly along a wide stretch of coastline. When the storm center passes, it causes a sharp rise in water level called the **surge.** This surge usually lasts about 2 to 5 hours; rises in sea level of 3 to 4 meters have been observed—usually slightly offset from the storm's center. Combined with extremely high waves generated by the storm, surges can be extremely destructive.

Following the storm, sea level continues to rise and fall as storm-caused oscillations pass. They are more-or-less free surface waves and have been termed the *wake* of the storm, like the wake left by the passage of a ship through the water. These resurgences can be quite dangerous, particularly because they are often not expected once the storm itself has subsided.

Storm surges can be predicted, based on wind speeds and direction, fetch, water depth, and shape of the shoreline. Other factors, such as currents, astronomical tides, and seiches set up by storms, complicate the calculations. Better observations of large storms and the availability of larger computers permitting more complicated calculations have combined to permit more accurate predictions of storm surges.

Storm surges have caused catastrophic flooding many times. In 1900 Galveston, Texas, was destroyed and about 6000 people were killed by a storm surge resulting from a hurricane. In 1969, Hurricane Camille, the second strongest recorded storm to hit the Gulf Coast,

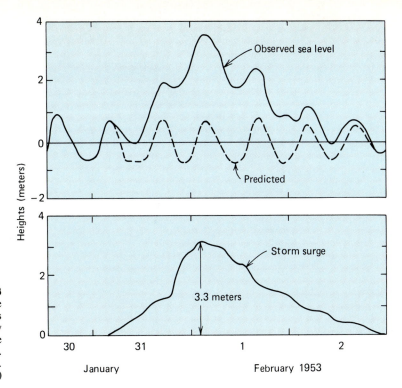

FIGURE 8-23
Storm surge of January–February 1953 in the North Sea on the Dutch coast. Sea-level changes resulting from the surge were estimated by subtracting the predicted tide level from the observed sea level. (Drawn from data in P. Groen. 1967. *The Waters of the Sea*. London: D. Van Nostrand Company, Ltd.)

caused $1.4 billion of damage; even with advance warning, it killed 256 people. A disastrous storm surge occurred in 1876 on the Bay of Bengal in the northern Indian Ocean, when 100,000 people were killed. In 1970, another storm surge hit the same area, killing an estimated half million people.

A northwest gale on 31 January–1 February 1953 blew across the North Sea with a fetch of 900 kilometers from Scotland to the Netherlands, causing sea level to rise more than 3 meters (Fig. 8-23). As a result of the storm surge combined with high tides and strong waves, waters broke through protective dikes and dunes, flooding low-lying areas on the Dutch and English coasts. Thousands of people were drowned.

WAVES ON BEACHES

Waves dominate beach processes. Currents and turbulence generated by waves stir up sediment, and **longshore currents** caused by the waves and tides transport sediment parallel to the coast (illustrated in Fig. 8-24). Transport generally takes place between the upper limit of wave advance on the beach and depths of about 15 meters. Large amounts of sand are transported in suspension. Relatively little sand is transported along the bottom.

Beaches change seasonally. During periods of low, long-period swell, sand is moved back onto the beach, usually building up the beach in height and width. Longshore bars migrate shoreward, filling in the troughs, and a new berm forms, usually at a level lower than the preceding one.

During periods of high, choppy waves (mainly in winter) beaches are cut back. The beach foreshore becomes more gently sloping, and a beach scarp forms as erosion proceeds. Strong longshore currents caused by the waves develop deep channels. Bars develop because of

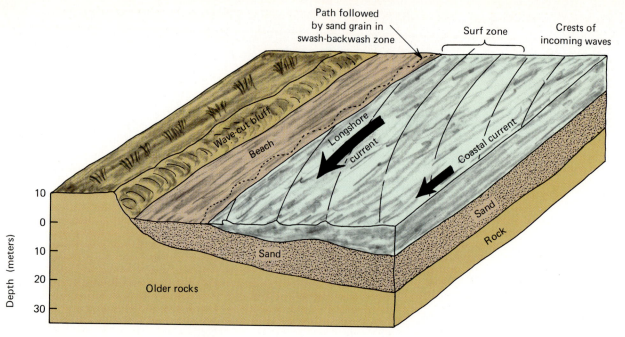

Path followed
by sand grain in
swash-backwash zone

Surf zone

Crests of
incoming waves

Longshore
current

Coastal current

Wave-cut bluff

Beach

Sand

Rock

Depth (meters)

10
0
10
20
30

Sand

Older rocks

FIGURE 8-24
Incoming waves are refracted and change directions as they enter shallow water. Striking the beach at an angle, they cause a longshore current in the surf zone that moves parallel to the beach.

offshore movements of sand from areas seaward of the breakers. Most of the sand removed from the beach is deposited nearby in the offshore zone to be moved back onto the beach during the next period of smaller waves.

Waves rarely approach a beach at right angles. Even though they are refracted on entering shallow water, so that crests are more nearly parallel to the coast, the process is rarely complete, and most waves approach the shore obliquely (Fig. 8-25). Wave energy acting parallel to the coast causes longshore currents that move in the same direction as the waves when they approach the coast. The current is strongest between the surf zone and the beach. The strongest currents are predicted to occur when the waves approach the shore from a 45° angle. This situation rarely happens; crests of most waves usually deviate less than 20° from being parallel to the beach when they strike the shoreline.

Each wave hitting the beach causes an uprush of a relatively thin sheet of water, or **swash,** onto the beach face. The water rises until all the energy of the oncoming wave is dissipated or until the water moved by the wave percolates downward into the sand. Any water remaining on the surface runs back down the slope of the beach face.

Waves rarely strike the beach head on (see Fig. 8-24) but rather at an angle. Thus the swash rises obliquely across the beach face. When the water with its entrained sediment runs back, it goes directly down the slope of the beach. Sand moving along the beach as a result of wave effects is called **littoral drift.** Its direction can change during a single day or over a season. This is a small-scale phenomenon, caused by waves hitting the beach.

Wave-induced sediment movements can be surprisingly fast—up to 25 meters per hour and 1 kilometer per day. A more typical rate would be 5 to 10 meters per day on the average.

When a wave breaks, there is substantial water movement. Water at the surface and along the bottom moves toward the beach, carrying with it materials floating or dragged along the bottom, which is why a beach acts as a convergence zone, collecting all sorts of debris as well as sediment.

FIGURE 8-25
Waves strike the reef at an angle on Angaur Island, Belau. Their approach at an angle causes longshore currents. (© Douglas Faulkner/Photo Researchers.)

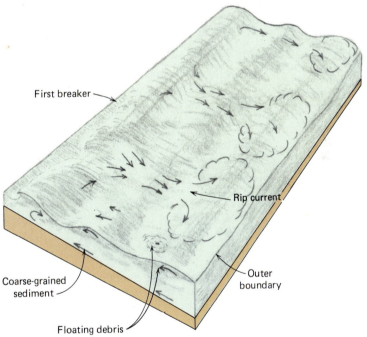

First breaker

Rip current

Coarse-grained sediment

Outer boundary

Floating debris

FIGURE 8-26
Schematic representation of breaking waves and associated water movements. Water depth at the outer boundary is approximately half the wavelength for storm waves. Note the regular spacing of the rip currents.

Rip currents are manifestations of water movements toward the beach (Fig. 8-26). After moving toward the beach in the breaking waves, water tends to flow parallel to the beach for short distances until it enters a narrow stream of return flow through the breaker zone—the **rip current.** In rip currents the most rapid flow is relatively narrow and has speeds of up to 1 meter per second until it reaches a distance of perhaps 300 meters from the coast.

FIGURE 8-27
Beach cusps at "Big Beach" near Makena, Maui, Hawaii. (© Carl Purcell/Photo Researchers.)

Seaward of the breaker zone, the current becomes more diffuse and spreads out, forming a *head* to the current. The water is caught up at this point in the general flow toward the beach. Rip currents and their associated water movements form a cell-like nearshore circulation system within the breaker zone.

Waves striking certain beaches cause **beach cusps,** which are rather uniformly spaced tapering ridges, with rounded embayments between them, as illustrated in Fig. 8-27. The regular spacing ranges from less than 1 meter to several tens of meters and is related to wave height. Higher waves are associated with wider spacing of cusps. There cusps form and re-form quickly, especially on fine-sand beaches.

SUMMARY

Ocean waves are disturbances of water surfaces caused by energy from winds, earthquakes, or volcanic explosions. Ideal waves are most easily studied. Mathematical techniques are often used to describe waves and to deal with their complexity.

Water parcels move in circular orbits as waves pass, but return to the same location in deep-water waves, where water depths exceed half of the wavelength. They are unaffected by the bottom. Shallow-water waves, where water depths are less than one-twentieth the wavelength, are affected by the bottom; water-parcel motions are essentially parallel to the bottom.

Large ocean waves can be formed by sudden disturbances such as large movements of the ocean bottom due to earthquakes or volcanic explosions. Free waves move independently away from their source. Forced waves remain under the influence of the force that caused them.

Waves are classified by size. The smallest are the capillary waves, also called ripples. These are round-crested, with V-shaped troughs. Capillary wavelength is less than 1.7 centimeters and wave period is less than 0.1 second; capillary forces dominate. Larger waves are most common, having periods of up to 5 seconds. Gravity is the dominant restoring force.

Most waves are generated by winds blowing across the ocean surface. Where waves are forming, the ocean surface is chaotic; this is called a sea. Ripples form first and then grow into larger waves as winds continue to put energy into the water surface. Outside the wave-generating area, the waves separate themselves according to size, forming swell, where waves are regular and smooth-crested. Longer waves travel faster than short ones. Most ocean waves are less than 6 meters high. The highest waves occur in the Pacific and in the South Atlantic.

Waves are altered when they enter shallow water. Waves change direction by refraction, which occurs when wave orbits drag on the bottom, moving slowest in shallow waters and faster in deeper waters. Waves can also be reflected to move in different directions. As the waters get shallower, waves eventually become unstable and break, forming breakers.

Waves can also form under the sea surface where there are density boundaries between layers. Seiches (standing waves) occur in basins when wave energy is reflected; the waveforms do not move but the water surface tilts in a regular manner.

Very strong, prolonged winds associated with storms can cause large, relatively slow-moving waves, called storm surges; such waves can cause flooding of low-lying coastal areas.

Energy can be extracted from waves by devices floating in the water and by fixed structures.

STUDY QUESTIONS

1. Draw an ideal wave. Label its parts.
2. Explain the differences between shallow- and deep-water waves.
3. List the causes of waves in the ocean. Where is each most important in causing waves in the open ocean.
4. Explain the differences between gravity and capillary waves.
5. Define sea and swell. Describe how each forms.
6. What limits the maximum size of waves in the ocean?
7. Describe tsunamis (seismic sea waves). How do they form?
8. What areas are most likely to experience seismic sea waves? Discuss the relation of these source areas to plate tectonic processes.
9. Draw a diagram showing a simple standing wave in a basin. How do standing waves differ from progressive waves?
10. Describe a storm surge. What causes storm surges?
11. Why are storm surges more likely from an extratropical storm than from a hurricane?
12. Describe a rip current. Discuss the forces that cause them.

SELECTED REFERENCES

Bascom, W. 1980. *Waves and Beaches: The Dynamics of the Ocean Surface,* rev. ed. Garden City, N.Y.: Doubleday Anchor Books. 366 pp. Elementary.

Clancey, E. P. 1968. *The Tides.* Garden City, N.Y.: Doubleday. 228 pp. Nontechnical.

Myles, D. 1986. *The Great Waves.* London: Robert Hale. 206 pp. Discusses catastrophic waves.

Russell, R. C. H., and D. M. MacMillan. 1954. *Waves and Tides.* London: Hutchinson. 348 pp. Elementary.

The flooding tide moving into the bay creates a tidal bore at Truro, Nova Scotia, Canada. (© G. Whiteley/Photo Researchers.)

CHAPTER 9

Tides

OBJECTIVES _____

1. To describe the general features and causes of ocean tides;

2. To describe tides and their behavior in open ocean basins;

3. To describe the behavior of tides in coastal waters;

4. To describe tidal currents, especially in coastal waters.

*T*ides are the pulse of the ocean. Their effects are felt most keenly in coastal areas where the periodic rise and fall of sea level, which alternately submerges and exposes the shallow ocean bottom, modulates plant and animal life and behavior. In this chapter, we discuss:

General features of tides;
Processes causing tides;
Tides in ocean basins;
Tidal currents, especially in the coastal ocean; and
Energy from tides.

TIDES

Tides are the periodic rise and fall of the sea surface. They are easily observed and measured. All that is required is a measuring pole attached to a post or stuck firmly in the bottom. At intervals—perhaps hourly—we record the height on the pole of the still-water surface. The height of the water surface plotted at each interval of time produces a **tidal curve.**

More elaborate installations are needed for continuous tidal observations. A simple mechanical *tidal station* is shown in Fig. 9-1(a). A basin with a restricted intake connects to the ocean. Thus the water level in the basin corresponds to the undisturbed sea level outside but is not disturbed by waves. A float on the water surface in the basin is connected to a marker which plots the tidal curve on a clock-driven, paper-covered drum. Modern tide gauges work automatically, and observations are recorded electronically for later computer processing and simultaneously transmitted by satellite to a central recording station [Fig. 9-1(b)].

TYPES OF TIDES

There are three types of tides in the ocean. They are differentiated by the number of high tides and low tides per day and their relative heights. Along most coasts, tidal curves have two high tides and two

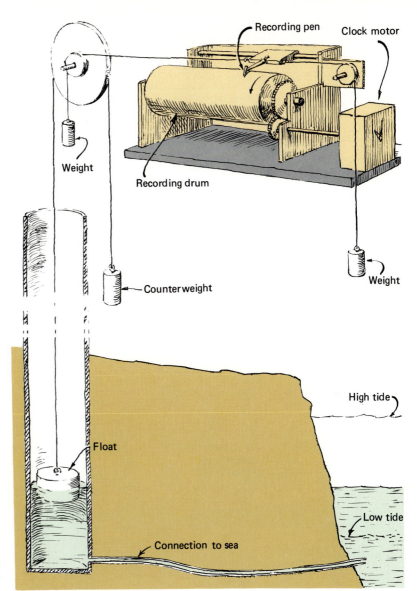

FIGURE 9-1

(a) Diagram of a simple mechanical tide gauge. (b) Modern tide gauges use acoustic techniques to determine sea level. A sound pulse is transmitted and reflected off the surface. The echo is detected by the sensor. The signal is processed at an installation ashore and transmitted by satellite to a central computer for further processing and data storage.

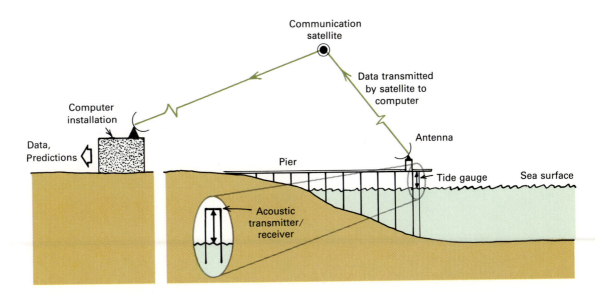

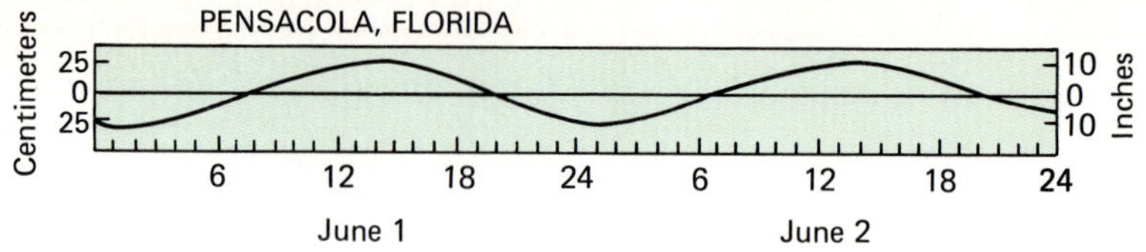

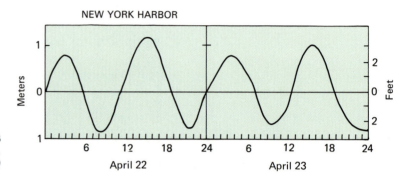

FIGURE 9-2
Tidal curve for a daily tide. (After Marmer. 1939.)

FIGURE 9-3
Tidal curve for a semidaily tide. (After Marmer. 1939.)

FIGURE 9-4
Tidal curves for mixed tides at Pacific ports. (After Marmer. 1939.)

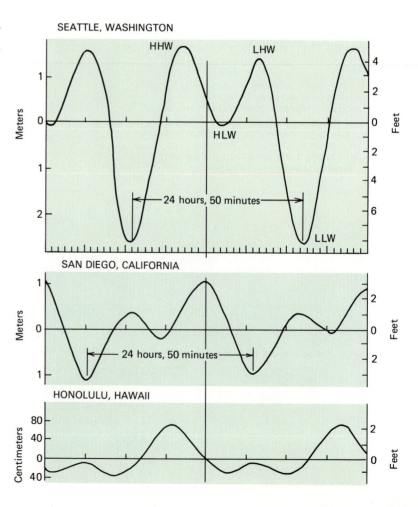

low tides per **tidal day**—24 hours, 50 minutes. This period corresponds to the time between successive passes of the Moon over any point on Earth. The time between successive high (or low) tides is known as the **tidal period.**

Some ocean areas, such as parts of the Gulf of Mexico, have only one high tide and one low tide each day (Fig. 9-2). These are called **daily** or **diurnal tides.**

When the two high and two low tides are approximately equal, they are called **semidaily** or **semidiurnal tides** (Fig. 9-3). Such tides are relatively easy to predict because high tides tend to occur at a known time after the Moon has crossed the longitude for that port. Tidal predictions for ports with semidaily tides have been made for centuries, based on the obvious relationship with the lunar cycle.

Still more complicated tidal curves show two high tides and two low tides per tidal day. The highs usually differ in height as do the low tides. These are called **mixed tides** and are shown in Fig. 9-4. The higher of the two high tides is called **higher high water** (abbreviated HHW); the lower is called **lower high water** (LHW). There are similar terms for the low tides—**lower low water** (LLW) and **higher low water** (HLW).

Mixed tides are more difficult to predict than semidaily tides. Timing of high and low tides stands is not simply related to the Moon's passage over the port.

From a record of only a few days' length, it is possible to determine the type of tide for any harbor (Fig. 9-5). (Typical tidal curves are shown in Fig. 9-6.) We can, for example, measure the **tidal range** (the difference between the highest and lowest tide levels) and the **daily**

FIGURE 9-5
Types of tides and spring tidal ranges (in meters) on North American coastlines. (After U.S. Navy Hydrographic Office. 1968.)

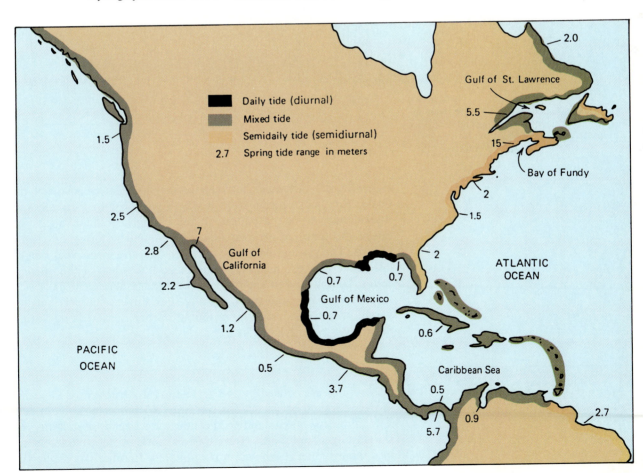

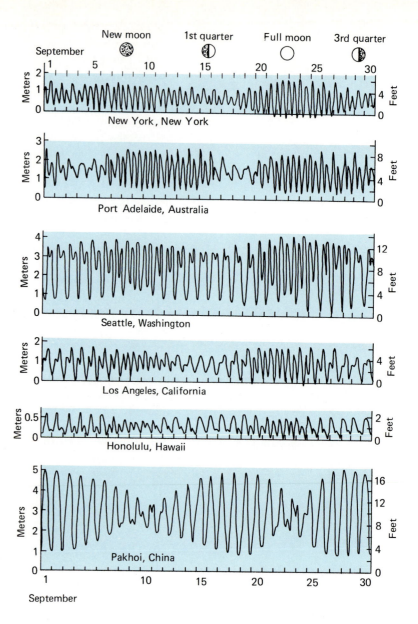

FIGURE 9-6
Tidal variations during a month. Note the variations in the time of the spring and neap tides relative to the new and full moons. (After Marmer. 1939.)

inequality (the difference between the heights of successive high or low tides). But tides change from week to week. With a record several weeks long, we see a pattern in the changes of tidal range. **Spring tides**—the times of greatest tidal range—occur during full and new Moons. The spring tidal range is larger than the **mean tidal range** (the difference between mean high and mean low tides) or the *mean daily range*. During the first and third quarters the tidal range usually is least; these are called **neap tides.** As Fig. 9-6 shows, there is substantial variation in the tides during a month. Other, less striking variations occur over periods of years.

TIDE-GENERATING FORCES

Ocean tides are caused by the gravitational attraction of the Sun and Moon acting on ocean waters. *Sir Isaac Newton* (1642–1727) laid the foundation for understanding the tides. He began by making several simplifying assumptions. He assumed a static ocean completely covering a nonrotating Earth with no continents. In other words, he started

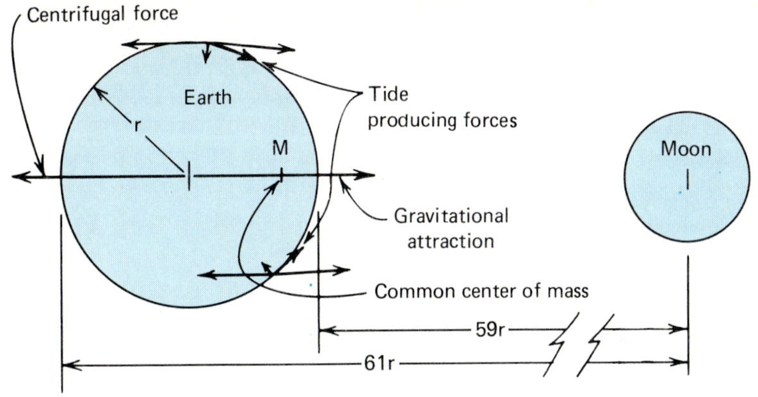

FIGURE 9-7
Schematic representation of the Earth-moon system revolving around their common center (*M*). Note that the side nearest the moon is about 59 Earth radii (*r*) away from the moon compared with 61 Earth radii on the side farthest from the moon. At the center of the Earth, the moon's gravitational attraction is balanced by centrifugal forces. Small unbalanced forces act on the Earth's surface to deform the ocean surface and cause the tides.

with the simplest case and then progressed to more complicated but more realistic cases.

Gravitational attraction pulls the Earth and Moon toward each other. Centrifugal forces, acting in the opposite direction, keep them apart, as illustrated in Fig. 9-7. The Earth and Moon thus behave like twin planets revolving about a common center, which, in turn, moves around the Sun. If the Earth and Moon were the same size, the center of revolution of the system would be located midway between them. The Moon, however, is only about $\frac{1}{82}$ the mass of the Earth. Consequently, the center of revolution of the Earth-Moon system is located within the Earth, about 4700 kilometers from the Earth's center. This situation is analogous to an adult and a small child on a seesaw. The adult must sit closer to the pivot than the child to achieve balance.

There are small but significant unbalanced forces in the system. Consider first the Moon's gravitational attraction on the Earth's surface. The force of gravity is inversely proportional (decreases as the distance increases) to the square of the distance separating two objects. Doubling the separation between the Earth and Moon reduces the force of gravity to one-fourth. A parcel of water located on the Earth's surface is only 59 earth radii away at a point nearest the Moon but 61 Earth radii away on the opposite side (see Fig. 9-7). Therefore, the Moon's gravitational attraction is greatest on the side nearest the Moon and least on the far side.

Centrifugal forces, however, are equal over the Earth's surface. On the side nearest the Moon, the Moon's gravitational attraction exceeds the centrifugal force. Thus water is pulled toward the Moon. On the opposite side of the Earth, the centrifugal forces are greater than the attraction of the Moon. There, too, a small force acts on the water, effectively pulling it away from the Earth. The Earth's gravity prevents the water in the two bulges from being pulled away from the Earth. These unbalanced forces on the Earth's surface, shown in Fig. 9-8, are the tide-generating forces associated with the Moon. A similar argument holds for the gravitational attraction of the Sun.

To understand how tides are created, consider these forces in more detail. Forces can be represented by *vectors* (shown in Fig. 9-9 as arrows) that point in the direction in which the forces act. The length of the arrow (the vector) corresponds to the relative strength of the force. Each vector can be resolved into two components. One component acts in a vertical direction, perpendicular to the Earth's surface. The other acts in a horizontal direction, parallel to the Earth's surface at that point (see Fig. 9-9). The forces act toward the Moon on one side of the Earth (and away from the Moon on the opposite side). The relative

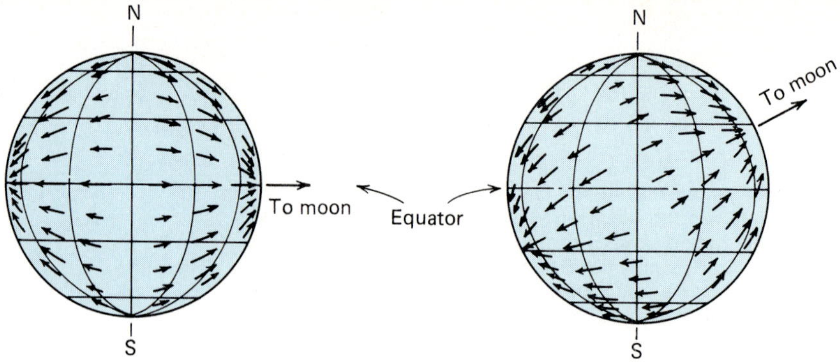

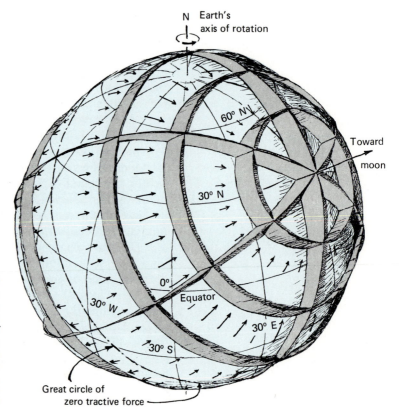

FIGURE 9-8
Horizontal component of the tide-producing forces acting on the ocean surface when the moon is in the plane of the Earth's equator (left) and when the moon is above the Earth's equator (right). Note that the tide-producing forces shift their orientation as the moon's position changes. (After U.S. Naval Oceanographic Office. 1958.)

FIGURE 9-9
Tide-generating forces predicted by the equilibrium-tide model when the moon is above the Earth's equator. The strength of the forces is shown by the length of the arrows. The height of the resulting tide or tidal bulge is shown by the height of the "fences." Note the variations in tidal height (along any parallel of latitude) that occur as the Earth rotates under the fences, which remain fixed in location with respect to the moon. (After Von Arx. 1962.)

strength of forces acting in a horizontal sense (parallel to the Earth's surface) varies over the Earth, as shown in Fig. 9-9).

Directly beneath the Moon and on the opposite side of the Earth, tide-generating forces act solely in a vertical direction. But these vertical components have little effect, for they are counteracted by the Earth's gravity, which is about 9 million times stronger. Horizontal components of the tide-generating force are also weak, but they are comparable in strength to other forces acting on the ocean's surface. It is thus these horizontal forces that cause the tides.

EQUILIBRIUM TIDE   Having considered the tide-generating forces, we can begin to predict tides. The **equilibrium tide theory** predicts some aspects of tides. For instance, it predicts that tidal bulges occur on the side of the Earth nearest the Moon and on the side opposite the Moon. As a result of the horizontal forces previously discussed, the water covering the Earth

228

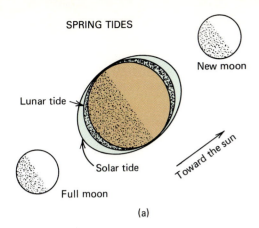

SPRING TIDES

New moon

Lunar tide

Solar tide

Toward the sun

Full moon

(a)

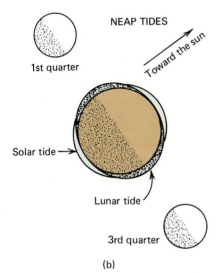

NEAP TIDES

1st quarter

Toward the sun

Solar tide

Lunar tide

3rd quarter

(b)

FIGURE 9-10
Relative positions of sun, moon, and Earth
during spring and neap tides.

forms an egg-shaped water envelope on our imaginary nonrotating water-covered Earth. The solid Earth also responds to these tide-generating forces and deforms slightly but much less than the ocean surface. There are also tides in the atmosphere.

If the Earth rotates inside its deformed watery covering, equilibrium tide theory explains successive high and low tides. When the Moon is in the plane of the Earth's equator, the tidal bulges of the equilibrium tide will also be centered on the equator. A tide gauge on the equator registers high tide when that point is directly under the Moon. After the Earth rotates 90°, the tide gauge registers low tide, in other words, it is now midway between the two tidal bulges. After the Earth rotates another 90°, the tide gauge is directly opposite the Moon and now registers high tide.

This simplified model explains semidaily tides with two equal high tides and two equal low tides per tidal day. Remember that the Earth rotates inside its slightly deformed water cover. The more-or-less egg-shaped deformed water surface, corresponding to the equilibrium tide, remains fixed in space, its position determined by the location of the Moon in our simplified case.

The Moon, however, does not maintain a fixed position relative to the Earth. It moves from 28.5° north of the equator to 28.5° south of the equator each month. When the Moon changes its position, so does the orientation of the tide-generating forces and the position of the equilibrium tide.

In order to demonstrate the effect of the Earth's rotation on the height of the tides, imagine a purely theoretical case of tide-generating forces on a water-covered Earth. (We identify points on the Earth's surface by their real names, but the conditions described do not necessarily apply at those points. Remember that the example illustrates conditions as they would be without continental barriers to water movements.) For instance, if the Moon were over the site of Miami, Florida, there would be a tidal bulge there and also one near the west coast of Australia. These bulges would be particularly well developed because of being in line with the tide-generating forces. After the Earth rotated 180°, these points would again experience high tide but a lower one because Miami and Australia would no longer be in line with the Moon. Tide-generating forces would now be greatest off northern Chile and in the Bay of Bengal–South China Sea area. (Using a globe will be helpful in demonstrating this point.) As this example shows, the equilibrium theory of tides explains semidaily tides and daily inequalities.

A similar analysis could be made for the Sun-Earth system to determine the Sun's influence on the tides. There are, however, several important differences. First, the greater distance of the Sun from the Earth (approximately 23,000 Earth radii) is only partially compensated for by its greater mass (330,000 Earth masses). Thus the tide-generating effect of the Sun is only about 47% as strong as that of the Moon. Second, solar tides have a period of about 12 hours rather than the 12-hour, 25-minute period of lunar tides. Finally, the Sun's position relative to the Earth's equator also changes—from 23.5° north to 23.5° south of the equator—but it requires a full year to make the complete cycle, in contrast to the monthly changes of the Moon's position.

The equilibrium theory also explains spring and neap tides. Spring tides occur every 2 weeks, usually within a few days of the new and full Moons (see Fig. 9-6). Considering their positions in space (shown in Fig. 9-10), we see that during the time of the full and new Moons, forces associated with the Sun and Moon act together, causing large tidal bulges and hence greater tidal ranges. During the first and third quarters of the Moon, tide-generating forces partially counteract each other by acting in different directions, resulting in the lowest tidal range—the neap tides.

DYNAMICAL THEORY OF TIDES

Since Newton's time, scientists have investigated tides by considering how the ocean responds to the tide-generating forces. A more realistic approach involves a dynamic rather than a static ocean. Like the equilibrium theory, it also requires simplifying assumptions in order to deal with complicated interactions. Among them is the neglect of vertical forces. This is not too serious because they are quite small compared with the Earth's gravitational attraction. Perhaps the most important difference between the dynamical treatment and the equilibrium theory is that the former considers the effects of a basin's shape and depth on tides. It also recognizes that tides involve water movements and must therefore include the Coriolis effect.

The dynamical theory treats tides as waves which can be resolved mathematically into several components and each treated separately. (We discussed this in the last chapter.) Tide-generating forces can also be resolved into **tidal constituents,** the most important being those due to the moon and the sun. Because of the changing positions of these bodies relative to the Earth, as many as 62 tidal constituents are used to make tidal predictions. The four principal ones usually account for about 70% of the tidal range.

The response of any bay or harbor—or an entire ocean basin—to each tidal constituent can be considered separately as a **partial tide.** The tide for any location thus can be calculated by combining partial tides (illustrated in Fig. 9-11), just as complicated waves in a sea can be reconstructed by combining several simple wave trains.

Study of tidal curves indicates how partial tides must be combined for a given port. Then tidal predictions are made by combining partial tides, using computers. Some of the earliest tide-predicting machines were simple mechanical or analog computers. More powerful computers permit the use of more sophisticated tidal models, requiring fewer simplifying assumptions.

Tides are very long period waves. Neglecting the Earth's curvature, we see that the two water bulges are the crests, and the intervening low areas are the troughs of a simple wave. Because of their immense size relative to the ocean basins, tides behave as shallow-water waves. Their wavelength, one-half of the Earth's circumference, is about 20,000 kilometers. The ocean basins average about 4 kilometers deep, thus $D/L = 4/20,000$. This is much smaller than the limit of $\frac{1}{20}$ for shallow-water waves. If tides were free waves, they would move at a speed of about 200 meters per second (720 kilometers per hour). But in order to "keep up with the Moon," the tides need to move around the Earth in 24 hours, 50 minutes. This means that they must move 1600 kilometers per hour at the equator.

For the tide waves to move fast enough at the equator to keep up with the Moon, the ocean would need to be about 22 kilometers deep. Because it is much shallower, the tidal bulges move as forced waves whose speed is determined by the movements of the Moon. The position of tidal bulges relative to the Moon is determined by a balance between the attraction of the Sun and Moon and frictional effects of the ocean bottom.

FIGURE 9-11
A daily and semidaily tide combine to produce a mixed tide. (After Marmer. 1926.)

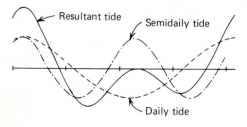

Resultant tide
Semidaily tide
Daily tide

Because the ocean is cut by the continents into north-south basins, it is impossible for the tide to move east-west across the Earth as a forced wave. Only around Antarctica can the tide move westward unimpeded around the Earth.

To see how the tide behaves, we will examine the tide in the Atlantic Ocean. After the tide enters the South Atlantic from the Southern Ocean, it moves northward. In the South Atlantic its course is relatively simple, although influenced by irregular boundaries and bottom topography. As it progresses northward, the wave from the Antarctic tide also interacts with the independent tide generated in the Atlantic.

Even if the Atlantic Ocean were completely isolated, it would still have tides. The tide in a basin like the Atlantic is somewhat similar to a standing wave, besides having some characteristics of a progressive wave. Every basin has a natural period for standing waves. If that natural period is near 12 hours, the standing-wave component of the tide will be well developed. If the period of the basin is much greater than or much less than 12 hours, the tide is less like a standing wave.

Each basin has a distinct tide. Tides in the North Atlantic are altered by reflections of tide waves from the complicated coastline. A standing wave is set up, but it is not the simple standing wave discussed previously. Since waters move over substantial distances and for long times, the wave is modified by the Coriolis effect. The result is a swirling motion such as we might get if we swirled water in a round-bottomed cup. Let us see why such a wave forms and how it behaves.

We start with a channel in the Northern Hemisphere and set water moving northward associated with a standing wave. Because of the Coriolis effect, water will be deflected toward the right, causing a tilted water surface, as shown in Fig. 9-12. When the water flows back, its surface will tilt in the opposite sense. The tilted wave surface rotates around the basin, once during each wave period. Near the center of such a system, called an **amphidrome system,** is a point where the water level does not change, the **amphidromic point.**

Figure 9-13 shows the movement of the high water associated with the principal lunar partial tide in the Atlantic. There we see a large amphidrome system in the North Atlantic. Other, smaller amphidrome systems (not shown on the map) occur in the English Channel and the North Sea. Points located near an amphidromic point for one partial tide will not be affected by that partial tide but will probably be affected by another. For example, an area near the amphidromic point for a semidaily tide may well have a daily-type tide.

Because of their dimensions (and hence their natural periods) ocean basins respond more readily to certain constituents of the tide-generating forces than to others. The Gulf of Mexico has a natural period of about 24 hours. Therefore, it responds more to the daily tidal constituents than to the semidaily constituents, and much of the Gulf has a daily tide (Fig. 9-5). The Atlantic Ocean, on the other hand, responds readily to the semidaily constituent of the tide-generating forces. Thus it tends to have a semidaily type of tide. Because of their size and shape, the Caribbean Sea and the Pacific and Indian oceans respond to both the daily and semidaily tidal forces. Thus they have mixed tides.

Each basin is affected by the tide in much the same way the Atlantic is affected by the tide moving around the Antarctic. A tidal

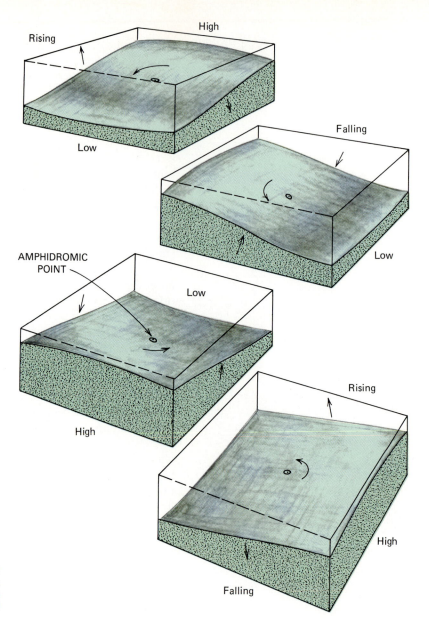

FIGURE 9-12
Amphidromic motion in a standing-wave tide in a basin in the Northern Hemisphere. (After Von Arx. 1962.)

wave advancing through the bay is reflected by the coast. If the basin has a natural resonance of the appropriate period, a standing wave may be excited as well. The tide in a bay, harbor, or sea is greatly influenced not only by the magnitude of the ocean tide at its mouth but also by the natural period of its basin and by the cross section of the opening through which the tide wave must pass. The small tidal range (less than 0.6 meter) of the Mediterranean Sea, for example, results from the small opening through the narrow Strait of Gibraltar into the Atlantic, which inhibits exchange of water during each tidal cycle. The small tidal range in several marginal seas of the Pacific can be similarly explained.

Where the natural period of a basin is near the tidal period, it is possible to set up a large standing wave, giving rise to exceptional tidal ranges. One example is the Bay of Fundy in the Canadian Maritime Provinces, whose natural period is apparently about 12 hours. Spring tides in the inner part of the bay have ranges of 15 meters (Fig. 9-14)

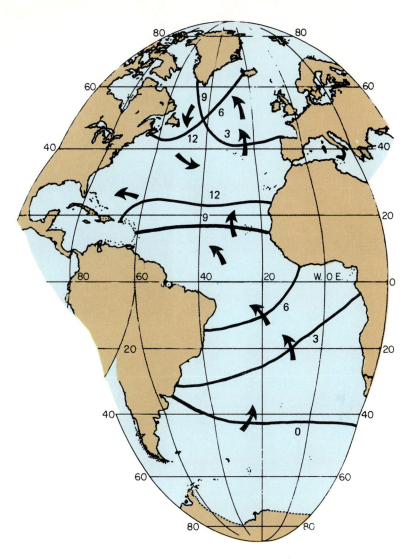

FIGURE 9-13
Locations of the high water of the principal lunar partial tide—the major tidal component caused by the moon. The figures on the dark lines indicate the number of hours since the moon crossed the latitude of Greenwich. (After Defant. 1958.)

because of an especially favorable situation. Large tidal ranges (about 13 meters) also occur on the Normandy coast of France and at the head of the Gulf of California (about 7 meters).

TIDAL CURRENTS

Tidal currents are horizontal water movements caused by the tides. But relationships between tides and tidal currents are not always obvious. Some seacoasts have no tidal currents, and a few have tidal currents but no tides.

Let us begin by examining currents caused by a tide which is a simple progressive wave. As we have seen, tides are shallow-water waves because of their extreme length and the relative shallowness of the ocean basins. In such a tide, the crest of the wave is high tide, and the trough is low tide. Orbital motions of water caused by the tide are ellipses, greatly flattened circles with their long axes parallel to the ocean bottom. In other words, most of the water motions associated with the tides consist of horizontal motions, with little vertical motion involved.

FIGURE 9-14
(a) High tide and (b) low tide at Halls Harbor, Nova Scotia on the Bay of Funday. (© Comstock/Photo Researchers.)

The familiar **reversing tidal currents,** in which water periodically flows in one direction for a while and then reverses to flow in the opposite direction, occur in restricted waters, such as harbors. Such tidal currents can be compared with water movement associated with the passage of a progressive wave (Fig. 9-15). As the wave crest moves into the harbor, the water flows in; this corresponds to the **flood current.** As the wave trough moves into the harbor, the water flows out of the harbor. This corresponds to an **ebb current.** Each time that the current changes directions a period of no current, known as **slack water,** intervenes.

Unfortunately, tides and tidal currents are rarely that simple. Progressive waves are reflected by the shore, so the tide observed in most coastal areas consists of several progressive waves moving in different directions, as well as the standing-wave component. Other complicating factors include frictional effects on the waves and non-tidal currents, such as river discharges. As a result, there is no simple relationship between high or low tides and the times of slack water or maximum currents. Just like the tides, tidal current predictions depend

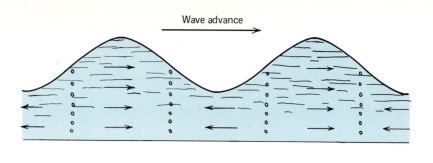

Wave advance

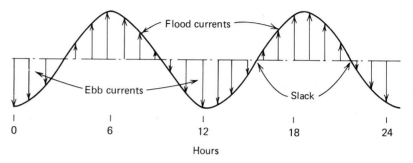

FIGURE 9-15
Relationship of tide and tidal currents in an idealized tide, consisting of a simple progressive wave tide.

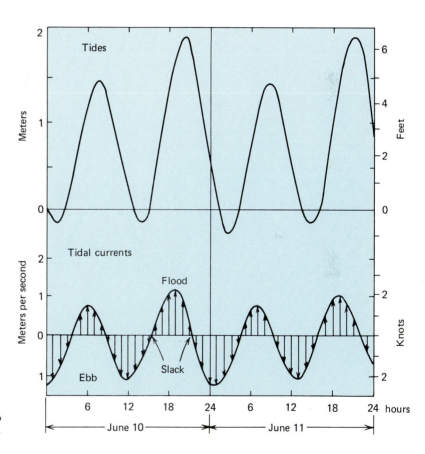

FIGURE 9-16
Tides and tidal currents, New York Harbor.

on the analysis or observations of tidal currents taken over a long period of time.

Tidal currents through a tidal period in New York Harbor are illustrated in Fig. 9-16. Beginning with slack water before high tide, the flood current increases until it reaches a maximum and then decreases again until it is slack water about an hour after high tide. After this slack, the current ebbs as the tide falls. The current again reaches a maximum and then decreases until about 2 hours after low tide, when it is slack water again. Elsewhere in New York Harbor, slightly different tidal current patterns may be observed. Tidal currents nearshore, for example, usually turn sooner than in the middle of the channel because of friction along the channel walls. Tugboat captains often take advantage of these nearshore tidal currents to avoid bucking strong midchannel currents.

Tidal currents are altered by winds or river runoff. River runoff, for example, can prolong and strengthen ebb currents because more water must move out of the harbor on the ebb than comes in on the flood. Also, different tides will have different tidal currents. In areas with large daily inequalities between successive high (or low) tides, there may be days with continuous ebb (or flood) currents that change in strength during the day.

The strength of a tidal current depends on the volume of water that must flow through an opening and the size of the opening. Thus it is not possible to predict the strength of the tidal current, given only the tidal range. The large tidal range in the Gulf of Maine, for instance, is accompanied by weak tidal currents. Conversely, Nantucket Sound has strong tidal currents but a small tidal range. At a given location, however, the relative strength of the tidal currents is generally proportional to the relative tidal range for that day. A spring tide, for instance, is usually accompanied by stronger tidal currents than a neap tide. In general, tidal currents are the strongest currents in coastal regions.

FIGURE 9-17
Average tidal currents during one tidal period at Nantucket Shoals off the Massachusetts coast. The arrows indicate the direction and speed (shown by arrow length) of the tidal currents for each hour during a tidal period. The outer ellipse indicates the movements of a log (tied by an elastic line) in the water in the absence of winds or other currents. (After Marmer. 1926).

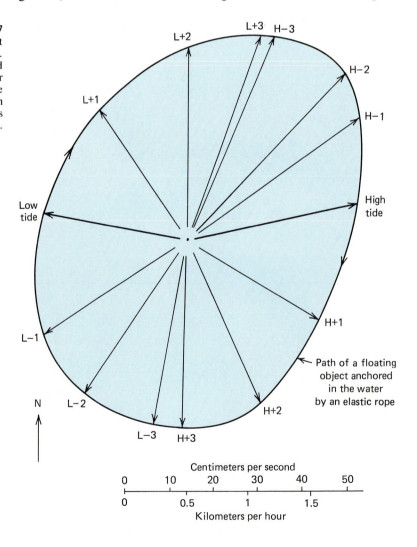

In the open ocean, **rotary tidal currents** exhibit patterns quite different from the reversing tidal currents in coastal areas. Rotary tidal currents continually change direction. Instead of slack-water periods with no current, such as we find in coastal areas, there are periods in which the current is at a minimum. In the Northern Hemisphere the current usually changes direction in a clockwise sense. Open-ocean tidal currents are generally weaker, typically about 30 centimeters per second (slightly more than 1 kilometer per hour), than in coastal waters.

A log anchored by an elastic tether near Nantucket Shoals (off the coast of Massachusetts) during a simple semidaily tide would move in a clockwise sense in an elliptical path, returning to its initial point after 12 hours, 25 minutes, as indicated in Fig. 9-17. The absence of wind or nontidal currents is again assumed.

In areas with mixed tides, with a substantial inequality, the current ellipse is more complicated (Fig. 9-18) because there are two different ellipses. A log placed in such a current pattern moves through two elliptical paths and returns to its original position after 24 hours, 50 minutes, assuming no winds or nontidal currents.

COASTAL-OCEAN TIDES

Tides in the coastal ocean affect estuaries and their tributaries. The tide enters an estuary as a progressive wave and travels upstream. In many estuaries the wave is gradually damped (reduced in height) by friction of the channel and opposing river flow. But in some estuaries the tide wave reaches the end of the estuary and is reflected back.

FIGURE 9-18
Average tidal currents during one tidal day at the entrance to San Francisco Bay. The radiating arrows show the direction and speed (shown by arrow length) of tidal currents for each hour of the tidal day. The outer ellipse indicates the movements of an object in the water (tied by an elastic band) in the absence of winds or other currents. (After Marmer. 1926.)

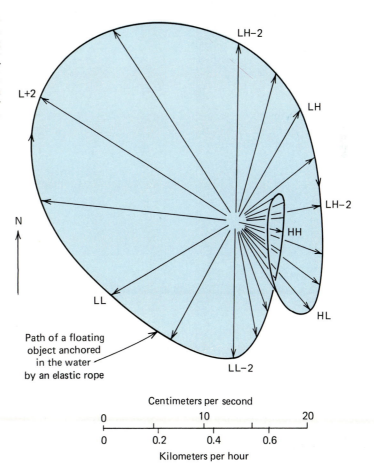

Chesapeake Bay (Maryland, Virginia) has relatively simple tidal currents associated with a tide that behaves primarily like a progressive wave. To follow the changing tidal currents, we begin when the waters are slack at the entrance, as shown in Fig. 9-19(a). From just inside the entrance, in the southern part of the bay, up to a second slack-water area, about midway up the bay, tidal currents are ebbing. In the northern part of the bay, beyond the second area of slack water, tidal currents are flooding, and we can see that areas of slack water separate sections with ebb and flood currents.

Two hours later both slack-water areas have moved into the bay about 120 kilometers, as shown in Fig. 9-19(b). By this time tidal currents are flooding at the bay entrance, and the pattern we observed previously has been displaced northward.

Four hours later the southernmost slack-water areas has reached the entrance to the Potomac River and flood-tidal currents at the mouth have diminished somewhat in strength, as shown in Fig. 9-19(c). A final look at Chesapeake Bay 6 hours after slack water before flood shows

FIGURE 9-19
Tidal currents in Chesapeake Bay show a progressive-wave tide in an estuary. Slack water before flood begins at the mouth of the bay; (b) two hours after flood begins; and (c) 4 hours after flood begins. Arrows indicate current direction, and the numbers give the maximum current velocities in kilometers per hour during spring tides. (After U.S. Naval Oceanographic Office. 1968.)

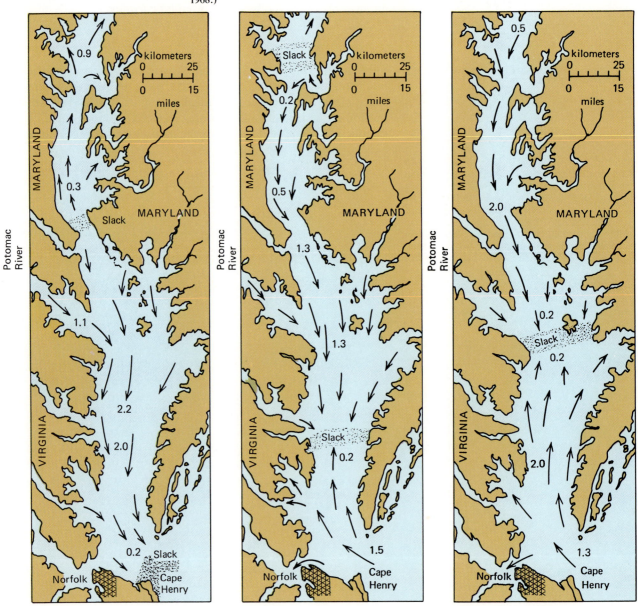

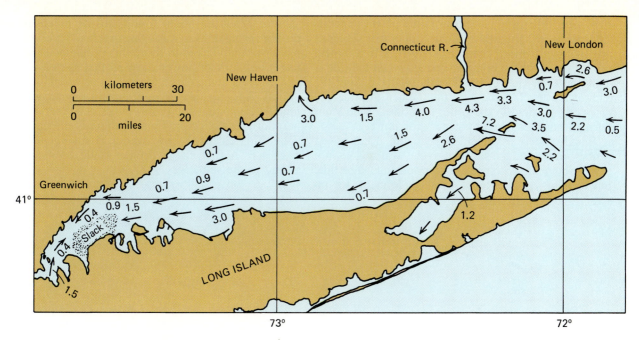

FIGURE 9-20
Tidal currents in Long Island Sound show a standing-wave tide during flood at the eastern entrance to the sound. Arrows indicate current direction, and the numbers give current speed in kilometers per hour. Note that the tide is flooding everywhere within the sound in contrast to the more complicated tidal current patterns in Chesapeake Bay. (Data from National Ocean Survey.)

that the area at the bay mouth is again experiencing slack water, but this time it is slack water before ebb. The arrows in the initial picture are now essentially reversed, with flood currents substituted for ebb currents.

Because of its narrowness, there is little tidal flow across most of Chesapeake Bay. Tidal currents therefore flow generally along the axis of the bay. In the wider parts, current patterns are complicated by flows across the bay.

Tidal currents in Long Island Sound are nearly the same over most of the sound at any time. Thus we say that the tide in Long Island Sound behaves like a standing wave. For instance, slack water occurs nearly simultaneously over the entire sound, and flood or ebb currents do also, although current strength varies between areas (Fig. 9-20). The area near the western end of the sound is an exception. An area of slack water separates the small area of opposing tidal currents that have advanced into Long Island Sound from the tidal system in New York Harbor, which connects with the sound through narrow channels.

The situation in which simultaneous slack water is followed by ebb or flood currents throughout the water body is typical of a standing-wave type of tide. This type of tide in Long Island Sound results in part from the natural period of the sound, which is apparently about 6 hours.

BOX 9-1

Energy from Tides

Tides are used to generate power in a few favorable coastal locations, primarily in France, Russia, and China. Three factors limit use of tidal power. First is the need for large tidal ranges, suitable topography, and the timing of power generation. The largest tidal ranges in the ocean are around 15 meters—for example, in the Bay of Fundy. Even with special turbines designed to work on the ebb and flood tides, the tidal range must exceed 5 meters to be useful. Such ranges are rare (Fig. B9-1-1), for the tidal range for most coasts is only about 2 meters. Furthermore, many potential sites are in remote areas where power transmission to urban and industrial centers would be expensive.

A second limiting factor for a tidal-power generating station is topography. Most tidal power schemes involve one or more dams (Fig. B9-1-2). In a typical system, gates in the dam are opened when the tide is high and then closed, keeping the water behind the dam at the level of the high tide. When the tide has dropped sufficiently, the water is allowed to flow out through turbines, thereby generating power.

Finally, there is a timing problem. Tidal-power generation is tied to the tidal cycle, which does not usually coincide with peak power demands. The tidal power plant on the Rance estuary (France), for in-stance, produces about four times as much power during spring tides as during neap tides. Several ideas have been advanced to solve this problem. One is to use a network of power lines so that the electricity can be used somewhere regardless of when it is generated. Another uses several dams to store water at high levels so that one basin serves as a reservoir and another as a collector. Finally, there is the option of generating electricity and storing it in some way for later use. A fuel like hydrogen could be made and stored in the same way.

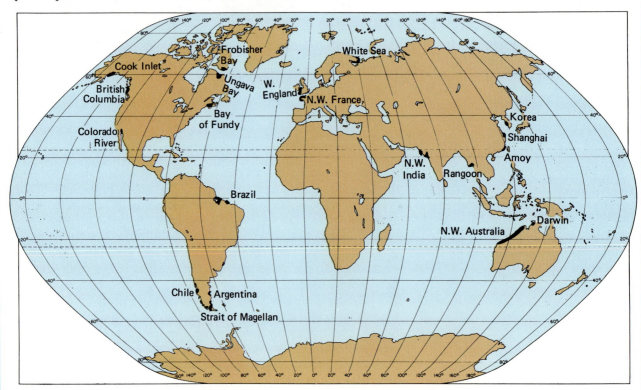

FIGURE B9-1-1
Areas where tidal heights exceed 5 meters.

FIGURE B9-1-2
A simple single-basin tidal-power installation.

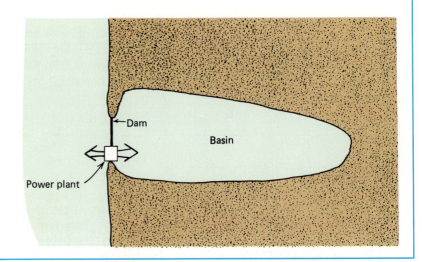

Tides are the periodic rise and fall of the sea surface due to the attraction of the Sun and Moon. The tidal period is the time between successive high (or low) tides. The tidal day is the time between successive transits of the Moon overhead, or 24 hours, 50 minutes.

There are three types of tidal curves—records of changing sea levels. Semidaily tides have two nearly equal high and two nearly equal low tides per tidal day. Daily tides have one high and one low tide per day. Mixed tides have two unequal high tides and two unequal low tides per tidal day.

Two different theoretical approaches have been used to explain how the Sun and Moon cause tides. The simplest, the equilibrium theory, assumes a static ocean completely covering a smooth Earth and considers only tide-generating forces. The gravitational attraction of the Moon (and Sun) combine with the centrifugal forces resulting from the revolution of the Moon and Earth together. This combination causes unbalanced forces on the Earth's surface that result in the tides. The horizontal component is the strongest (vertical forces are ignored). The result is a distorted ocean surface with two bulges: one on the side nearest the Moon, the other on the side opposite the Moon. Tides are thus very long waves (wavelength half the earth's circumference) whose location is determined by the Moon's position with respect to the Earth. In other words, tides are forced waves, unlike wind waves, which are free waves.

The dynamical theory includes the responses of the ocean due to the different shapes of the ocean basins. This theoretical approach is primarily mathe- matical and treats tides as very long waves. Tidal observations are analyzed mathematically into constituents which are used to make predictions. Each constituent can be related to the effects of the Sun and the Moon. Tidal predictions are calculated using these techniques.

Each basin responds differently to tide-generating forces. The natural periods of some basins favor standing waves. Because of the Coriolis effect, such waves form amphidromic systems in which the wave rotates about a fixed location where there is no tidal effect. In nearly enclosed basins, factors controlling tidal effects include the range of ocean tides at the entrance to the basin, the basin's natural period, and the size of the ocean entrance.

Tidal currents are horizontal water movements caused by tides. Such currents may be viewed as the water movements caused by the passage of tides which are essentially shallow-water waves. Near coasts, tidal currents reverse—flood currents flowing toward the coast, ebb currents flowing away from the coast, separated by slack water when the currents stop and reverse. Open-ocean tidal currents flow continuously with particles moving in elliptical orbits; these are called rotary currents.

In Chesapeake Bay the tide is a progressive wave, moving up the bay. In Long Island Sound the tide is primarily a standing wave, so high tide or low tide occurs at the same time throughout the sound. Tidal currents are altered by river flows and winds, making them difficult to predict. Tides have been used to generate electrical power in France and the USSR, where they have large tidal ranges.

STUDY QUESTIONS

1. Show diagrammatically the three types of tidal curves. Label the principal features of each.

2. Define tidal range; daily inequality; spring tide; neap tide.

3. Describe the equilibrium theory of tidal generation. What are the major assumptions? What features of tides does the theory fail to explain?

4. Explain the dynamical theory of the tides. How does this theory differ from the equilibrium theory of tidal generation?

5. Show (diagrammatically) how tides and tidal currents are related in the simplest case.

6. Explain the difference between rotary and reversing tidal currents. Draw a simple diagram showing water flows in each.

7. Discuss the difference between progressive-wave and standing-wave types of tides.

8. Explain why the tide acts like a forced shallow-water wave in the deep ocean.

9. Draw a diagram showing the relationship between the stage of the tide and tidal currents in a coastal plain estuary.

10. Draw a diagram showing an installation for obtaining power from tides. Contrast this with the installation for obtaining power from waves.

SELECTED REFERENCES

BRIN, A. 1981. *Energy and the Oceans*. Westbury House, Surrey, England. 133 pp. Discusses tidal energy.

DARWIN, G. H. 1962. *The Tides and Kindred Phenomena in the Solar System*. W. H. Freeman, San Francisco, Calif. 378 pp. Reprint of a classic.

DEFANT, A. 1958. *Ebb and Flow*. University of Michigan Press, Ann Arbor, Mich. 121 pp. Descriptive and mathematical treatment.

RUSSELL, R. C. H., AND D. M. MACMILLAN. 1954. *Waves and Tides*. Hutchinson, London, 348 pp. Elementary.

Distributions of waters rich in plant pigments (shown in red) demonstrate the complex interactions among currents, river discharges, and complex topography. The waters of the coastal ocean contain less pigment (shown in blues) than the nearshore waters. (Courtesy NASA.)

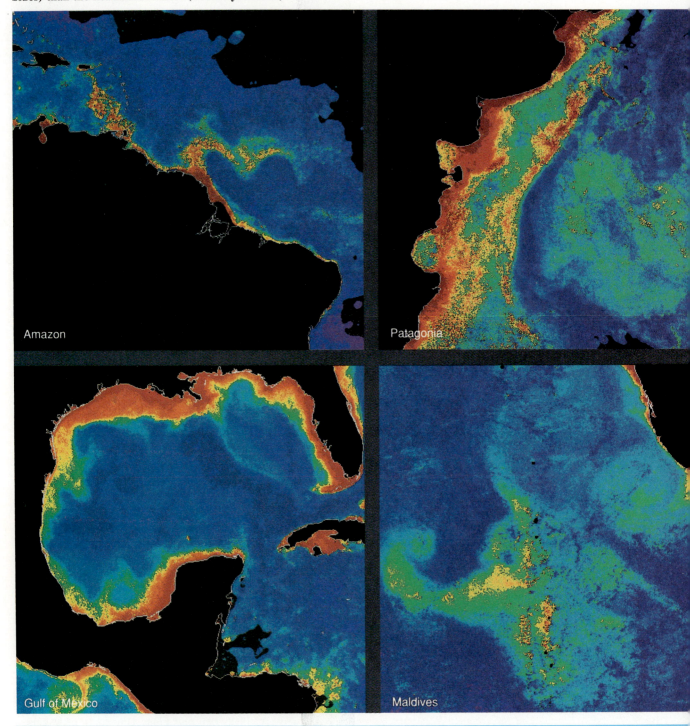

Amazon

Patagonia

Gulf of Mexico

Maldives

Coastal Ocean

OBJECTIVES

1. To describe water movements in coastal oceans;

2. To explain the behavior of water temperature and salinity in coastal oceans;

3. To describe estuaries and estuarine circulation;

4. To explain processes affecting beaches and coastlines;

5. To describe the principal features of the coastal ocean;

6. To explain processes affecting deltas and other wetlands.

*T*he coastal ocean lies on the continental margin. Where the continental margin is narrow, as in the Pacific, coastal ocean processes extend farther seaward than the continental shelf. The shallow bottom and complex shorelines make the coastal ocean far more complicated than the open ocean. Many water bodies connected to the coastal ocean are partially isolated, and therefore each has its own characteristics.

In this chapter we discuss:

Temperature and salinity in coastal ocean waters;
Origins of estuaries, fjords, and lagoons;
Water movements in estuaries and fjords;
Processes affecting beaches and coastlines; and
Processes affecting deltas and wetlands.

COASTAL CURRENTS

The coastal ocean is strongly influenced by the presence of the continental margin under it. **Coastal currents** generally parallel the shore. They are primarily driven by winds (especially storms) and by river discharges. Such currents can form, disappear, or change flow direction within a matter of a few hours to a day. **Fronts** (Fig. 10-1) usually separate the coastal ocean from offshore waters.

Coastal currents are strongest when freshwater runoff is large and winds are strong. Winds often control the currents. For example, surface waters can be blown shoreward and held there by winds, depressing the pycnocline. A sloping sea surface results, creating geostrophic currents that parallel the coast.

Coastal circulation is bounded seaward by boundary currents. A frontal zone near the edge of the shelf often marks the outer limit of coastal circulation (see Fig. 10-1). Along many coasts, nearshore currents often set in the opposite direction to boundary currents offshore. When winds diminish or freshwater discharge is low, coastal currents weaken or disappear. Such conditions are especially common in summer (gentle winds, little river discharge), when coastal currents can

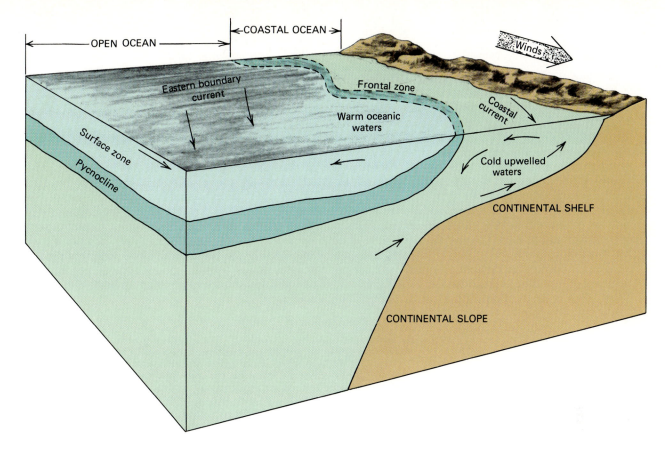

FIGURE 10-1
Coastal-ocean circulation in an eastern boundary current region with wind-induced upwelling.

virtually disappear. With the onset of winter storms, coastal currents are reestablished within a day or two.

Water discharged by rivers moves across coastal currents to mix into the open ocean. Because currents tend to parallel the coast, transfer of freshwater across the coastal ocean is slow. In other words, the **residence time** of freshwater is long. For example, between Nantucket Shoals off Cape Cod, and Cape Hatteras, North Carolina, the freshwater necessary to account for the lowered salinity is equivalent to about $2\frac{1}{2}$ years of discharge from the region's rivers. For comparison, in the Bay of Fundy and in Delaware Bay it is only about 3 months and about a year in Chesapeake Bay.

UPWELLING

When winds move surface waters away from the coast, subsurface waters move upward to replace them, as seen in Fig. 10-1. This wind-induced upwelling is common along the eastern sides of ocean basins, in eastern boundary current areas. Coastal upwelling occurs along the California-Oregon coast and the western coasts of South America (Fig. 10-2, p. 249) and Africa.

During upwelling, deeper, colder waters are brought to the surface from below the pycnocline, usually from depths of 100 to 200 meters. The cold surface or near-surface waters occur near the coast (see Fig. 10-1). Upwelling is most noticeable along arid, subtropical coasts, such as Southern California. There little rainfall or river runoff is present to form a low-salinity surface layer to inhibit upwelled waters from coming to the surface. Near rivers upwelling occurs when winds move the low-salinity surface layer seaward, and cold, upwelled waters are exposed at the surface.

BOX 10-1

Upwelling at Capes

Upwelling is often especially intense at capes. The cold waters and high productivity make them preferred locations for birds and for fishermen. Until recently, the reasons for this association were unknown.

Satellite images of sea-surface temperatures show complex features located near capes. These large complex structures, called "jets" and "squirts", extend seaward from the capes. Not only is upwelling intensified, but upwelled waters are moved seaward into the offshore coastal ocean. Experiments and satellite observations have now established the reasons for this association.

When a current encounters a submerged ridge associated with a cape or head land, it is deflected seaward. This is easily seen off the coasts of Peru and Northeast Africa (Fig. B10-1-1). This deflection causes strong upwelling on the downcurrent side of the ridge, also easily seen in the upwelled cold waters.

When the current bends back toward the coast (downstream of the ridge), it forms a sharp bend. The head of this feature is unstable and forms pinched-off eddies which move offshore away from the coast. (These are eastern boundary current analogues to the rings spun off by western boundary currents.) The wave in the current persists and can move downstream 50 to 150 kilometers depending on the size and shape of the ridge. In this way, the shoreline and coastal ocean bottom molds coastal currents into complicated shapes easily seen by Earth-orbiting satellites.

FIGURE B10-1-1

The waters off Peru (left) and Northwest Africa (right) are among the most productive in the world due to the upwelling that occurs there. Note the association between the capes and the chlorophyll-rich upwelling zones (shown in reds and yellows). Also note that the upwelling is displaced downcurrent from the capes. The current comes from the south (bottom) off Peru and from the north off Northwest Africa. (Courtesy NASA.)

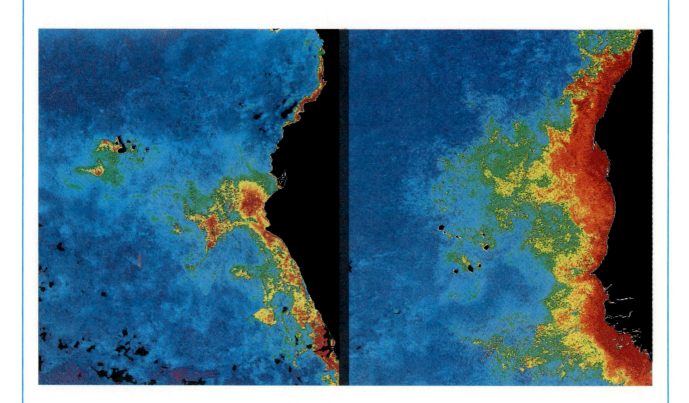

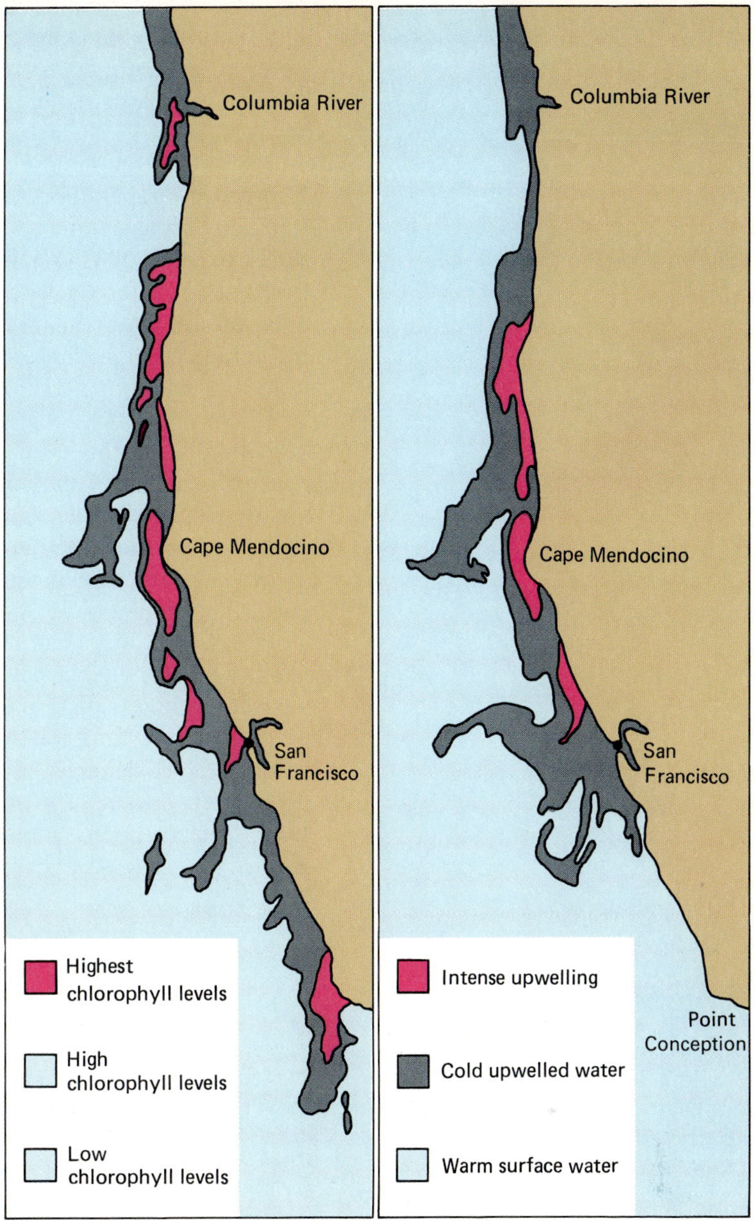
Columbia River
Cape Mendocino
San Francisco
Highest chlorophyll levels
High chlorophyll levels
Low chlorophyll levels
Columbia River
Cape Mendocino
San Francisco
Point Conception
Intense upwelling
Cold upwelled water
Warm surface water

FIGURE 10-2
Sea surface temperatures (right) on July 7, 1981, along the Pacific coast of the United States. Note the complicated patterns of upwelling associated with the major capes along the coast. (Date courtesy M. Abbott and P. Zion, NASA.)

TEMPERATURE AND SALINITY

Large temperature and salinity changes occur in shallow, partially isolated coastal waters. Winds blowing from a continent over the coastal ocean affect water temperatures and salinities to a marked degree. On the U.S. Atlantic Coast, winds coming from the continent are much warmer than the ocean in summer and much colder in winter. They strongly influence coastal water temperatures. Dry winds from continents cause evaporation when they blow across coastal waters. In contrast, winds coming from the ocean onto a continent, as in the northwestern United States during winter months, cause heavy rainfall and low salinities.

Salinity extremes occur in coastal waters, usually in isolated embayments. In the eastern Mediterranean and northern Red Sea, for example, evaporation is high, and surface-water salinities exceed the average for their latitude by more than 4‰.

Evaporation occurs from all ocean surfaces. But near shore fresh water discharged by rivers is mixed back into the ocean. For this reason, the lowest salinities also occur in coastal waters near large rivers.

Extreme temperatures also occur in surface coastal waters. Surface-water temperatures exceed 40°C in the Arabian Gulf and in the Red Sea during summer, whereas water temperature in the open ocean rarely exceeds 30°C. Again, water in such enclosed seas is restricted from mixing laterally with a larger body of water, in contrast to the open ocean, where heat is distributed by surface currents. Secondly, the limited fetch in small ocean areas limits wave action, thereby inhibiting vertical mixing.

Lowest surface-water temperatures are controlled by freezing seawater, at about −2°C. In winter, sea ice forms in high-latitude coastal areas, particularly in shallow bays and lagoons. There salinities are low due to river discharge, and cooling is rapid because of the large surface areas and small volumes of water involved.

Marked seasonal temperature differences occur in the coastal ocean, particularly at midlatitudes. Coastal waters are mixed by storms. Cold waters can mix all the way to the bottom during winter. Offshore waters never get as cold as continental shelf waters because surface cooling is distributed throughout a thicker water column.

In spring, surface waters are warmed, and the depth of the warmed layer deepens. As the summer progresses, the warm mixed layer deepens and a pronounced thermocline develops. During autumn, surface layers cool by radiation, evaporation, and mixing with deeper waters, caused primarily by storms. By the time winter begins,

BOX 10-2

Rising Sea Level

Sea level has been rising since the glaciers began retreating about 18,000 years ago, as water from melting glaciers returns to the ocean. Sea-level rise is also caused by the expansion of sea waters as they become warmer.

At present, sea level is rising between 1.5 and 2.0 millimeters per year (6 to 8 inches per century). While the rise is imperceptible over short periods, it is significant over decades and centuries. Rates of sea level rise also may be increasing. This is especially likely to happen if global warming of the atmosphere occurs as a result of the greenhouse effect. Rates of sea-level rise also vary locally due to sinking or rising of the land due to mountain-building, compaction of sediments, or even withdrawal of groundwater, oil, or gas. In some areas, Scandinavia for instance, the land is rising due to the disappearance of the glaciers and the removal of their weight from the crust.

Rising sea level is most critical in low-lying areas such as the Atlantic coast of North America. Beaches, barrier islands, and shorelines move slowly landward as sea level rises. This is especially bad news to owners of waterfront property. The beach

survives these movements; the houses may not. Often shorelines move quickly during major storms. The present south shore of Long Island formed within a few hours as a hurricane passed over the area in 1938. The New Jersey shore was eroded severely during the Ash Wednesday northeaster in 1962.

To protect coastal properties and communities against erosion of beaches and flooding from storms, sea walls are often constructed. While expensive to build and maintain, sea walls do provide some protection. Another technique is to supply sand to beaches to build them up. This usually lasts for several years before more sand must be brought in. This approach is used in many resort communities where the beaches are essential to their tourist business.

Rising sea level causes other problems in coastal areas. Saltwater can intrude into rock strata used to produce water, which can destroy local water supplies. There is also a risk of flooding of sites containing hazardous wastes, many of which are situated in low-lying areas subject to flooding.

CHAPTER TEN   COASTAL OCEAN

FIGURE 10-3
Daily variations in near-surface water temperatures. At midnight (a) the surface zone is nearly isothermal. Cooling continues through the night, so that by dawn (b) the surface layer is cooler than the waters beneath. After the sun comes up, the surface layer is warmed. By noon (c) the surface layer is distinctly warmer than water immediately below the surface. Warming continues until later afternoon, when surface temperatures are highest. Later, the surface layer begins to cool and is slightly cooler at dusk (d). During the night, the surface zone cools until it is again isothermal around midnight. Daily temperature changes are usually confined to the upper 10 meters or less. (After E. C. Lafond. 1954.)

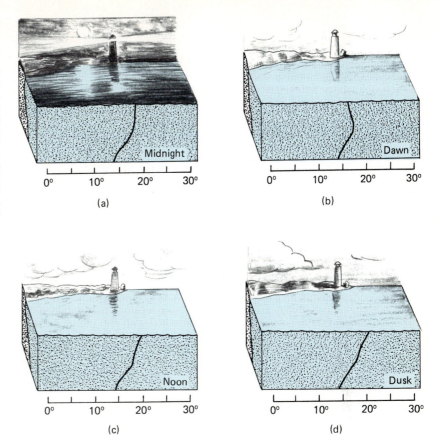

coastal waters are thoroughly mixed, with almost no temperature or salinity differences between surface and bottom waters.

Warming and cooling of surface waters during the day can be detected fairly readily, especially in areas protected from winds and waves. Surface-water temperatures are highest in midafternoon and lowest at dawn, as indicated in Fig. 10-3.

ESTUARIES, FJORDS, AND LAGOONS

The nature of coastal embayments is determined by whether the shoreline is sinking (on a stable or Atlantic-type margin) or rising (Pacific-type margin) compared to rate of sea-level rise. On stable (or sinking) coasts, **estuaries**—semienclosed embayments—formed during the past 18,000 years as the sea rose to its present level. The sea has been near its present level for only about 3000 years. Coastal features are still adjusting as sea level continues to rise at the rate of about 15 centimeters per century.

Another complicating factor is glacial erosion of high-latitude coastlines. On the Canadian Maritime Provinces, New England, and northwest U.S. coasts, glaciers removed the soil cover and carved deep valleys. When these valleys were flooded, they formed **fjords** (Fig. 10-4). The continental margin was partly destroyed, leaving a narrow, irregular continental shelf. These regions have remained almost unaltered for thousands of years since the sea reached its present level.

In the U.S. mid-Atlantic region, south of the glaciers during the last advance of the glaciers, there are many **coastal plain estuaries**

FIGURE 10-4
The coast from New York City (upper right corner) to southern Virginia (bottom center) has several large estuaries formed as the rising sea level flooded river valleys. Such coastlines are typical of stable continental margins. The coast between the large estuaries has nearly continuous barrier islands with lagoons between them and the mainland. (Courtesy NASA.)

filling the drowned mouths of ancient rivers. Coastal plain estuaries are typically broad, and they gradually deepen seaward. Some, such as the Hudson or Columbia rivers, have associated submarine canyons extending across the former coastal plain, now the submerged continental shelf.

A **lagoon** is a wide, shallow estuarine system. Flow of water between estuary and coastal ocean is usually restricted by a **barrier beach** offshore, generally paralleling the shoreline. Barrier beaches enclosing a lagoon are interrupted by narrow inlets through which tidal currents move water in and out of the lagoon (Fig. 10-5).

On many low-lying coasts, barrier beaches separate lagoons from the coastal ocean. Often a single stretch of beach isolates several lagoons, which communicate with the coastal ocean through narrow inlets. Such barrier islands are common along the U.S. Gulf Coast and much of the Atlantic coast, up to Long Island.

Where mountain building is active, and the coast is rising, estuaries or bays **(tectonic estuaries)** are cut off from the ocean by young mountain ranges. On such a coast, there are few estuaries or lagoons. The Pacific coast of North America is an example.

ESTUARINE CIRCULATION

Mixing of fresh and saltwater in estuaries and in the coastal ocean results in a long-term average flow seaward in the surface layers and a net landward flow along the bottom. This is called an **estuarine circulation.** Let us see how it develops.

In an ideal estuary (no friction, no tides, no winds), low-salinity river water flows in at the head of the bay (Fig. 10-6) and spreads out over the seawater beyond because it is less dense (Fig. 10-7). There is a more-or-less horizontal pycnocline zone with a marked density discon-

FIGURE 10-6

A small inlet on Ellesmere Island in the Canadian Arctic was cut by a glacier, a remnant of which (the large white mass) can be seen at the center left). The icebergs floating in the water come from such glaciers. Note the small floes of sea ice in the foreground. (Photograph courtesy National Film Board of Canada.)

FIGURE 10-7

Pamlico Sound, North Carolina, is formed by the barrier islands (light-colored narrow strips) which partially isolate drowned river valleys from the coastal ocean. Note the plumes of turbid water coming out of the inlets and the cloudlike water masses moving along the coast. The small white masses on the right are clouds. (Photograph courtesy NASA.)

(a)

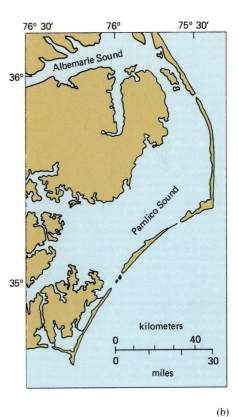

(b)

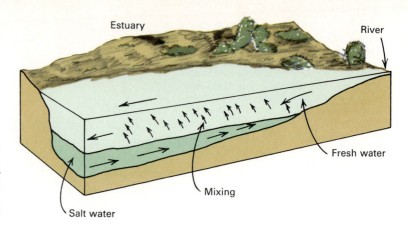

FIGURE 10-8
A simple salt-wedge estuary showing the two-layered structure, which a landward flow of the salty subsurface layer and seaward flow in the less saline surface layer. (After D. W. Pritchard. 1955.)

tinuity separating the two layers. The subsurface saltwater in such an estuary forms a wedge with its thin end pointed upstream; so in its simplest form this is called a **salt-wedge estuary** (Fig. 10-8).

This simplified picture is modified by several factors. First, friction between seaward-moving fresh water and the underlying seawater causes currents, as shown in Fig. 10-8. Water is entrained (dragged up from below) and mixed with the surface layer. This newly mixed water cannot reenter the lower layer because its reduced salinity makes it less dense than seawater. Consequently, ocean water flows into the estuary along the bottom to replace that drawn seaward by surface flows. Each volume of fresh water ultimately mixes with several volumes of seawater, and surface water salinities increase in a seaward direction. Landward flow along the bottom is much greater in volume than the flow of river water into the estuary (Fig. 10-9).

Simplified salt-wedge stratification, such as described, is observed only where river flow is large and tidal range is low. Salt-wedge estuaries are found where a large river discharges through a relatively narrow channel, such as the mouths of the Mississippi or the Columbia rivers during floods. Forces associated with large river flows are much stronger than the tides, so they dominate the estuarine circulation. A nearly ideal salt-wedge estuary is formed, one in which relatively little saltwater is mixed into the upper layer. Nearly fresh water is discharged into the coastal ocean, where it mixes with ocean water over the continental shelf.

As fresh water discharge into an estuary decreases, tidal effects become more important. An estuary may act like a salt-wedge estuary during floods. During low-flow periods, the same estuary is usually strongly influenced by tides and tidal currents. Then it departs substantially from the pattern of a simple salt-wedge stratification and becomes a *moderately stratified estuary* (Fig. 10-10).

FIGURE 10-9
Variation in salinity and current speeds with depth in a salt-wedge estuary.

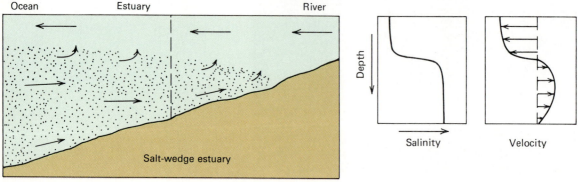

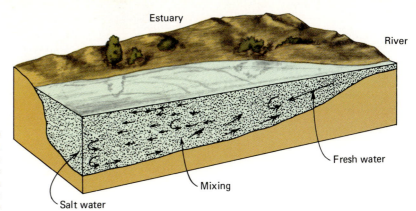

FIGURE 10-10
Water movements in a moderately stratified estuary. There is a net landward movement in the subsurface layers and a seaward flow in the surface layers, although the difference between the layers is not as pronounced as in a salt-wedge estuary. (After D. W. Pritchard. 1955.)

Estuary

River

Fresh water

Mixing

Salt water

Tidal currents cause mixing between layers so that the waters become only moderately stratified (Fig. 10-11). Tidal flow causes **turbulence** (random movements) throughout the water column, which, in turn, increases mixing. Consequently, more saltwater is transferred from the subsurface to the surface layer. Some fresh water from the surface also mixes downward, and salinity decreases in a landward direction in both the surface and subsurface layers. As in the salt-wedge estuary, however, there is a net landward flow in the subsurface layer, replacing saltwater lost from the system. A net seaward flow in the surface layer removes both fresh water and saltwater from the estuary.

Consider the volume of flow in the two layers. River water has almost no salt as it enters the head of the estuary, whereas seawater moving in along the bottom has a salinity of 33‰. A mixture of equal volumes of fresh and saltwater has a salinity of 16.5‰; By the time surface-water salinity has reached 30‰, 1 volume of river water has mixed with 10 volumes of seawater. In other words, at this point, landward flow in the subsurface layer is 10 times the volume of river water discharged at the head of the estuary. The volume of surface water moving seaward in the surface layer will be 11 times the original river discharge (Fig. 10-12, p. 254).

Tides dominate currents in most estuaries. Thus to observe estuarine circulation, it is necessary to measure currents over many tidal cycles. When averaged over many tidal cycles, tidal currents cancel one another, leaving a nontidal or estuarine circulation. Long-term average currents in the surface layers indicate a net flow landward in the deeper layers.

Where tidal effects are relatively strong, waters in the estuary are

FIGURE 10-11
Variation in salinity and current speeds in a moderately stratified estuary.

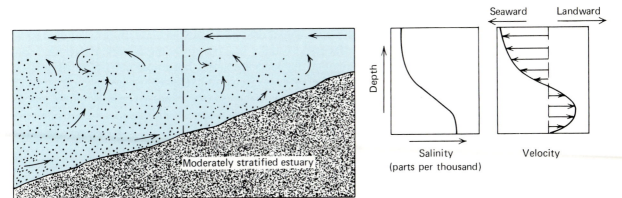

Moderately stratified estuary

Seaward   Landward

Depth

Salinity
(parts per thousand)

Velocity

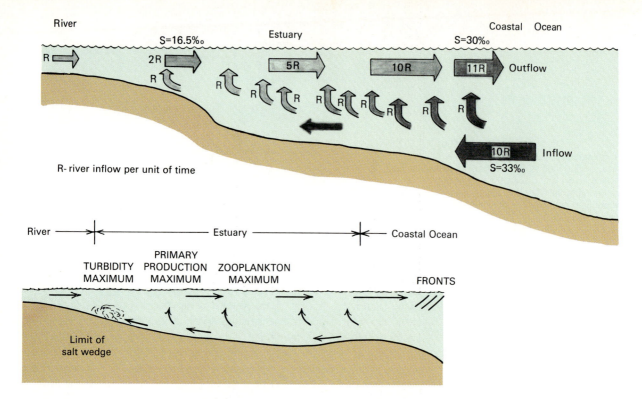

FIGURE 10-12

(a) In an estuarine system, an amount of river water (*R*) draws 10 times as much seawater upward from below and carries it downstream toward the ocean. To replace this outflow, 10*R* amount of seawater moves into the estuary. Thus the effect of estuarine circulation is to draw subsurface, high-salinity water into the surface layer, essentially an upwelling process. (b) Due to estuarine circulation, various processes occur in different parts of the estuary. Sediment is trapped in the turbidity maximum. Growth of microscopic plants (primary production) occurs where the waters are clearer, after losing their sediment. These plants are eaten by microscopic animals (zooplankton). Fronts near the mouth of the estuary separate the estuarine circulation from the coastal ocean.

less stratified. In an estuary with small river flow but large tides and tidal currents, the waters may be mixed almost completely from top to bottom. These are called *well-mixed estuaries*.

The tidal rise and fall of sea level controls river discharge into the ocean. Most fresh water is discharged on the ebb tide and enters the ocean as a series of pulses. River discharge forms cloudlike parcels of low-salinity water, which, when turbid with suspended sediment, are easily recognizable from the air (Fig. 10-13). Each tends to overrun the previous one, forming a complex of irregular fronts (Fig. 10-14). As the cloudlike masses move away from the river mouth, waves and tidal currents continue the mixing over the continental shelf. Boundaries are obscured, and eventually a large lens or *plume* of low-salinity water forms. Such plumes then move with coastal ocean currents.

Because of their great depths and irregular bottom topography, fjords have complicated circulation systems. Many fjords have a **sill** or submerged ridge at the entrance deposited by the glacier that cut the fjord. This sill cuts off most of the deeper water from communication with the adjacent ocean, as can be seen in Fig. 10-15.

Fresh water flowing into a fjord forms a low-salinity surface layer that moves seaward, in a typical estuarine circulation. This layer, however, often extends no more than a few tens of meters below the surface and usually involves only the waters above the top of the sill.

The deeper waters of a fjord may be almost completely isolated from the surface circulation. However, they are strongly affected by conditions outside the estuary. Since river flow, tides, and winds have little effect on this deep water, its circulation is controlled primarily by density differences. If the water at or slightly above the sill outside the estuary becomes more dense, it can flow into the fjord. This displaces the deeper waters in the fjord, which then move out slowly. Strong winds can affect the deep waters by setting up seiches (standing waves), which cause the waters to move back and forth.

FIGURE 10-13
Small estuaries along the coast of South Carolina discharge turbid waters into the coastal ocean. (Photograph courtesy NASA.)

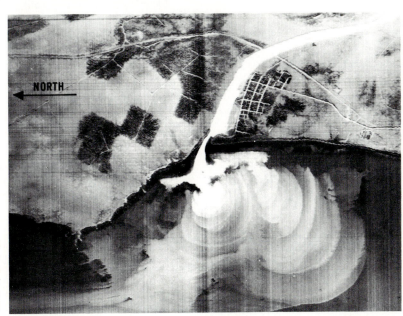

NORTH

FIGURE 10-14
Fresh water from the Quinault River discharges into the northeast Pacific Ocean. Differences in water temperature outline the cloudlike water masses of low-salinity, cold water mixing with coastal waters. (Photograph courtesy NASA.)

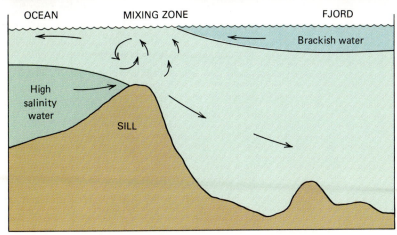

OCEAN   MIXING ZONE   FJORD

Brackish water

High salinity water

SILL

FIGURE 10-15
Circulation in a simple fjordlike estuary. Note the water mass formed by mixing surface and subsurface waters while flowing over the sill. (After M. Waldichuk. 1957.)

MARGINAL SEAS

Several large ocean areas near or between continental areas are partially isolated from the open ocean. Because of their isolation, waters in several of these seas are either diluted by large rainfall or river discharge, especially in high latitudes. Such marginal seas exhibit an estuarinelike circulation. Low-salinity waters flow out at the surface and are replaced by denser waters flowing in along the bottom. Circulation of surface waters is driven primarily by the winds.

The *Black Sea* is a marginal sea with an estuarinelike circulation. It receives the discharge of several large rivers, including the Danube River, which drains Central Europe, and the Don, which drains central Asia. High-salinity water ($S = 38.52‰$) flows into the Black Sea from the Mediterranean Sea along the bottom of the Bosporus, a narrow channel with a shallow sill. Low-salinity water ($S = 17.2‰$) from the Black Sea flows out in the surface layers. This two-layered flow was discovered by fishermen whose nets were dragged in the opposite direction from the surface currents. In short, the Black Sea and the Bosporus have an estuarine circulation.

The Black Sea is unusual in that waters below the pycnocline are isolated from contact with the atmosphere by the strong density contrast between surface and subsurface waters. Flows into the very large basin are relatively small that it takes approximately 2500 years to replace all the water. During that time, dissolved oxygen in the bottom waters is used up, so that none occurs below 150 to 200 m. Instead, anaerobic bacteria obtain their oxygen by decomposing sulfates in seawater, releasing hydrogen sulfide (H_2S) as a byproduct. Most marine organisms cannot live in the presence of hydrogen sulfide, so only bacteria survive in the deep waters of the Black Sea. Consequently, organic matter produced in the surface waters is not decomposed, and the sediments are rich in organic carbon.

When the partially isolated sea lies in an arid climate, surface waters are usually strongly evaporated. If there are no large rivers flowing into the basin, the sea exhibits the opposite type of vertical circulation, sometimes called an antiestuarine circulation. Low-density waters flow in at the surface, and denser, usually more saline, waters flow out at depth.

The *Mediterranean Sea* is an example of a basin with strong evaporation. It is connected to the North Atlantic by exchanges of waters through the Strait of Gibraltar, to the Black Sea through the Bosporus, and to the Indian Ocean through the Suez Canal, a shallow dredged canal through which little water flows. The Aswan Dam in Egypt cuts off the flow of the Nile into the Mediterranean. No other river discharges are large enough to make up for the large evaporation. As a result, surface waters in the Mediterranean are strongly evaporated, and salinities are greater than $39‰$ in the eastern end. These highly saline waters are strongly cooled in winter, causing deep convective circulation along the southern French coast, in the southern Adriatic Sea (off southeastern Italy), and south of Greece. These highly saline waters ($S = 38.4$ to $39.0‰$, $T = 12.7°$ to $14.5°C$) fill the deep basins.

Highly saline waters flow out of the Mediterranean through the Strait of Gibraltar. These waters flow down the continental slope until they reach a comparable density level, around 1000 m. There they flow out into the North Atlantic. This flow into the North Atlantic can be traced across the entire basin. The salty waters from the Mediterranean cause the North Atlantic to have the highest salinities of any of the major ocean basins.

The *Red Sea* is another marginal sea where high-salinity subsurface waters form. The Red Sea is a narrow basin, connecting to the Indian Ocean through a narrow opening at its southern end. There are no large rivers discharging into it, so that the desert climate results in some of the highest salinities (greater than 40‰) observed anywhere in the ocean. During winter, these highly saline waters are chilled at the northern end of the sea, filling the deep basins. These highly saline waters flow out into the northern Indian Ocean. Circulation is controlled by winds which shift seasonally, a part of the monsoon circulation. These highly saline waters can also be recognized below the pycnocline in the Indian Ocean.

The bottom of the Red Sea also has hydrothermal circulation in the newly formed ocean crust. These discharges of hot (up to 60°C) highly saline waters (dissolved salt contents equivalent to 257‰) fill the deepest holes in the basin. These highly saline brines are extremely dense and do not readily mix with overlying waters.

COASTAL PROCESSES

The **shoreline**—where land, air, and sea meet—is the most dynamic part of the ocean. It is shaped by tides, winds, waves, changing sea level, and humans. The shore, as shown in Fig. 10-16, extends from the lowest tide level to the highest point on land reached by wave-transported sand.

Processes affecting shores act for periods of time ranging from a few seconds to tens of thousands of years. Among the most obvious are waves, which break, rush up the beach face, and then retreat in a matter of seconds. Along open coasts there are few days without waves to move materials on the beach, dissipating energy originally imparted to the sea surface by winds.

Usually these forces act most effectively under extreme conditions, such as storms or exceptionally high tides. Violent storms or exceptional tides strike most coasts every 10 to 100 years. The intervals involved are short in the billion-year history of a continent. Storms of such strength have major effects on coasts, particularly beaches.

Storm surges—relatively sudden large changes in sea level resulting from strong winds blowing across a shallow and partially enclosed body of water—also cause large changes in the shoreline and the coastal zone landward because of the flooding of normally protected

FIGURE 10-16
Profile of a typical beach and adjacent coast.

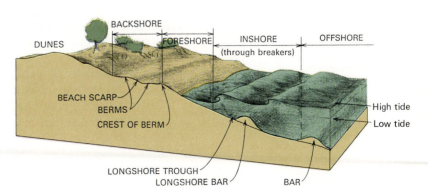

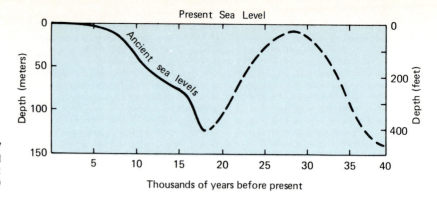

FIGURE 10-17
Changes in sea level caused by growth and melting of continental glaciers during the most recent glacial stage. (After J. R. Curray. 1965.)

areas. Although major storms come only every few decades or even centuries, the changes they make remain visible for many decades.

The most profound influence on coasts as we now see them was the repeated advance and retreat of continental glaciers and resulting sea-level changes (Fig. 10-17). When sea level stood at its lowest, about 18,000 years ago, the shore was near the present continental shelf break. As glaciers waned and water was released, the shoreline moved back across the continental shelf until it reached its present level about 3000 years ago. If the remaining ice sheets (Antarctica, Greenland) melt, the sea surface may eventually stand as high as 50 meters above its present level. Today's shoreline would be submerged and a complex of submerged beach ridges and lagoon or estuarine deposits would mark its former location.

COASTLINES

Most of the world's coastlines exhibit signs of rising sea level. They have been modified by wave action and by deposition of sand and gravel to form beaches or mud to form deltas and marshes (Fig. 10-18).

Some coastlines are formed mainly by terrestrial processes. The ocean has not been at its present level long enough to reshape the land features along all its margin. An example of such a coastline is the drowned river valleys of the U.S. East Coast.

These former river valleys are virtually intact although largely underwater and somewhat modified by erosion and deposition. In glacially carved regions movements of the great ice sheets created deep valleys with U-shaped cross sections. Now filled by seawater, the U-shaped bottoms are concealed, but the valleys form fjords in Canada, Alaska, and Chile.

Spectacular coastline features are formed by volcanoes. On the island of Hawaii lavas flow into the sea. In other areas the volcanoes have been worn down (Fig. 10-19, p. 262).

Marine processes or marine organisms shape many coastlines. They cut in rocks or sediments soft enough so that even the limited time of the present sea-level stand has been sufficient to cut back the coast. Where bluffs of unconsolidated sand and gravel rise above the shore, they can be cut back by waves to straighten the coastline, as illustrated in Fig. 10-20, p. 262. Materials derived from the erosion of bluffs are deposited near shore or form small beaches or **baymouth bars** across nearby bays and inlets. Sand is moved along coasts by longshore currents, forming **barrier islands** and **spits** that separate inlets and bays from the ocean.

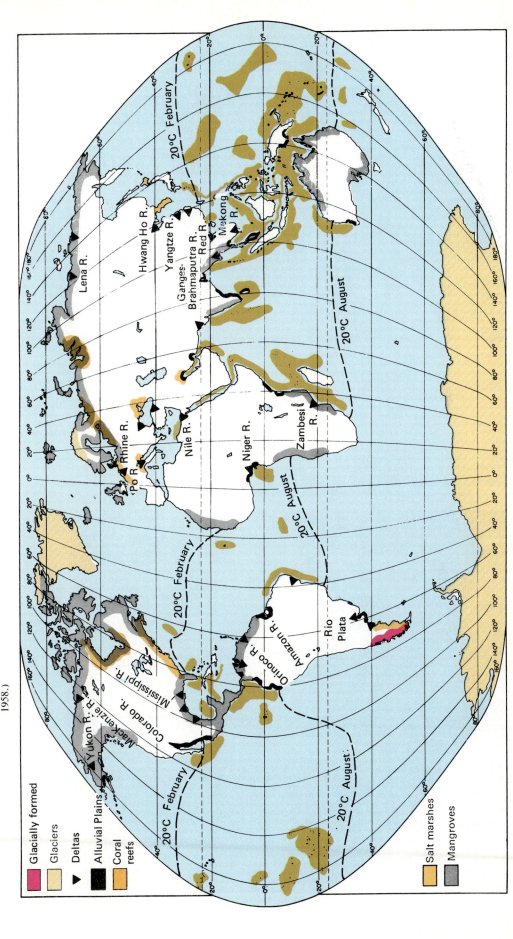

FIGURE 10-18
Coastlines of the world. (After J. T. McGill. 1958.)

Glacially formed

Glaciers

Deltas

Alluvial Plains

Coral reefs

Salt marshes

Mangroves

Mackenzie R.
Yukon R.
Colorado R.
Mississippi R.
Orinoco R.
Amazon R.
Rio Plata

Rhine R.
Po R.
Nile R.
Niger R.
Zambesi R.

Lena R.
Hwang Ho R.
Yangtze R.
Ganges-Brahmaputra R.
Red R.
Mekong R.

20°C February
20°C August

FIGURE 10-19

Coastline at Waikiki Beach on Oahu, Hawaii. Diamond Head in the background is the remnant of an extinct volcano. The original volcanic features of the island have been altered by erosion and by humans. (Photograph courtesy Hawaiian Visitors Bureau.)

FIGURE 10-20

The shoreline of northern Long Island, a glacial deposit of sand and gravel, has been modified by wave erosion and sediment transport. Sand and gravel eroded from the points of land have been transported and deposited to fill in the irregularities in the original shoreline. (Photograph courtesy National Ocean Survey.)

Beaches are familiar shore features (Fig. 10-21). Sand beaches, barrier islands, and bays border the U.S. Atlantic Coast from New York to Key West, Florida. On the Gulf Coast region, barrier islands and lagoons dominate. On mountainous and recently glaciated coasts, beaches are usually less extensive, being generally restricted to low-lying areas between rocky headlands.

A **beach** is a sediment deposit in motion. Movements of sand grains in the surf may be obvious while the rest of the beach appears quite stable. But, during major storms, large segments of beaches are moved. Higher or more protected parts may move only during exceptionally powerful storms.

To most of us, the word beach means a beach composed of sand grains (diameters between 0.062 and 2 millimeters). In tropical and subtropical areas where silicate rocks are rare or absent, beaches are made from broken carbonate shells and skeletons of marine organisms. Such beaches are often white or slightly pink.

Not all beaches are made of sand. Where wave and current action is especially vigorous, sand may be washed away faster than it is being brought in, leaving behind gravel or boulders. Still other beaches consist of mixtures of gravel and sand where wave action is not strong enough to remove all the sand.

Beaches are accumulations of locally abundant materials not immediately removed by waves, currents, or winds. Beach sands and gravels can be derived from erosion of glacial deposits, originally containing unsorted gravels, sands, and clays. Only the gravels and sands remain on the beaches. Silt and clay-sized particles are usually washed out of beach areas by even weak waves or tidal currents. Fine-grained sediments tend to accumulate in areas with little wave action or tidal currents. Thus they occur on continental shelves at depths below about 30 meters or in lagoons, bays, or tidal marshes.

Beaches typically form near sediment sources—at the base of

FIGURE 10-21

A barrier island isolates a lagoon from the adjacent coastal ocean. Fine-grained muds accumulate in the lagoon while sands eroded from the headlands are moved along the beaches by wave and longshore currents. (After Judson, Kauffman, and Leet, 1987.)

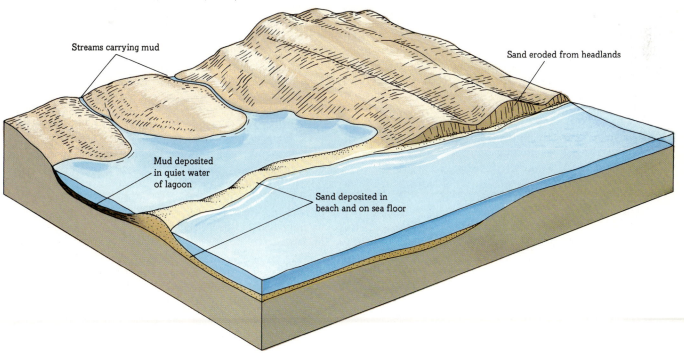

Streams carrying mud

Sand eroded from headlands

Mud deposited in quiet water of lagoon

Sand deposited in beach and on sea floor

BOX 10-3

Waste Discharges

Wastes discharged to streams and estuaries eventually reach the ocean. In the ocean, wastes can cause problems, in some cases killing organisms and in others rendering them unfit for human consumption.

Most disposal systems discharging into the ocean are designed to dilute wastes flowing into the receiving waters. If the systems are properly designed and working, waste concentrations near the discharge points are relatively low. Later movements and mixing by currents dilutes them further. Offshore waste disposal from large pipes discharging below the permanent thermocline off Southern California has been used for many years to dispose of sewage wastes from Los Angeles and San Diego. Around the United Kingdom, offshore disposal of sewage wastes is widely practiced. There strong tidal currents quickly dilute the wastes.

Where properly designed and operated, discharges of sewage wastes have relatively little long-lasting effect on water quality and on pelagic organisms which can swim out of the waste-containing waters. For example, wastes discharged below the pycnocline are isolated from the ocean surface, where they might come in contact with people.

Elimination of discharges of untreated wastes has greatly improved water quality in the previously polluted Thames River near London, so that salmon can now be caught there. Water quality in the upper Potomac River near Washington, D.C. has also improved due to more treatment of sewage-plant discharges.

Long-continued waste disposal operations can cause drastic changes in the condition of the bottom and of organisms which cannot move away from the waste materials. Among these are changes in the physical and chemical characteristics of the bottom. For example, areas on the continental shelf used by the New York metropolitan area for disposal of sewage solids for many decades were covered by soft and easily eroded silts instead of the hard sand bottom that originally covered the area. Consequently, organisms that require hard sand bottoms, such as lobsters, were excluded. Only pollution-tolerant organisms such as fast-growing worms were able to survive in the waste-contaminated areas.

Many waste materials entering the ocean are either associated with particles or quickly become attached to solids. Waste solids usually settle out near the discharge points. Thus nearby waste deposits frequently contain high concentrations of carbon, bacteria, and viruses as well as oils and waste chemicals. Such materials are also abundant in deposits dredged from urban harbors. Often, contaminated dredged materials are placed in diked areas rather than dumped into estuaries or onto continental shelves.

When the circulation near a contaminated bottom is sluggish, consumption of dissolved oxygen by these deposits can cause depletion of dissolved oxygen in overlying waters. This happened in the summer of 1976 in the New York Bight. When the dissolved oxygen was used up, benthic organisms died. Hydrogen sulfide was eventually widespread in near-bottom waters. Clams and bottom-living fish worth many millions of dollars were killed. Similar depletion of dissolved oxygen in the near-bottom water of Chesapeake Bay has resulted in greatly increased incidences of dieoffs of bottom organisms during warm summer weather.

Bacteria, viruses, and possibly fungi in contaminated sediment deposits can also damage animals which live on the bottom. In the New York waste disposal area, bottom-living fish often have large sores, and eventually their fins are eaten away. This is called **fin rot.** Similar conditions are observed in lobsters living in the waste disposal area. They have deposits in their gills, and parts of the carapace is often eaten away. Such conditions are observed only in the most contaminated areas. Their causes are not known.

Perhaps the most widespread effect of waste disposal in coastal waters is in making marine animals unfit for human consumption. Filter-feeding organisms concentrate bacteria and viruses as well as various particle-associated materials. Thus such organisms (clams, oysters) cannot be harvested or sold. Large areas of highly productive clam and oyster beds are now closed to commercial harvest because of the unacceptable quality of the overlying waters.

A tragic example of human exposure to industrial wastes from eating shellfish occurred in Minamata, Japan. There mercury wastes were discharged between 1953 and 1960 into coastal waters near shellfish beds which were harvested for food. The mercury was changed by bacterial action into a form that was concentrated by shellfish. Those villagers living nearby who ate large quantities of shellfish developed severe mercury poisoning and some died. The effects of the poisoning were most serious in pregnant women and children. This illness was called **Minamata disease.**

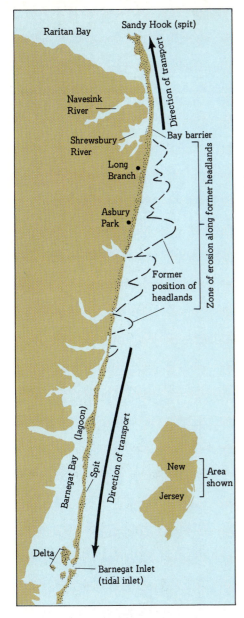

FIGURE 10-22
Barrier beaches along the New Jersey shoreline were formed by sand moving north and south from an area of bluffs near Long Branch and Asbury Park. (After Judson, Kaufmann, and Leet, 1987.)

cliffs or near river mouths. Sediment is moved onto beaches by waves and currents, replacing materials either moved out into deeper water or transported along the coast.

On the North Atlantic Coast of the United States virtually no river-borne sediment escapes the estuaries to enter the ocean, so Atlantic Coast beaches are formed primarily from erosion of nearby cliffs (Fig. 10-22) or from sands deposited offshore during times of lower sea level. Beaches often form where transport of sand along the coast is interrupted by an obstruction, such as a headland.

Not all beaches occur on coastal plains. If the land along a coast is low lying and slopes gently toward the ocean, sand deposits form parallel to the coast and a short distance offshore. If submerged, they are called **longshore bars.** Where large enough, they form barrier islands, which typically have a shallow lagoon or bay between the island and the mainland. Such barrier beaches, formed by the onshore movements of sands and the longshore movements of currents, are the most common type of beach occurring along low-lying coastlines around the world. When connected to the mainland, usually at some **headland** (a point of land that juts out from the coast), they are known as barrier spits. At small indentations, a barrier of sediment may build completely across the mouth of a bay—then the barrier is known as a baymouth bar (Fig. 10-23, p. 266). Several kinds of beaches can occur within a single bay.

Offshore from beaches are usually submerged low sand ridges on the ocean bottom, called longshore bars. These run parallel to the shore, generally submerged in a few meters of water (Fig. 10-24, p. 265). At extreme low tides the tops of these bars may be exposed. Separating the longshore bar from the beach proper is the longshore trough, which normally remains filled with water, even at low tide. Bars can usually be identified, even when submerged, by the waves breaking on them. On many coasts several sets of bars can be spotted from the lines of breakers offshore.

As we leave the offshore portion of the beach and come onto the beach proper, there is a sandy area dipping gently seaward, the **low tide terrace.** This part of the beach is exposed at low tide and submerged at high tide. A small **scarp** or vertical face often occurs at or near the upper limit of the low tide terrace. It is commonly left by a recent cycle of more intensive wave action that has caused erosion into the beach profile formed during a preceding cycle. The seaward-dipping portion of the beach, collectively known as the **foreshore,** leads up to the **berm crest** or **berm,** the highest part of the beach. Several berms may be present on a beach at any time. Berm crests usually form during storms and represent the effective upper limit of wave action during that storm. As a rule, the highest berm on a beach is formed during winter storms and is referred to as the winter berm.

The berm slopes gently downward toward the base of the cliffs or **dunes** behind the beach. These sands are usually not moved often, as indicated by the large trees that grow there. Where a beach is not backed by cliffs, dunes are often formed from beach sands blown by the prevailing onshore winds, as illustrated in Fig. 10-24. When not actively gaining or losing sediment, dunes may be colonized by salt-tolerant plants or trees, to become stabilized through time; this process is shown in Fig. 10-25, p. 267. Dunes protect low-lying lands behind them.

Sediment budgets are useful in determining major sources and losses (sinks) of sand along a beach. The major sand sources are usu-

FIGURE 10-23

A baymouth bar formed by longshore currents moving sediment northward along the Oregon coast at the mouth of the Siuslaw River. The light-colored spit is an area of active sand dune migration with no vegetation. Darker areas on the spit are depressions between dunes; some have small lakes in them. Dark-colored areas landward of the spit are heavily vegetated, which stabilizes sand deposits and inhibits their movement by winds. (Photograph courtesy U.S. Geological Survey.)

ally either rivers or eroding sea cliffs [Fig. 10-26(a), p. 266]. Major losses of sand to the beach occur through longshore transport out of a particular segment, offshore transport (down submarine canyons), and losses due to wind transport to form sand dunes or to deposit it in marshes behind the beach.

In Southern California the coast is divided into four cells [shown in Fig. 10-26(a)]. Each cell consists of a river (or rivers) providing sand, littoral drift along the shore, and a submarine canyon that comes in

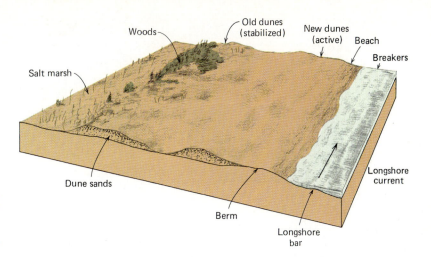

FIGURE 10-24
Typical sand beach with active dunes, stabilized dunes, and salt marshes. Such a beach-dune complex is found at many locations on all coasts. (After D. L. Larsen. 1969.)

FIGURE 10-25
Beach grasses stabilize small dunes on an island made of dredged materials. (© Dan Guravich/Photo Researchers.)

close to the shore to trap the sand flow, carrying it offshore [Fig. 10-26(b), p. 268]. Each cell is separated from adjoining cells by a stretch of rocky coast devoid of large beaches because the sand has moved down a submarine canyon into water too deep to be moved by waves.

Inlets and lagoons behind barrier islands also trap sediment moving along the coast. Sand is stirred up and put in suspension by waves and then moved into the inlets or lagoons by flood-tide currents. When they slacken or when sediment encounters the dense vegetation of the tidal flats, it settles out. Without resuspension by waves, ebb-tide currents cannot move this sediment back out of the inlet. The resulting sediment deposits form wetlands, which we discuss in a later section in this chapter.

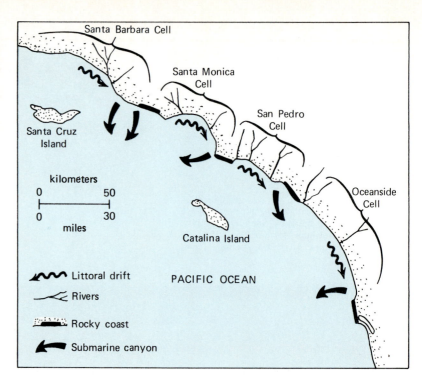

FIGURE 10-26
The Southern California coast includes four littoral cells (a). Beach sands move from the sources (primarily rivers) to the submarine canyons, where sand is transported offshore out of reach of the beach processes. (After D. L. Inman and J. D. Frautschy. 1966.) (b) Beach sands move down a submarine canyon in Baja California, Mexico, forming a sandfall. (Photograph courtesy Scripps Institution of Oceanography, University of California.)

BOX 10-4

Oil Spills

Large oil spills are among the most dramatic forms of ocean pollution (Fig. B10-4-1). As larger quantities of oil are transported to the industrialized countries, this form of ocean pollution is likely to increase in importance. Spills usually occur when tankers are wrecked, often during storms. The spilled oil is widely dispersed by waves and tidal currents and deposited on nearby beaches and occasionally on reefs or in wetlands. The technology for dealing with oil spills is still primitive. The first line of defense is to deploy barriers to retard the spread of oil slicks onto marinas or beaches or into sensitive areas, such as wetlands. Any oil trapped by the barriers is pumped into containers for disposal. Oil slicks are also removed by using special skimmer boats. All these techniques are best suited to deal with small spills. There is no technology suitable to deal with large spills at sea.

Once oil has reached an area, extensive efforts are usually made to retrieve it. First, the oil is normally taken up, using straw or some other inexpensive material, which is then physically removed to be burned or buried. Beach sands coated with oil are then removed and buried.

FIGURE B10-4-1
The Tanker *Exxon Valdez* unloads oil to the *Exxon Baton Rouge* after running aground near Valdez, Alaska. (Photograph courtesy Hazardous Materials Branch, NOAA.)

A large spill occurred when the *Exxon Valdez* (Fig. B10-4-2) went aground near Valdez, Alaska, on March 24, 1989. More than 10 million gallons of crude oil was spilled and moved by currents (Fig. B10-4-3). Much of the oil went onto beaches (Fig. B10-4-4) where it was removed by steam and hot water. Sea birds and sea otters were affected.

The behavior of spilled oil depends on its composition, water temperatures, and environmental conditions where it is spilled. For example, in a small spill on Cape Cod, the oil went into a marsh and was buried. The refined oil involved was much more toxic than crude oil. Furthermore, the lack of dissolved oxygen in the sediments slowed the oil's decomposition. Effects of this spill were visible for years. The effects of most crude oil spills, however, are not recognizable a few years later.

In most cases, benthic (bottom-dwelling) organisms are seriously affected, since they cannot move away from oil-affected areas. Seabirds and marine mammals are also likely to be affected.

FIGURE B10-4-2
Crude oil on the water is moved by winds and currents after spilling from *Exxon Valdez*. (Photograph courtesy Hazardous Materials Branch, NOAA.)

FIGURE B10-4-3
Crude oil coats rocks and sands on beach. (Photograph courtesy Hazardous Materials Branch, NOAA.)

FIGURE B10-4-4
Sea otter recovers after being cleaned of crude oil. (Photograph courtesy Hazardous Materials Branch, NOAA.)

Deltas are huge sediment deposits at the mouths of streams where they enter the ocean or large lakes (Fig. 10-27). Sediment carried by streams is usually deposited in estuaries or at the river mouth. For many small streams their sediment load is a small, rivermouth sandbar, moved by the tides and waves. Most of this sediment is eventually resuspended by wave action and moved along the coast by longshore or tidal currents.

Some rivers, however, carry far more sediment than can be dispersed along the adjacent coast. In this case, sediment usually is deposited at the river mouth, forming a delta. A well-known example is the Mississippi (as shown in Fig. 10-28).

The first step in delta formation (illustrated in Fig. 10-29, p. 272) is the filling of an estuary with sediment. The Mississippi, for instance, probably filled its estuary soon after sea level reached its present position. Rivers with large estuaries and moderate sediment loads are still filling their estuaries. Until the estuary is completely filled, little or no sediment can be deposited at the river mouth, and no delta forms.

As a delta forms, the river builds channels across it called **distributaries,** as shown in Fig. 10-30, p. 272, through which water flows on its way to the ocean. Often a series of these channels extend across the delta as long, radiating, and often branching fingers. An active distributary continually builds its mouth farther seaward until the distance to the sea is so great that the river flow can no longer maintain that channel. At this time the river shifts course, often during a flood, and the flow cuts a new channel through a different set of distributaries to reach the ocean and the whole process begins again. The Mississippi Delta has several abandoned distributaries, each with its own subdelta, forming a complex *lobate delta* (Fig. 10-28). Distributary abandonment is not always sudden but may occur gradually as one channel becomes too shallow to carry a large amount of water and another gradually receives more of the flow and becomes enlarged.

FIGURE 10-27
The Nile Delta was formed by sediments transported by the Nile River and deposited when the river entered the ocean. The delta is easily seen because of the vegetation in contrast to the desert around it. (Photograph courtesy NASA.)

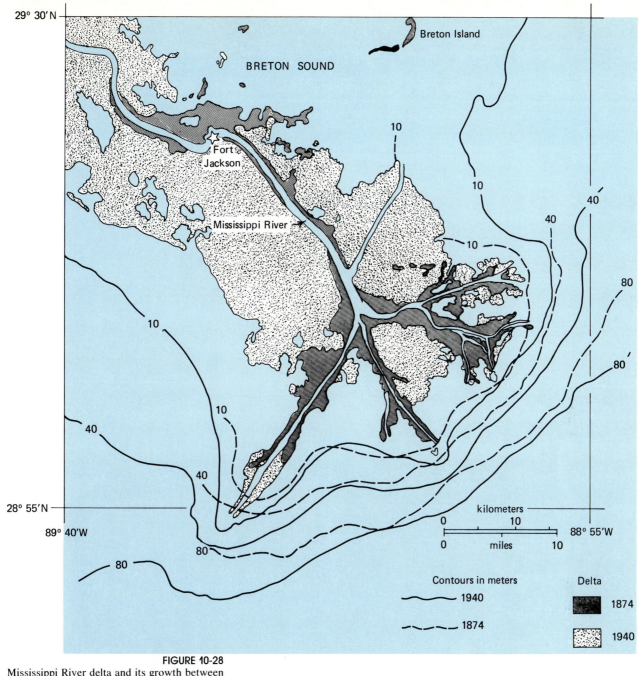

FIGURE 10-28
Mississippi River delta and its growth between 1874 and 1940. Note the "bird-foot" shape resulting from each distributary building seaward with only occasional flooding between the distributaries. The delta is now eroding in many of the areas built in the 1940s.

WETLANDS

Salt marshes—also called **wetlands**—are low-lying areas that are submerged at high tides but protected from direct wave attack. Their surfaces are generally overgrown by salt-tolerant plants and look very much like grass-covered meadows (Fig. 10-31, p. 273).

The size and shape of a marsh is determined by the general out-

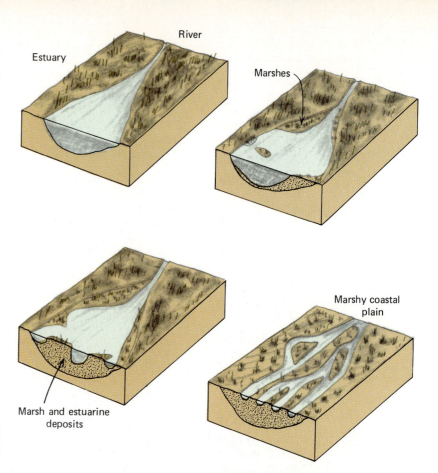

FIGURE 10-29
Stages in the development of a small estuary and its filling by sediments, converting it into a salt marsh.

FIGURE 10-30
Distributaries of the Suwanee River, Florida, showing the tidal marsh and cypress swamps lying between them. Note the meandering tidal creeks in the marsh. A road and dredged channel are the straight light-colored features in the upper right. (Photograph courtesy U.S. Geological Survey.)

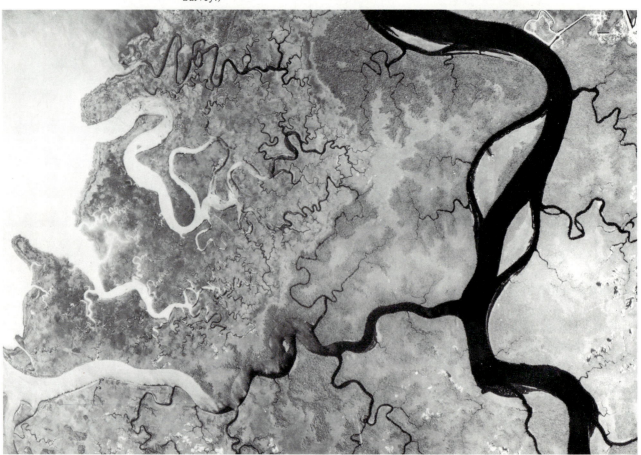

FIGURE 10-31
A wetland near Chesapeake Bay. (Photograph courtesy Michael J. Reber.)

FIGURE 10-32
A tidal creek in a Chesapeake Bay wetland. (Photograph courtesy Michael J. Reber.)

line of the depression in which it forms; vertical extent is controlled by the tidal range. The upper limit is generally controlled by the spring tides, which is the highest level to which ocean water can transport sediment. Most marshes are nearly flat-topped banks of sand or mixtures of sand and silt. The bank tops, known as **tidal flats,** are commonly exposed at low tide and submerged at high tide (Fig. 10-32).

Cutting through tidal flats are meandering channels, through which seawater enters as the tide rises and drains from the marsh as the tide falls. The largest and deepest channels contain water even at low tide and plants rarely grow in them. The bottom material is generally shifting sand or gravel because strong tidal currents resuspend the finer-grained materials, to be deposited on the flats. These large channels connect with smaller ones (Fig. 10-30) that are often empty at low tide.

Fine-grained sediment is transported into marshes by currents. As rising water moves out into the smaller channels and then onto the tidal flats, current velocities decrease. At some point current velocities are too low to keep sediment in suspension, so particles settle out.

FIGURE 10-33

Mangroves growing out into the water at Ten
Thousand Islands, Florida. (Photograph courtesy
Florida News Bureau.)

When the tide goes out, current velocities are inadequate to resuspend the sediment grains, which therefore remain where they settled out of the water. Plants on tidal flats also retain sediment and so marshes trap fine-grained sediment.

In tropical areas **mangroves** (Fig. 10-33) dominate marshes. These large treelike plants have extensive root systems which form dense thickets that shelter marine and land animals. Some organisms are especially adapted to survive in this environment. The mangrove oyster, for instance, attaches itself to roots and branches that are exposed at low tide. Mangrove roots trap sediment and organic matter and eventually the swamp is filled in—to be replaced by a low-lying tropical forest.

SUMMARY

The coastal ocean lies on the continental shelf. It is bounded on one side by the shoreline and on the other by open-ocean currents. Coastal currents parallel the shore and are driven by winds and density distribu-

tions resulting from river discharges. Upwelling occurs when winds blow surface waters offshore, and subsurface waters rise to replace them.

Temperature and salinity variations in coastal-ocean waters are greater than in the open ocean. These changes occur daily and seasonally. Salinity changes are greatest near the mouths of large rivers. Highest salinities occur in semiisolated basins in arid midlatitude areas.

Estuaries formed in the past 18,000 years as glaciers melted and sea level rose to its present position, about 3000 years ago. Where glacially eroded mountain valleys were flooded, they formed fjords. Flooded river valleys formed coastal plain estuaries. Lagoons formed when shallow coastal ocean areas were partially isolated from the ocean by barrier islands.

Estuarine circulation is a two-way current pattern. Low-salinity waters flow seaward, and landward-flowing high-salinity waters coming in from the ocean flow along the bottom. Separation between flows is sharpest where river flow is large and tidal range is small; this is the case in salt-wedge estuaries. Where river flow is small and tidal ranges are large, estuaries are well mixed; the volume of water flowing seaward is much larger than the river flow into the estuary. Discharges from estuaries are partially controlled by tides.

Marginal seas are coastal ocean areas which are partially isolated by land from the open ocean. The Black Sea receives the discharge of several large rivers and has an estuarine circulation. Waters below the pycnocline are isolated. Deep waters are devoid of dissolved oxygen and contain hydrogen sulfide. The Mediterranean Sea has strong evaporation, so it has an

antiestuarine circulation. Atlantic waters flow in at the surface; warm saline Mediterranean waters flow out at depth. The Red Sea is a long, narrow basin formed by the rifting of Africa. It, too, forms a warm saline water mass which discharges below the surface.

Shorelines are dynamic parts of the coastal ocean. Processes affecting shorelines act over relatively short periods of time. Major storms often have a lasting impact on coastlines. Sea-level rise since the last glacial stage has shaped many shoreline features, including flooded river valleys. Many coastlines are formed by marine processes which modify shorelines by erosion and sediment deposition.

Beaches are especially dynamic, since they are deposits of unconsolidated sediments. Variations in sea level due to tides and variable wave conditions change beaches rapidly. Sands are moved by winds to form dunes behind beaches. Sands and gravels eroded from headlands are moved along beaches by waves and then intercepted by submarine canyons to move offshore to depths below wave action. Inlets and wetlands also trap sands, moving along beaches.

Deltas are formed by river-borne sediment deposited at river mouths. During floods, rivers overflow their banks to deposit sediment between the distributaries, which builds up the delta. Flat-lying wetlands are covered with salt-tolerant plans, which trap and retain the sediments. In temperate midlatitudes, grasses dominate; mangroves dominate in the tropics.

STUDY QUESTIONS

1. Explain why water temperature and salinity variations are greater in coastal ocean areas than in the open ocean.

2. What causes coastal currents?

3. Describe net nontidal estuarine circulation. What causes an estuarine circulation?

4. Describe the difference between a salt-wedge and a well-mixed estuary.

5. Describe how an estuary changes into a wetland.

6. Draw a profile of a beach. Label the major features.

7. Discuss the effect of continued sea-level rise on shoreline features.

8. What causes sediment particles to move along a beach?

9. Draw a cross section of a continental margin and show the relationship to the deep-ocean floor for an Atlantic- and a Pacific-type margin.

SELECTED REFERENCES

BASCOM, W. 1980. *Waves and Beaches: The Dynamics of the Ocean Surface,* rev. ed. Garden City, N.Y.: Doubleday Anchor Books. 366 pp. Elementary.

BIRD, E. C. F. 1969. *Coasts.* Cambridge, Mass.: MIT Press. 246 pp. Ocean effects on coastlines; elementary.

KAUFMAN, W., AND O. PILKEY. 1979. *The Beaches are Moving.* Anchor Press, Garden City, N.Y.: Doubleday. 326 pp. Elementary, popular account of effects of rising sea level.

SHEPARD, F. P., AND H. R. WANLESS. 1971. *Our Changing Shorelines.* New York: McGraw-Hill. Study of U.S. coastlines.

Dinoflagellates are major producers of organic matter in the ocean. (© D. P. Wilson/Science Source/Photo Researchers.)

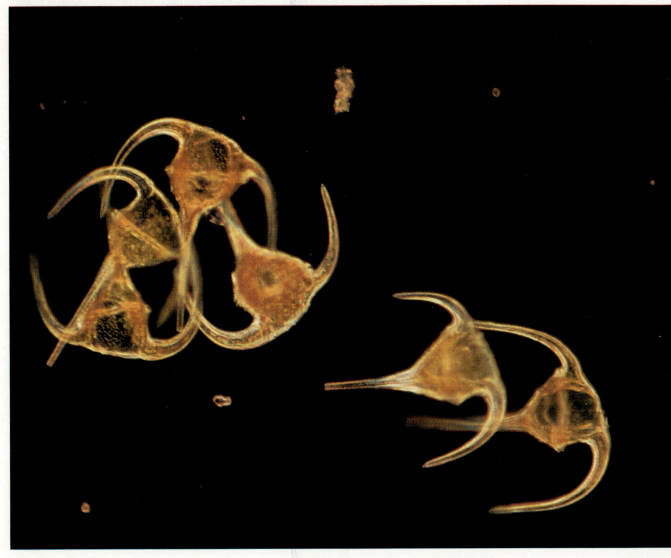

CHAPTER 11

Biochemical Processes

OBJECTIVES

1. To understand the principal features of oceanic ecosystems;

2. To understand primary and secondary production in the ocean;

3. To explain how organisms affect the composition of seawater;

4. To describe and explain the distribution of primary production in the ocean.

Many processes act together to control abundances and distributions of life in the ocean. Some are similar to those we know on land, but others are quite different. In this chapter, we examine the ocean as an enormous biochemical system in which energy, primarily from the sun, is converted into living matter and how this affects seawater. In the next three chapters, we focus on different parts of the ocean to see how energy is produced and used by organisms and communities of organisms.

In this chapter we examine:

How energy is transformed into organic matter;

How energy is used by living organisms;

Processes limiting production; and

Effects of life processes on abundances of various constituents in seawater.

ECOSYSTEMS

Oceanic plants and animals depend on one another to provide the conditions and materials that make life possible in the ocean. Organisms exchange matter and energy with each other and with the waters around them. The simplest system is called an **ecosystem** (Fig. 11-1). It includes **autotrophic** organisms (usually plants or bacteria) that produce food from inorganic substances. Plants use energy from the sun, in a process called **photosynthesis.** Some bacteria use energy stored in compounds such as methane (a component of natural gas) or hydrogen sulfide in a process called **chemosynthesis.** (Both of these food-making processes are discussed in following sections.)

Heterotrophic organisms (animals and bacteria) eat this food, thereby obtaining needed energy. **Decomposers** (primarily bacteria, fungi, and other organisms) break down tissues after an organism's death. This recycles the chemical constituents needed to make tissues or skeletons. This chapter is primarily concerned with the cycle: *production, utilization, decomposition,* and *recycling.*

The first part of the ecosystem we consider deals with feeding relationships among organisms. Here we define the concept of **trophic**

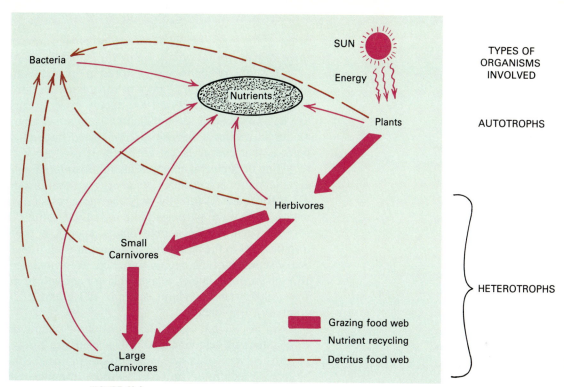

FIGURE 11-1
A simple ecosystem includes organisms, an energy source, and the environment in which the organisms live. Chemical constituents are recycled within an ecosystem. Energy is not recycled and flows only one way.

level, or simply put—who eats whom. Plants are the first trophic level. They are eaten by **herbivores,** the second trophic level. **Carnivores** eat herbivores, making them the third trophic level. Higher-level carnivores eat lower-level ones.

Outside this neat arrangement are the **omnivores,** who eat both plants and animals. (Humans are an example.) At every stage, **detritus** or waste products are released to the waters. These then support another complex of organisms, called **detritivores.** The concept of trophic levels does not work well with detritivores. For one reason, we do not understand that system very well.

In this chapter we are concerned with flows of energy and materials within ecosystems. This is an example of the application of **ecology**—the study of organisms in their environment.

FOOD CHAINS AND FOOD WEBS

Relationships among primary producers and consumers in ecosystems are usually complex. For simplicity, we begin by considering the simplest possible relationship. In a **food chain,** organisms at each level are eaten only by organisms of the next higher trophic level. One example is a *grazing food chain,* consisting of the following (Fig. 11-1):

Primary producers → herbivores → carnivores

Detritus (particles of organic matter) is produced at all levels in food chains. These particles sink and supply food (and energy) to all depths in the ocean and to a great variety of organisms. (We discuss the deep ocean in the next two chapters.) This supply of organic matter supports the *detritus food chain,* based on dead organic matter:

Organic matter → bacteria, fungi → microscopic animals

Many of the organisms in detritus food chains are extremely small, and these organisms are poorly known.

Essentially all the energy stored in organic matter is consumed by organisms. Only a small amount of organic matter falls to the ocean bottom and is finally buried too deeply in sediments for organisms to use it. (As we shall see in Chapter 15, a small amount of the organic matter in sediment deposits is converted into oil and gas.) The key point is that energy is not recycled. It is a one-way flow, unlike the recycling of chemical constituents in seawater. Each time a particle of organic matter is consumed, its energy content is reduced as organisms use it for food. Eventually only the most resistant organic matter remains, so that the particles have little food value.

At each step in a food chain, energy is transferred from one trophic level to the next. Energy is lost at each transfer because organisms use much of the energy they take in for maintenance. Only about 10% of the food intake is available for growth at each trophic level. Thus 1000 grams of plants support 100 grams of herbivores and 10 grams of carnivores.

So far, we have considered only food chains. But as pointed out earlier, feeding relationships in the ocean are rarely simple. Most organisms eat more than one type of food. In turn, each organism is eaten by many predators. The result is more complicated feeding relationships, called **food webs.**

To quantify the amount of organic material in an ecosystem, we use the concept of **biomass.** Biomass is the amount of plant or animal material, expressed as the weight of organic carbon, per volume of seawater or area of ocean bottom. It is similar to the expression used for **production.** Production expresses how much is being produced, while biomass indicates how much is available to other organisms to eat. For example, an organism may have a high rate of production but use most of the energy quickly, storing little—for example, a hummingbird. In that case, production may be high, but the biomass is low.

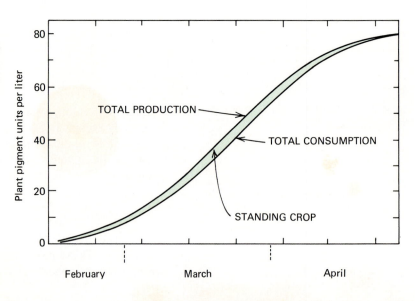

FIGURE 11-2
The standing crop is only a small part of the total plant production. At any time, only about 3% of the total production was in the standing crop in this example. The rest was eaten by grazing herbivores or was lost through other processes, such as sinking. (After Raymont. 1963.)

A slow-growing whale is the opposite case. It grows slowly for a very long time, so that the biomass is large even though the productivity may be small. Another way of expressing this relationship is the **standing crop,** which is the difference between the amount produced and the amount consumed (Fig. 11-2).

PHOTOSYNTHESIS

Organic matter is produced in sunlit, near-surface waters by photosynthesis. In most of the ocean, photosynthesis is carried out primarily by minute, one-celled plants called **phytoplankton.** Around the ocean margins, where the waters are shallow enough for the bottom to receive enough sunlight, larger plants, primarily algae, also photosynthesize.

Plants (and some bacteria) can produce organic matter because they contain **chlorophyll** or other light-absorbing pigments. These pigments capture energy from sunlight, which they use to combine dissolved carbon dioxide with water, forming carbohydrates, energy-rich compounds consisting of carbon, hydrogen, and oxygen, as shown below:

$$\longrightarrow$$

PHOTOSYNTHESIS
(chlorophyll)

$$6CO_2 \ + \ 6H_2O \ + \ \text{light} \ \leftrightarrow \ C_6H_{12}O_6 \ + \ 6O_2$$

Carbon   water   energy   carbohydrate   oxygen
dioxide

RESPIRATION

$$\longleftarrow$$

The process is reversed during **respiration** when oxygen is taken up and carbohydrates are decomposed to obtain energy. Heterotrophic organisms (animals and many bacteria) use this process to obtain energy by breaking down energy-containing compounds (such as carbohydrates and fats) synthesized by autotrophic organisms. Plants also respire, using some energy captured during photosynthesis.

CHEMOSYNTHESIS

Some bacteria produce organic matter without sunlight—a process called **chemosynthesis.** Instead of using sunlight, these bacteria extract energy from hydrogen sulfide, methane, metals, or even hydrogen gas to make new organic matter (Fig. 11-3). Chemosynthesis is common in dark, oxygen-deficient environments such as sediment deposits in marshes.

FIGURE 11-3
Comparison of photosynthesis and chemosynthesis, showing energy sources for each. Photosynthesis depends on energy from sunlight. Chemosynthesis is independent of sunlight, deriving energy from oxidizing reduced compounds, such as methane, metals, or sulfur compounds.

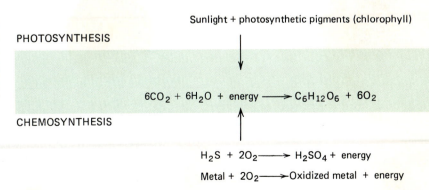

Organisms living on the deep-ocean bottom near vents depend on chemosynthesis for their food. Too little food sinks from the surface to feed the many organisms living near vents. Vent waters supply abundant sulfides and metals, which are the raw materials required by these specialized bacteria. Vents can supply food as long as they remain active.

Chemosynthetic bacteria live in or on their host organisms. In worms, the bacteria live in specialized organs. In clams, the bacteria live on the gills. The food they produce is absorbed through the gill surfaces much as is dissolved oxygen. These bacteria thus provide food directly to the hosts, while waste products from the host provide nutrients to the bacteria. Vent organisms have special blood proteins to transport sulfides to the bacteria without poisoning themselves. This relationship is an example of **symbiosis,** in which two organisms grow together to their mutual benefit. (We will see this again when we discuss corals.)

Bacteria also grow in mats on the bottom near the vents where snails and other grazing organisms can feed on them. In addition, filter-feeding organisms filter out bacteria living in the water.

Chemosynthesis also occurs where there are discharges of methane- or hydrogen sulfide–bearing waters. Such waters are expelled from sediments subducted at trenches or by groundwaters discharging on continental slopes. Communities of chemosynthetic organisms may be common on the deep-ocean bottom. We still know little about them.

PRIMARY PRODUCTION

To understand life in the ocean, it is important to know how much production occurs and where. You could answer such a question for your yard by weighing the grass clippings over a year. Production in the ocean is more difficult to measure.

One way is to label the newly formed organic matter (called **primary production**) in a water sample. Radioactive carbon-14 is commonly used. It is injected into several closed containers that are exposed to light intensities similar to those that organisms experience at various depths below the sea surface. After a set time, the samples are filtered and the radioactivity determined for the organisms caught by the filters. This technique measures *gross primary production,* the quantity of inorganic radioactive carbon made into organic matter. To measure **net primary production** (the amount available to other organisms), one must correct for any carbon used up (usually 10 to 50%) by respiration of plants and animals in the container while the measurement was being made. (The respired carbon would remain in the water and not on the filter.) Such measurements provide only an estimate of production at the time of the observation. Estimating production over a season or a year requires many observations. Since these are rarely available, total production must be estimated using available measurements.

SECONDARY AND TERTIARY PRODUCTION

So far, we have discussed processes involved in primary production. Before examining what limits them and determines their distribution, we need first to consider what happens to energy-rich organic matter produced by the primary producers.

Every organism requires energy to maintain itself. This is called

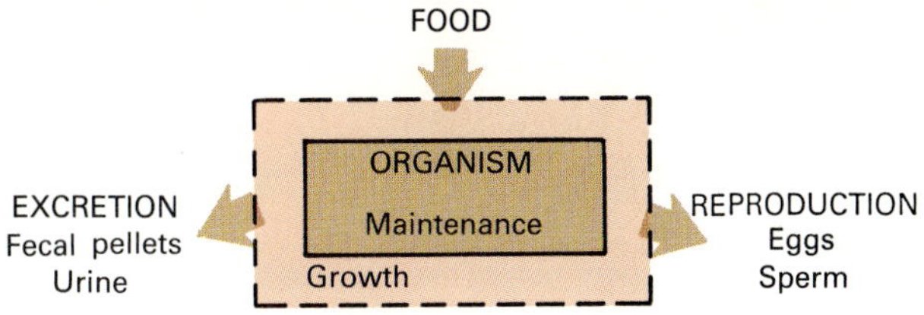

FIGURE 11-4
Most of the food taken in by an organism is used in maintaining itself and for growth. Some is excreted and some is used for reproduction.

EXCRETION
Fecal pellets
Urine

ORGANISM
Maintenance
Growth

REPRODUCTION
Eggs
Sperm

maintenance (Fig. 11-4), and it takes a large fraction of the energy available to an organism. Even plants respire carbon dioxide as a result of the processes of building and repairing their tissues.

Some of the energy available is lost immediately due to sloppy eating on the part of herbivores. This loss may amount to as much as one-third. The herbivores then use much of the available energy for their own maintenance. In addition, they excrete some as urine or fecal pellets (which become part of the detritus). Finally, a large amount of the energy may be used in reproduction. The eggs and sperm produced require large amounts of energy. Thus only a small amount of energy is available for growth of new tissues. This new growth is called *secondary production* in herbivores, *tertiary production* in first-level carnivores, and so on up through the food chain.

LIGHT LIMITATION

Availability of light controls plant growth and phytoplankton distributions in the ocean. As previously indicated, most food for marine organisms is produced by phytoplankton through photosynthesis in sunlit, near-surface waters. While most production occurs near the surface, respiration is nearly independent of depth. Thus at some depth, the instantaneous rate of photosynthesis equals the instantaneous rate of respiration; this is the **compensation depth** (Fig. 11-5).

The **photic zone,** where plants can produce enough to survive, is above the **critical depth,** where total photosynthesis equals total respiration. Below this depth, plants respire more than they produce and therefore die. The zone below the critical depth is called the **aphotic**

FIGURE 11-5
Production is highest near the ocean surface and diminishes with depth. Respiration is constant with depth. The compensation depth is where instantaneous production equals instantaneous respiration. At the critical depth, total respiration equals total production.

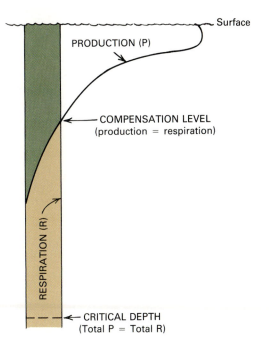

Surface

PRODUCTION (P)

COMPENSATION LEVEL
(production = respiration)

RESPIRATION (R)

CRITICAL DEPTH
(Total P = Total R)

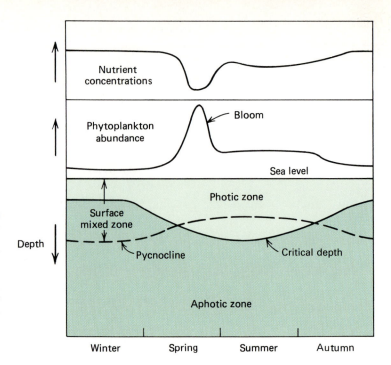

FIGURE 11-6
In winter, there is little light in the surface waters, and the pycnocline is relatively shallow. Thus phytoplankton are readily mixed out of the photic zone, there is little phytoplankton growth, and populations are small. In spring and summer, a shallow pycnocline keeps the phytoplankton in the photic zone. A bloom occurs because of the abundance of nutrients in the photic zone and ends when the nutrient supply is exhausted. Continued phytoplankton production depends on recycling of nutrients to the surface zone.

zone. Light intensity at the critical depth is about 1% of the incoming light at the surface.

To see how the availability of light affects phytoplankton abundances seasonally, we will examine the relationships in the midlatitudes. In winter, low light levels cause the critical depth to occur at relatively shallow depths (Fig. 11-6). At the same time, the surface mixed layer (bounded by the top of the pycnocline) is much deeper because of cooling of surface waters and also mixing by storm waves. Therefore, plants are frequently mixed below the critical depth and do not remain in the photic zone long enough for photosynthesis to exceed respiration. Thus only a few cells survive in surface waters over winter.

During spring, light intensities increase (Fig. 11-6), causing the critical depth to deepen. At the same time, surface waters begin to warm, resulting is a relatively shallow mixed layer. As a result, fewer phytoplankters are mixed below the photic zone and therefore remain in the photic zone long enough for photosynthesis to exceed respiration. (Remember that phytoplankton is the population composed of individual phytoplankters.) Thus there is net photosynthetic growth. This results in a rapid increase in the phytoplankton, which is called a **bloom.** The numbers of cells can double in a day or two. (We discuss the role of nutrients in the next section.)

The cycle reverses itself in fall and winter. As light intensities drop, the critical depth is shallower. Mixing due to cooling and wave activity cause a deeper mixed layer, and phytoplankton production drops to low winter levels.

NUTRIENT LIMITATION

Scarcity of essential substances necessary for plant growth—called **nutrients**—also limits population sizes. As plants grow, they take elements needed to make tissues and skeletons from surrounding waters. Many constituents are available in abundance in seawater, such as

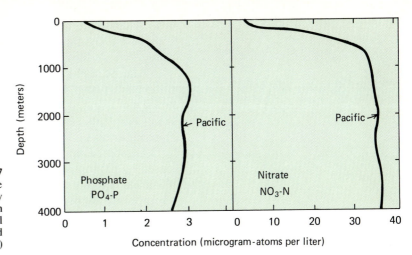

FIGURE 11-7
Near-surface waters are depleted of phosphate and nitrogen compounds due to uptake by phytoplankton. These data were taken between 30°N and 30°S, where there is little seasonal variation in productivity. (After Sverdrup and others. 1942.)

carbon, oxygen, and sulfur. Others, such as nitrogen compounds and phosphate, are scarce. Scarcity of nutrients limits growth for many organisms, even when there is enough sunlight.

As plants and animals die, about 95% of the nutrients contained in the tissues are released by bacterial decomposition in surface waters and are quickly taken up by growing organisms. The remaining nutrients are released when tissues decompose and skeletons dissolve below the surface zone, where there is usually too little light for photosynthesis. Nutrients released in the aphotic zone accumulate there (Fig. 11-7), increasing nutrient concentration. These deep-sea nutrients move with subsurface currents and are returned, often decades or centuries later, to surface waters. In the tropics, where light levels remain high throughout the year, phytoplankton growth continuously removes nutrients from the waters. Thus there is no opportunity for them to build up seasonally as they do where insufficient light limits growth in winter.

In the midlatitude example of Fig. 11-6, the spring bloom occurred because of the abundance of nutrients in surface waters when winter ended and light was no longer limiting. The bloom ended when the supply of nutrients in the surface layer was exhausted, even though light levels remained high. Later in the summer as phytoplankton died and decomposed, the nutrients released supported additional phytoplankton growth. Small blooms occur when strong winds in storms mix nutrients into surface waters while light intensities are still high.

Another factor limiting phytoplankton growth is the presence or absence of *organic trace constituents,* such as vitamins and hormones produced by bacteria. Such substances are essential for the growth of certain plants or animals. Recently upwelled waters are sometimes relatively barren of plant life even though they contain high nutrient levels because they lack some trace constituent. Some organic compounds combine with potentially toxic metals, such as copper, detoxifying the waters so that plants can grow in them.

PHOSPHOROUS AND NITROGEN CYCLES

Phosphorous and nitrogen compounds are necessary for phytoplankton growth. Because of differences in their chemical behavior, phosphorous and nitrogen compounds are released in different ways after an organism's death. While both compounds are primarily recycled in near-surface waters, they are also released below the pycnocline when some particles sink and decompose. Thus both types of compounds are more abundant in deep waters than in near-surface waters, where they are removed by plant growth.

Marine plants can survive short periods of low nutrient concentrations. If plants experience high nutrient concentrations, they can hoard more in their tissues than are immediately needed for growth. In this way, plants can grow for several generations by using reserves of stored nutrients.

Dissolved and particulate organic phosphorus compounds released by phytoplankton and animals are used by bacteria and phytoplankton. Many bacteria are eaten directly by one-celled animals, so phosphorus compounds can be taken up directly.

Nitrogen cycles more slowly because it must be released and then chemically transformed before it can be taken up by plants. Zooplankton, however, excrete nitrogen compounds, such as urea and ammonia, which can be taken up directly by phytoplankton.

Nitrogen compounds are also formed from atmospheric nitrogen by organisms living in open-ocean surface waters. Such a source of newly fixed nitrogen compounds increases the production of phytoplankton over the levels supported by recycled nitrogen compounds released through decomposition.

In addition to the recycling of these compounds, there is a slow resupply of nutrients by waters as they move upward through the pycnocline. As they do, they carry nutrients released below the photic zone back into the photic zone. This supply of nutrients balances the removal of nutrients by particles when they fall into the deep ocean. The amount of phytoplankton production supported by upwelled nutrients is called **new production.**

DISSOLVED ORGANIC MATTER

Most organic matter in the ocean is dead, occurring either as dissolved compounds or as small particles (Table 11-1). Nonliving organic matter is generally most abundant in near-surface waters. Dissolved organic matter remains in the ocean for very long times—many thousands of years. It is roughly equal to all living matter on Earth.

There are many sources of dissolved organic matter. Decomposition of dead plant and animal matter is one source. Up to 50% of an organism's weight dissolves in seawater as bacteria decomposes it. Continued bacterial action releases organic matter into the water as tissues are broken down.

Secretion of organic compounds by living plants is also important (Fig. 11-8). Usually less than 10% of the carbon assimilated during photosynthesis is released to the water as dissolved organic compounds. But under stressful conditions, such as unusually high light levels, 50% or more of the photosynthesized carbon compounds may be released. Excretion by animals is yet another source.

Dissolved organic matter enters food webs primarily through tiny bacteria, less than 0.6 micrometer across—too small to be seen by ordinary light microscopes. These bacteria, which live floating in seawater, can utilize dissolved organic matter. They are eaten by flagellates and ciliates (see Fig. 11-8), which in turn may be eaten by zooplankton, but no one is yet certain of that. If they are not eaten by zooplankton, this enormous reservoir of dissolved organic matter and

TABLE 11-1
Relative Abundance of Various Forms of Organic Matter in Seawater

FORM OF ORGANIC MATTER	RELATIVE ABUNDANCE (percent)
Dissolved organic matter	95
Particulate organic matter (nonliving)	5
Phytoplankton	0.1
Zooplankton	0.01
Fishes	0.0001

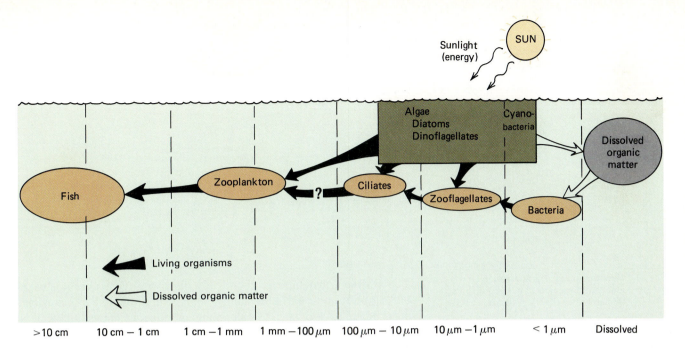

Sunlight
(energy)

SUN

Algae
Diatoms
Dinoflagellates

Cyano-
bacteria

Dissolved
organic
matter

Fish

Zooplankton

?

Ciliates

Zooflagellates

Bacteria

Living organisms

Dissolved organic matter

| >10 cm | 10 cm – 1 cm | 1 cm –1 mm | 1 mm –100 μm | 100 μm – 10 μm | 10 μm –1 μm | < 1 μm | Dissolved |

FIGURE 11-8
Dissolved organic matter, released by phytoplankton and decomposing organisms, is used by bacteria, which are eaten by flagellates and ciliates. These organisms may be eaten by zooplankton and enter the grazing food chain.

the organisms living on it apparently do not contribute to the zooplankton—fish food webs exploited by humans. Instead, their role may be in recycling nutrients for use by other organisms.

PARTICLES

Particles dispersed in seawater influence its chemical and biological behavior. The total amount of particles in the ocean is about 10,000 million tons (10^{16} grams). Particles are most abundant in near-surface, coastal-ocean waters but relatively rare in deep-ocean waters. Particles are important because they are the source of food for animals living on the bottom and a major source of sediments, which we discuss in Chapter 15.

Particles are also food for suspension-feeding organisms in the water. Bacteria on particles supply organic matter as well as needed trace organic compounds, much like our vitamins. Particles also help remove toxic metals and organic compounds from the water.

Biological particles in the ocean are also relatively large, ranging from 1 micrometer to 1 millimeter. (For comparison, remember that a human hair is about 100 micrometers, or 0.1 millimeter, in diameter.) Shell fragments, fecal pellets, and pieces of tissue released by zooplankton are most abundant, especially in upwelling areas. Such particles constitute up to 70% of the particulate matter in the ocean. Particulate organic carbon constitutes about 25% of all oceanic particulate matter.

Particle concentrations are lowest in areas of high productivity because they are removed by filter-feeding organisms. Large fecal pellets formed by these organisms sink rapidly (a few hundred meters per day) and are little altered by biological or chemical processes. Smaller particles sink more slowly and are more likely to decompose before reaching the bottom.

Particles are usually destroyed in two stages. First, they are broken up, either mechanically by grazing or by chemical dissolution.

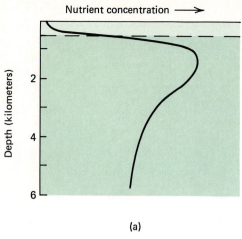

Nutrient concentration ⟶

(a)

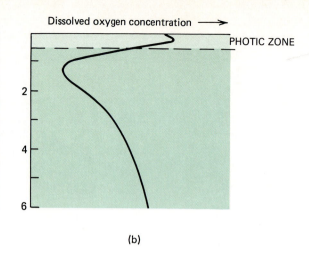

Dissolved oxygen concentration ⟶

PHOTIC ZONE

(b)

FIGURE 11-9
Variation in nutrient and dissolved oxygen concentrations typical of the North Pacific Ocean. (After Sverdrup and others. 1942.)

Then smaller fragments are chemically dissolved. Dissolution is especially important for particles smaller than 10 micrometers. As particles sink, some dissolve, such as calcareous or siliceous shells and skeletal parts. These dissolving particles release nutrients, silica, and metals, thus changing the chemical composition of deep-ocean waters and changing the composition of sediment deposits.

DISSOLVED OXYGEN

We turn now to chemical processes affected by the biological processes we have discussed. First we consider dissolved oxygen and then trace elements.

Oxygen dissolved in seawater participates in both biological and chemical processes. Most of the oxygen dissolved in seawater comes from the atmosphere through the sea surface. (Photosynthesis in near-surface waters also produces oxygen, but we will ignore that for this discussion.)

Dissolved oxygen and nutrient concentrations are inversely related to each other (Fig. 11-9). Where nutrient concentrations are high, dissolved oxygen concentrations are low and vice versa.

Cold, oxygen-rich water masses that form in Arctic and Antarctic regions supply dissolved oxygen to the deep ocean. Where such water masses form, dissolved oxygen concentrations are nearly constant throughout the water column (see Fig. 11-10; note the curve for Greenland).

Dissolved oxygen is consumed at all depths in the ocean. Below the surface zone, dissolved oxygen can come only from subsurface waters. The longer a water parcel is isolated from the surface, the lower its dissolved oxygen concentration. Thus dissolved oxygen concentrations are highest where the water mass is formed and lowest where water masses have been isolated from the atmosphere the longest. (Note the low dissolved oxygen concentrations in the middepths of the Pacific, as shown in Fig. 11-10.)

Oxygen consumption in the deep ocean is slow. Low temperatures and high pressures reduce metabolic rates of deep-water organisms. Also, scarcity of food keeps populations sparse. As waters slowly move along the bottom, their dissolved oxygen is used up. The patterns of change in dissolved oxygen concentrations are used to indicate the directions of deep-ocean water mass movements (see Fig. 7-16).

FIGURE 11-10
High concentrations of dissolved oxygen near the surface are due to exchanges with the atmosphere. High concentrations near the bottom are caused by cold waters from the high latitudes supplying dissolved oxygen. The North Atlantic near Greenland is where cold, oxygen-rich waters are injected into the deep Atlantic basin. Low dissolved oxygen levels at middepths are due to consumption of dissolved oxygen by decomposing organic matter as its sinks to the bottom. (After Sverdrup and others. 1942.)

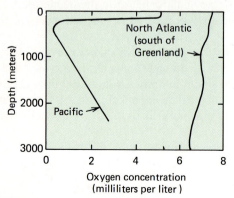

North Atlantic (south of Greenland) ⟶

Pacific ⟶

Oxygen concentration (milliliters per liter)

In estuaries and fjords, dissolved oxygen in bottom waters may be totally used up in the deepest parts of the basins, where water movements are sluggish. **Anaerobic bacteria** can live in such environments, since they do not require dissolved oxygen. They obtain oxygen by breaking down oxygen-containing compounds in seawater. They first use nitrate (NO_3^{2-}) and then nitrite (NO_2^-). Since there is little nitrate and less nitrite in ocean water, these are quickly used up. Other bacteria then break down sulfate (SO_4^{2-}) to obtain oxygen and release hydrogen sulfide (H_2S) as a byproduct, causing the familiar rotten-egg smell. Hydrogen sulfide occurs in sediments and in bottom waters of isolated basins, such as the deep waters of the Black Sea.

TRACE ELEMENTS

Chemical elements occurring in very low concentrations in seawater are usually highly reactive and involved in biological processes. These are called **nonconservative elements.** (Their concentrations are not obviously related to seawater salinity as are the conservative constituents. Recall that concentrations of **conservative constituents** are affected only by physical processes, such as mixing.) Very few elements occur at trace concentrations and yet display conservative behavior (not obviously involved in chemical or biological reactions).

Many nonconservative elements involved in biological processes exhibit distributions in seawater similar to nutrients [see Fig. 11-11(a)]. In other words, these elements are least abundant in near-surface waters and most abundant in the deep ocean. The explanation is simple: Growing organisms remove them from surface waters. Later, when the organisms die and decompose, these elements are released to subsurface waters by decomposition of particles. A few trace elements substitute directly for nutrient substances, such as arsenic and selenium, which substitute for phosphate, and germanium for silicon.

Distributions of some trace elements are controlled by how they enter the ocean [Fig. 11-11(b)]. For instance, manganese and cobalt are most abundant in near-surface waters, near land. These elements come from oxygen-deficient (reducing) environments in wetlands and from carbon-rich sediments on continental margins. Formation of manganese- and cobalt-rich crusts or nodules removes them from near-bottom waters.

FIGURE 11-11
Variation with depth in concentrations of dissolved constituents in seawater. (a) Some elements behave like nutrients. (b) Other distributions are controlled by the sources of that element, such as manganese and cobalt, which come into the ocean from oxygen-deficient sediment deposits on continental shelves and slopes. (c) Some distributions are controlled by the removal of the elements on particles.

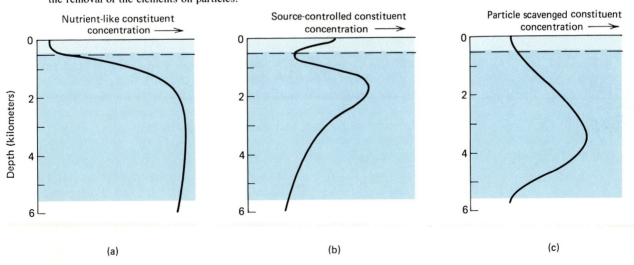

(a)

(b)

(c)

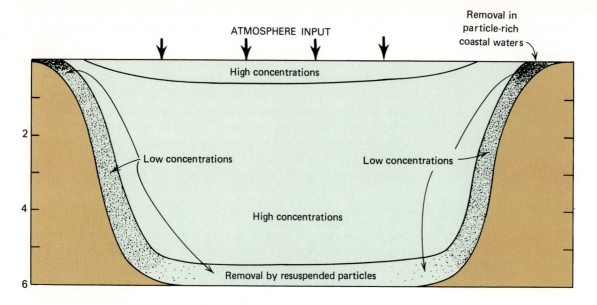

ATMOSPHERE INPUT

Removal in particle-rich coastal waters

High concentrations

Low concentrations   Low concentrations

High concentrations

Removal by resuspended particles

FIGURE 11-12
Constituents removed by particles show high concentrations in the central waters and low concentrations in waters near the bottom. Highest concentrations occur at the surface when the constituent is introduced by atmospheric fallout. (After D. W. Spencer, M. P. Bacon, and P. G. Brewer. 1981. Models of the distribution of ^{210}Pb in a section across the North Equatorial Atlantic Ocean. *J. Mar. Res.* 139:119–38.)

Finally, removal from seawater through chemical reactions with particles [Fig. 11-11(c)] controls distributions of some trace elements, such as lead and copper. Concentrations of trace elements removed by particle interactions are lowest where particles are abundant. Conversely, their concentrations are highest in particle-free waters such as near-surface open-ocean waters in the centers of gyres, far from land (Fig. 11-12).

UPWELLING

We have discussed the processes causing upwelling. Now we examine why upwelling areas are so rich in marine life.

Major upwelling areas occur on the eastern ocean margins along the equator and around Antarctica. First we examine coastal upwelling. As you recall, the pycnocline is nearest the surface in eastern boundary areas and deepest on the west. Because the pycnocline is relatively shallow, upwelled waters from depths of 100 to 200 meters come from below the pycnocline. (If the pycnocline were much deeper, upwelled waters would come from the surface zone and contain few nutrients. This is what happens during El Niños.) They are rich in nutrients, usually low in dissolved oxygen, and colder than surface waters. All these features identify recently upwelled waters.

Upwelling (Fig. 11-13) resembles estuarine circulation, whose productivity we have already discussed. Landward-flowing upwelling waters receive particles sinking out of seaward-flowing surface waters. Thus decomposition of particles takes place in the upwelling waters, and nutrients are quickly recycled back into surface waters. Also, organisms, larvae, and spores sink into the upwelling waters to seed them. Newly upwelled waters often contain few phytoplankton. After a day or so, cells in the upwelled waters have reproduced. First to appear are small forms (flagellates), followed by larger, chain-forming diatoms. In upwelling areas, organisms are 50 to 100 times more abundant than in nearby nutrient-poor surface waters. In some upwelling areas, the waters contain so many plants and animals that filters on cooling-water intakes for ships' engines quickly clog.

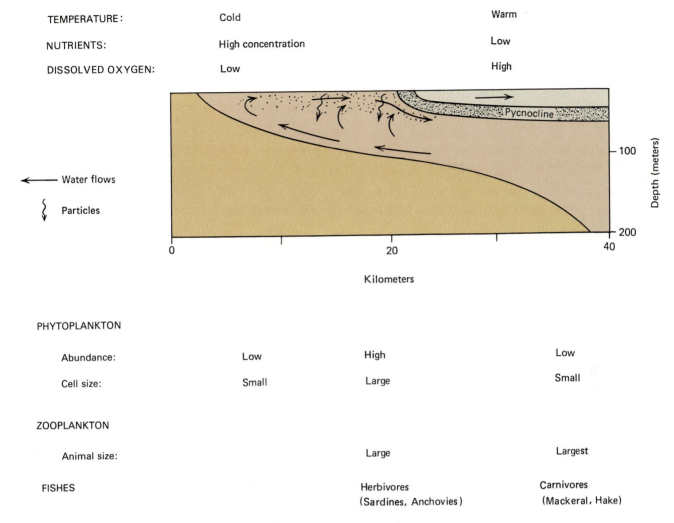

UPWELLED WATERS   SURFACE WATERS

TEMPERATURE:	Cold	Warm
NUTRIENTS:	High concentration	Low
DISSOLVED OXYGEN:	Low	High

← Water flows

⟨ Particles

PHYTOPLANKTON

Abundance:	Low	High	Low
Cell size:	Small	Large	Small

ZOOPLANKTON

Animal size:		Large	Largest

FISHES		Herbivores (Sardines, Anchovies)	Carnivores (Mackeral, Hake)

FIGURE 11-13
Schematic representation of the abundance and distribution of nutrients and marine organisms in an upwelling area.

Fish and other animals take advantage of the abundance of food. Filter-feeding fish (sardines, anchovies) live in areas of greatest algal production. Farther downstream, other herbivores (copepods) feed on the algae. Carnivorous fish (mackerel) then feed on the herbivores. The largest zooplankton (euphausiids) live near the edges of continental shelves along with larger fish (hake) which feed on them. Little energy is lost in these short, simple food webs. Recall that open-ocean food webs are complicated. Thus much energy is lost in the many transfers among trophic levels, and the total efficiency is much lower than in upwelling systems. Whalers learned to exploit the high productivity of upwelling areas by hunting whales downcurrent from them.

Both the equator and the Antarctic region are examples of divergences. In both cases, surface waters flow away from an area and are replaced by subsurface waters. This brings nutrients into the photic zone, making the areas highly productive. Production continues throughout the year along the equator. In Antarctica, light limits production except during the southern summer.

BOX 11-1

Food from the Sea

Culturing marine organisms, called **aquaculture,** is a promising way to increase food production from the sea (Fig. B11-1-1). Since many bottom-dwelling organisms do not move around, they are readily grown for human consumption. For example, oysters and clams can be taken while young from areas where they settled and moved to more favorable areas to fatten for the market. Another common technique is to attach oysters or mussels to ropes and suspend them beneath floats (Fig. B11-1-2). There they can mature away from predators, such as starfish.

Algae are also cultured, especially in Japan. There, a red alga (seaweed), called *nori,* is the principal crop. It is used as a protein supplement in foods. In the early 1980s, 30,000 farms employed nearly 80,000 persons to produce a crop worth about $1 billion.

Scarcity of domesticated marine plants and animals and ignorance of their diseases limits aquaculture. Also, disputes over legal status and conflicting uses of potential growing areas make it difficult to prevent losses or crop damage. Despite such problems, aquaculture is an important source of high-cost foods, such as oysters, mussels, and clams. It is expected to expand in countries where laws and labor costs are favorable.

New animals are being investigated for aquaculture. An example is the giant clam, *Tridacna,* which grows in tropical waters. (This is the so-called killer clam of underwater adventure movies. The name came from the fear that a diver could be trapped when such a clam, which can grow to be a meter across, closed its shell. There is no evidence that this ever happened.)

The animal has photosynthetic algae in its brightly colored mantle tissues, which it exposes to sunlight in shallow waters, typically on reefs. The photosynthetic algae provide food to the clam and it in turn provides nutrients to the algae. The result is a fast-growing animal whose mantle and abductor muscle (used to close the shell) can be eaten. The shells themselves are highly prized. The animal grows to marketable size within a few years.

Each *Tridacna* functions as both male and female. It first releases sperm to the surrounding waters and then changes sex to release millions of eggs. *Tridacna* is being studied for possible culturing in the tropical Pacific.

FIGURE B11-1-1
Oysters are grown in containers suspended in the water at an oyster farm in Washington. (© Bruce W. Heinemann, The Stock Market/Photo Researchers.)

FIGURE B11-1-2
Organisms are grown in mesh containers supported by ropes which hang down from rafts.

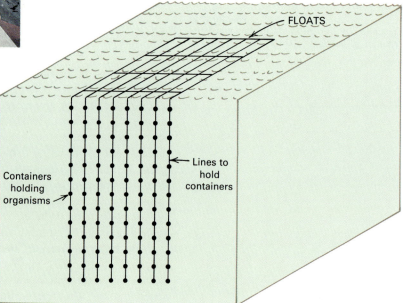

BOX 11-2

Marine Biotechnology

Efforts are now underway to improve the yields of mariculture, using *biotechnology*—the manipulation of organisms (often bacteria) to obtain useful products or to change the characteristics of organisms (Fig. B11-2-1). This is faster and more controllable than selective breeding which has been used with domesticated animals over thousands of years.

One technique in biotechnology is to insert a gene into a bacterium to make it produce large quantities of a substance controlled by the particular gene involved. When isolated and purified, the product can be used for many purposes.

This technique has been used to make growth hormone to treat fishes, such as trout and striped bass. The fish can be made to eat and grow twice as fast as animals in the wild. This may be used with many different fishes in the future.

Still another approach is to inject new genetic material directly into fish eggs. Some of the genic material is incorporated by the developing egg. This results in a transgenic animal—one that does not occur in nature. Such fishes will likely be grown first in controlled enclosures for mariculture to prevent their accidental release to the ocean.

FIGURE B11-2-1
World aquaculture production for 1971–78, excluding sport, bait, ornamental fish, and pearls. (From J. H. Ryther, 1981. Mariculture, ocean ranching, and other culture-based fisheries. *Bioscience* 31(3):223–230.)

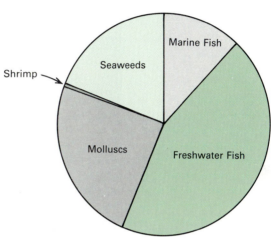

DISTRIBUTION OF PRODUCTION

Figure 11-14, p. 294 shows the distribution of primary production of surface waters and during the spring bloom in the North Atlantic (Fig. 11-15, p. 295). Most of the ocean's organic material is produced (Fig. 11-16) in open-ocean waters, while most major fisheries are located in coastal waters, particularly in upwelling areas (Fig. 11-17). This apparent discrepancy arises from differences in the relative efficiences of open-ocean and coastal food chains.

Open-ocean food webs are typically long, involving many energy transfers. Only the smallest phytoplankters (less than 0.005 millimeter) grow in the nutrient-poor, open-ocean waters. They, therefore, can be consumed only by very small herbivores.

These tiny animals are preyed on by animals in the 1-millimeter size range and they, in turn, by secondary carnivores that are about ten times as large, or about a centimeter long. (Examples of these organisms are described in Chapter 12.) In many cases, one or two larger invertebrate animals or fishes form additional links in an open-ocean food chain before it reaches a carnivore, such as mackerel or tuna.

Energy is lost at each transfer between trophic levels. Open-ocean predatory animals grow more slowly and expend more energy hunting for food than do coastal-ocean predators. Transfer efficiencies average about 10% in open-ocean food chains.

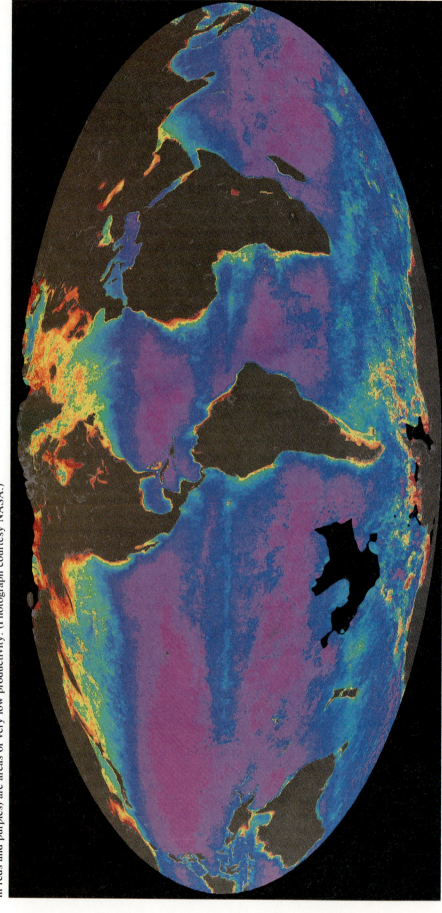

FIGURE 11-14

The distribution of productivity on land and in the ocean as seen by the Coastal Zone Color Scanner between November 1978 and June 1981. Note the bands of high productivity (shown by light blues) along the equator and the very high productivity in the coastal regions (shown by reds and yellows). The centers of the gyres (shown in reds and purples) are areas of very low productivity. (Photograph courtesy NASA.)

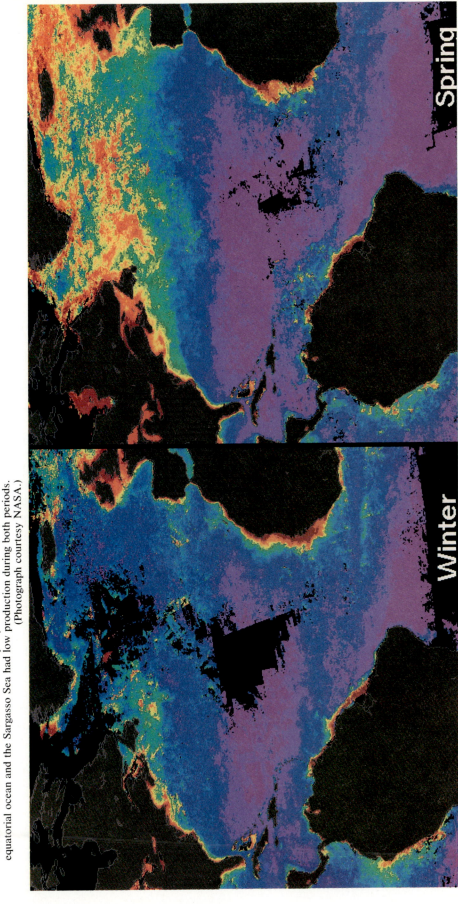

FIGURE 11-15

In winter 1979, high productivity (shown by reds and yellows) was confined to coastal ocean areas, especially around Iceland (upper center) and around the British Isles where tidal mixing was strong. During spring 1979, the spring bloom occurred over much of the northern Atlantic. Note the spotty distribution of the bloom. The equatorial ocean and the Sargasso Sea had low production during both periods. (Photograph courtesy NASA.)

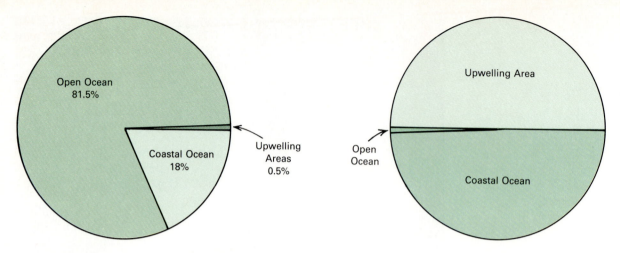

FIGURE 11-16
Estimated primary production of different ocean areas (in percent). Total primary production of carbon is estimated to be approximately 20 million metric tons per year. (Data from Ryther. 1969.)

FIGURE 11-17
Estimated fish production in different ocean areas. (After Ryther. 1969.)

In contrast, upwelling areas are characterized by a large proportion of phytoplankton (diatoms) that aggregate into clumps or long chains. They can be eaten directly by fishes or by small herbivorous zooplankton. In any case, highly productive areas commonly support short food chains of rapidly growing organisms where transfer efficiencies may be 20% or higher. Little energy is wasted in hunting, and the yield of fish to humans is much higher for the same amount of primary production. About half the world's fish catch comes from upwelling areas (Fig. 11-16).

Coastal areas without upwelling have intermediate productivities. Large phytoplankton are often present, especially during blooms. The density of even very small phytoplankton, however, is higher in coastal waters than in the open ocean, and long food chains undoubtedly exist along with very short ones, depending on local conditions. Transfer efficiencies in coastal ocean waters average around 15%.

SUMMARY

An ecosystem includes autotrophic organisms that produce food, heterotrophic organisms that consume food to obtain energy, and decomposers that break down tissues, recycling essential compounds to sustain new growth. Trophic levels in an ecosystem describe who eats whom. Plants, the first trophic level, are eaten by herbivores, the second trophic level. Carnivores, the third trophic level, eat herbivores. Energy transfers between trophic levels are about 10% efficient. Biomass is the quantity of plants or animals per unit volume of seawater or area of sea surface.

Photosynthesis is the process by which chlorophyll-containing organisms make food out of inorganic substances using energy from sunlight and releasing oxygen. Respiration is the opposite process. Organisms combine food and oxygen to obtain energy for building tissues or other processes. Chemosynthesis is a process in which bacteria use reduced compounds to obtain energy to make food in the absence of sunlight.

Primary production is the organic matter produced by plants. Secondary production is the organic matter produced by animals. Standing crop is the amount of organic matter per unit area (or volume) at a given time.

Light is essential for photosynthesis. Light is available in the photic zone to sustain photosynthesis above the compensation depth (light level about 1% of the surface value). In the aphotic zone below, there is too little light for net growth because plants use more energy to sustain themselves than they can produce through photosynthesis. Availability of sufficient light together with a stable water column to permit plants to remain in the photic zone leads to rapid growth of the plants, called a bloom.

Substances scarce in seawater but essential for growth are called nutrients. Lack of these substances—usually compounds of nitrogen and phosphorus—limits plant growth even with enough sunlight. When plants die and decompose, they release nutrients. Many organisms sink below the base of the surface mixed layer and decompose in the subsurface waters. Nutrients depleted by plant growth in surface waters are not immediately recycled through decomposition. As a result, nutrient concentrations removed from surface waters build up in subsurface waters. Nutrients are returned to surface waters by mixing or by upwelling of waters through the pycnocline. Other substances, called trace organic constituents, are also necessary for plant growth.

Most organic matter in the ocean is dead and occurs dissolved in seawater. Some organic matter is incorporated in growing plants. Most is released when the plant dies and decomposes. Fecal pellets of filter-feeding organisms form rapidly sinking particles.

Oxygen dissolves in seawater at the surface and is produced by photosynthesis. Dissolved oxygen levels are highest where water masses form in high latitudes. Dissolved oxygen is used at all depths as organic matter is decomposed. Dissolved oxygen concentrations are normally the opposite of nutrient concentrations—high where nutrient levels are low

and vice versa. Changes in dissolved oxygen levels indicates movements of subsurface waters. In isolated ocean areas where dissolved oxygen is used up, organisms form hydrogen sulfide by breaking down sulfate to obtain oxygen for their metabolic processes.

Many elements dissolved in seawater react in biological or chemical processes. These are called nonconservative elements. Conservative elements are not involved in such processes and are affected only by physical processes. Still other elements are removed by chemical reactions with particles suspended in seawater.

Biological productivity is especially high in upwelling areas, usually near the eastern boundaries of ocean basins and along the equator. The central portions of the ocean have low production. Short food chains in upwelling areas make energy transfers more efficient than in the open ocean, where food chains are long. Upwelling areas are especially productive because their circulation patterns retain nutrients and also seed the upwelling water with larvae of organisms that grow in the surface waters.

STUDY QUESTIONS

1. Describe an ecosystem.
2. What factors limit production of organic matter in the ocean? Where is each factor most important?
3. Describe the processes that control the distributions of phosphorus and nitrogen compounds in the open ocean.
4. Describe photosynthesis.
5. Describe chemosynthesis. Tell how it differs from chemosynthesis.
6. On an outline map of the world show the major upwelling areas.
7. Describe the distribution of dissolved oxygen in the three major ocean basins. Explain the differences.

8. Discuss the factors limiting the amount of fish that can be caught in the ocean.
9. Discuss food webs. How do they differ when dominated by diatoms and by bacteria?
10. Explain the different depth distributions of trace elements in the ocean.
11. Describe the processes controlling productivity in upwelling areas.
12. Discuss how particles can affect the abundances and distributions of trace elements in the ocean.
13. Why is half the world's fish production taken from upwelling zones?

SELECTED REFERENCES

Austin, B. 1988. *Marine Microbiology.* Cambridge: Cambridge University Press. 227 pp.

Buchsbaum, R., M. Buchsbaum, J. Pearse, and V. Pearse. 1987. *Animals without Backbones,* 3d ed. Chicago: University of Chicago Press. 572 pp. A classic.

Cushing, D. H., and J. J. Walsh. 1976. *The Ecology of the Seas.* Philadelphia: Saunders. 467 pp. Advanced.

Longhurst, A., and D. Pauly. 1987. *Ecology of Tropical Oceans.* London: Academic Press. 407 pp.

Nybakken, J. W. 1988. *Marine Biology: An Ecological Approach,* 2d ed. New York: Harper & Row. 514 pp. Elementary.

Parsons, T. R., M. Takahashi, and B. Hargrave. 1985. *Biological Oceanographic Processes,* 3d ed. Oxford: Pergamon Press. 344 pp. Reviews biochemical processes.

Valiela, I. 1984. *Marine Ecological Processes.* New York: Springer-Verlag. 546 pp.

Radiolarians, one-celled animals, are prominent members of oceanic zooplankton communities. (© M. I. Walter/Photo Researchers.)

CHAPTER 12

Plankton

OBJECTIVES _____

1. To describe the life environment for open-ocean planktonic organisms;
2. To describe the role of microorganisms in the ocean.

*I*n the open ocean, organisms must float or continually swim. Life is constantly in motion, responding to small-scale turbulence or to large-scale eddies and currents. Those organisms that cannot swim fast enough are moved around by the waters. We call such organisms **plankton.** In this chapter, we see how such organisms have adapted to this environment, which is so unfamiliar to us who live attached to the land. In particular, we are concerned with:

The open-ocean environment for organisms;
Strategies of competing for resources;
Adaptations for open-ocean life; and
Distributions of planktonic organisms.

MARINE ENVIRONMENTS

Living conditions for marine organisms are markedly different from those on land. On land we live at the bottom of the atmosphere, which only birds, some microbes, and a few spiders inhabit. Even the tallest trees extend only a few tens of meters above the land surface. The open ocean, in contrast, is a three-dimensional world where life exists at all depths—from sunlit surface waters to the dark ocean bottom. In the open ocean, subtle changes in temperature, salinity, or light levels are the boundaries. In this chapter, we are concerned with the adaptations organisms make to live in this environment.

There are three *oceanic life styles: drifting* (**pelagic** *or planktonic*), *swimming* (*nektonic*), or *attached* (**benthic**). The drifters, called **plankton,** are weak swimmers and easily carried by currents. They are usually small and include **bacteria, phytoplankton** (one-celled plants, Fig. 12-1), and **zooplankton** (planktonic animals, Fig. 12-2). **Nekton** are strong swimmers and include fishes, squid, and whales. **Benthos** or bottom-dwelling organisms, include large plants that grow in shallow waters and bottom-dwelling animals at all depths.

Here we discuss planktonic organisms that live in the open ocean, primarily in the surface zone, and their adaptations to that environ-

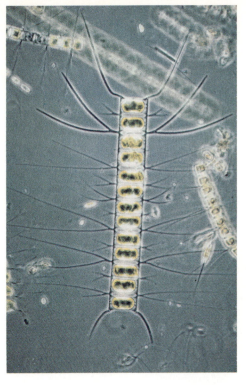

FIGURE 12-1
A chain-forming diatom. (Photograph courtesy NASA, J. Yoder.)

FIGURE 12-2
Planktonic larval form of a star fish. These larvae go through several stages before they finally metamorphose and settle to the bottom as a star fish. The larvae swim by means of cilia. (Photograph courtesy M. J. Youngbluth, Harbor Oceanographic Institution.)

ment. In the next chapter, we consider organisms that are strong swimmers and can move independently of currents. In Chapter 14, we discuss organisms and communities that live on the bottom.

PLANKTONIC LIFE

Seawater density (1.025 grams per cubic centimeter) is much greater than the density of air (0.0012 gram per cubic centimeter). Because the density of seawater is near the density of living tissues, marine organisms do not require heavy skeletons needed to support both plants and animals on land. The waters around them provide the needed support. Indeed, many marine organisms have no skeletons and consist largely of water.

One problem unique to planktonic life is **sinking.** Maintaining the proper depth is critical to survival, since most open-ocean organisms live in a particular layer (or depth) that meets their light and temperature needs. An organism that is less dense than its surrounding water rises. Those denser sink unless they swim or can somehow maintain buoyancy. *Drag,* or resistance to sinking, determines an object's rate of sinking. (You experience drag on your hand when you stick it out the window of a moving car.) Drag is greater for large objects than for small ones. By altering its density or drag, an organism can change its sinking rate. Also, objects sink more slowly in cold than in warm waters. Some organisms living in warm waters grow long spines or have other adaptations to retard their sinking in warm waters. Other organisms incorporate air or fats in their tissues to decrease their density.

Most marine organisms are coldblooded. In other words, their internal temperatures are the same as the surrounding waters. Each species can tolerate temperatures within a limited range. Coldblooded organisms are more tolerant of low than of high temperatures. Low temperatures usually cause quiescence, during which organisms require less food and less oxygen than at higher temperatures. Extreme heat increases metabolic rates and food consumption and may cause death.

Most marine organisms have body fluids with salt contents similar to seawater. Thus they do not need to protect themselves from losing (or gaining) water as land organisms must. Where salinity is variable, as in estuaries or in the coastal ocean, organisms must be able to maintain the salt balance of their blood to survive.

All organisms seek to avoid being eaten by their predators. Planktonic organisms use several *defensive strategies*. *Small size* is perhaps the most common. This has many advantages for plankton, as discussed in the next section. The average marine organism that we catch in nets is about the size of a mosquito.

A second defensive strategy is *transparency*. Transparent or translucent organisms are difficult for predators to see. This strategy is especially common among the gelatinous plankton that we discuss later.

Schooling (Fig. 12-3) is a third strategy. By remaining in schools of many hundreds or thousands of other organisms of the same species and usually of the same size, individuals are likely to escape if the school is attacked. Also, schooling confuses predators, further protecting individuals. Many fishes, squids, and shrimplike organisms exhibit schooling behavior.

Organisms also *migrate vertically* (see Fig. 12-4). They stay in

FIGURE 12-3
Schooling of herring fishes provides protection by confusing predators. (Photograph by C. Arneson.)

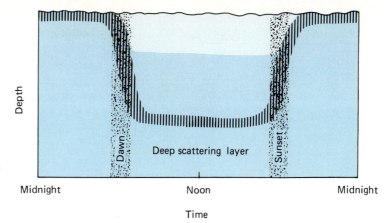

FIGURE 12-4
Organisms migrate to the surface at night to feed. During the day, they return to the darkness of the aphotic zone, where they are more difficult to see. These organisms form the "deep scattering layer" that is detected by echo sounders.

(Diagram labels: Depth; Dawn; Deep scattering layer; Sunset; Midnight; Noon; Midnight; Time)

FIGURE 12-5
A dark red shrimp lives in the dim light around 1000 meters below the surface. The dark red color is common among deep-sea organisms, apparently protection against predators being able to see it. (Photograph by B. Robison, MBARI.)

FIGURE 12-6
Hatchet fish are common in the mid-depths of the ocean. Bioluminuscent organs on the bottom of the fish provide enough light to obscure the shadow the fish would otherwise make in the dim light coming down from the surface. This is called counter shading. (Photograph courtesy B. Robison, MBARI.)

dark subsurface water during the day. At night, they swim to the surface layers to feed. As day brightens, they return to the safety of the dark, where predators who hunt by sight cannot readily find them. Some organisms are known to migrate several hundred meters daily.

Organisms also use *color* to blend into the background. Organisms near the surface are often bluish to match the color of the sea surface. Deeper in the ocean, where light is very dim, shrimp (Fig. 12-5) are often deep red. (Remember that red light is absorbed very near the surface, so that red shrimp are invisible.) In the deepest ocean, the organisms are gray or black.

Many fishes use *countershading,* in which they have light-emitting organs on their undersides (Fig. 12-6). This permits them to eliminate the shadow that a predator would see from below when looking toward the light coming down from the sea surface.

Still another defensive strategy is to increase the size of the organism. This is one of the advantages that gelatinous organisms have. They contain large amounts of water in their tissues, so that a large organism may contain only a small amount of organic matter. Another approach is to grow spines. A spiny organism is more difficult to swallow than one without spines.

Open-ocean plants are mostly microscopic floating one-celled organisms (Fig. 12-7) called **phytoplankton.** These tiny plants produce food for open-ocean organisms.

Small plankton—called *ultraplankton* (including bacteria)—are less than 0.005 millimeter in diameter. Next largest are *microphytoplankton* (0.07 to 1.0 millimeter), about the same size as many zooplankton organisms (microscopic planktonic animals). (Remember that a human hair is about 0.1 millimeter, or 100 micrometers, in diameter.)

Very small plankton (less than 50 micrometers) are important in coastal- and open-ocean equatorial waters, where they make up about 50 to 80% of the standing crop. In coastal waters, larger microphytoplankton are more abundant than in the open ocean, but very small plankton called *nannoplankton* still dominate primary production. Where currents are strong or upwelling replenishes nutrients, larger forms are apparently favored. In open-ocean waters, there is less turbulence to return phytoplankton to the surface layers. Here the swimming abilities of many smaller plankters provide an advantage over larger forms, which slowly sink out of the photic zone.

Net plankton, (so called because they are easily caught by biologists' nets) have large standing crops at higher latitudes. Diatoms [Fig. 12-7(a)] are usually most important. Division rates can exceed one per day; therefore, populations can increase 500 to 2000 times over the winter "seed crops." These minute plankters cannot ingest large diatoms. When zooplankters grow and reproduce in response to abundant food, their grazing quickly reduces phytoplankton biomass.

Diatoms have protruding threads and form long chains, especially in nutrient-rich waters. They grow by division, the cells becoming smaller with each generation (Fig. 12-8, p. 306). When cells decrease to a critical size, both halves of the old shell are discarded. Then the naked cell doubles or triples in size before a new set of shells forms.

Diatoms (and many other organisms) form resting spores under unfavorable conditions. They can remain alive as spores for long periods, even years. After death, the glasslike shells of diatoms sink to the ocean bottom, where they dissolve or are buried in sediment deposits. (More about this in Chapter 15.)

Dinoflagellates are second in abundance to diatoms [Fig. 12-7(b)]. These organisms also resemble one-celled animals because many are heterotrophic. Some can live on dissolved or particulate organic matter absorbed or ingested from seawater, and many can tolerate low nutrient concentrations. Dinoflagellate blooms can exceed diatom production. This may result from a scarcity of silicon, which limits diatom growth but does not affect dinoflagellates. Changing light intensity also affects successions of species.

Armored dinoflagellates have tiny cellulose plates covering the cell. Usually two flagella provide propulsion. These, combined with the organism's sensitivity to light, permit it to swim to its preferred light level. **Coccolithophores** [Fig. 12-7(c)], another major group of flagellates, have coatings of tiny calcareous plates. They are important primary producers in the ocean and major contributors to sediment deposits. Less common are *silicoflagellates* [Fig. 12-7(d)] and other flagellated nannoplankton.

For such plants, small size has several advantages. They rely on diffusion to supply nutrients and remove wastes from the cells. Thus large surface area relative to body mass facilitates exchanges of dis-

FIGURE 12-7
Major types of plankton. (a) Diatoms include rod-shaped, spool-shaped, and pillbox-shaped organisms. The longest ones measure about 80 micrometers. (b) Dinoflagellates are about 400 micrometers long. (Photograph courtesy J. M. Sieburth and University Park Press, Baltimore, Md.) (c) The coccolithophore measures about 10 micrometers in diameter. Note the calcareous plates forming its shell. (Photograph courtesy Susumu Honjo, WHOI.) (d) The silicoflagellate shell (containing debris) is about 30 micrometers in diameter, including spines. (Photograph courtesy J. M. Sieburth and University Park Press, Baltimore, Md.)

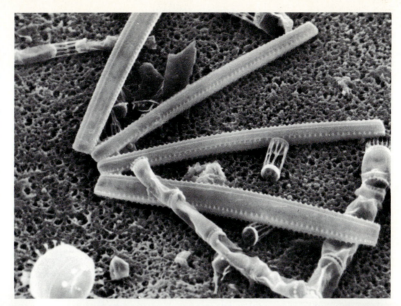

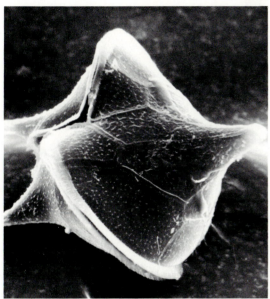

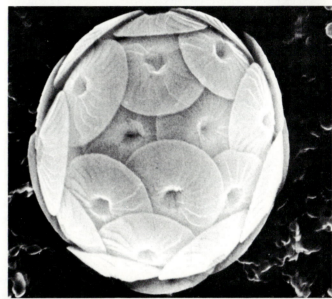

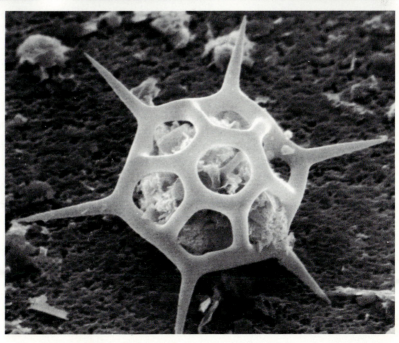

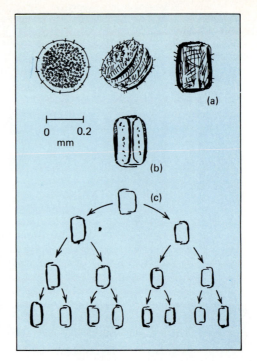

FIGURE 12-8
Reproduction in diatoms. This is a "pillbox shell" (a) with pores for exchanging nutrients and metabolic products with the surrounding waters. When a diatom grows large enough to divide, the "lid" and the "box" separate (b) and each gets about one-half the cell contents. A new box is secreted over the exposed tissue, so that the original shell half becomes the new lid. Daughter cells become progressively smaller.

solved substances between cells and the waters around them. On land or in shallow waters, air and water move past plants, bringing nutrients to the roots and carrying away wastes.

To maintain themselves in the photic zone, some phytoplankton can move relative to their surroundings. These have tiny, whiplike flagella to propel them through the water by lashing motions. Others sink slowly and depend on turbulence and upwelling currents in the water to return them to the photic zone.

NANNOPLANKTON AND BACTERIA

Nannoplankton (less than 50 micrometers in diameter) are widespread throughout the ocean; accounting for up to 80% of the standing crop. Organisms smaller than 10 micrometers are the dominant producers and consumers of organic matter in the ocean. These tiny organisms include heterotrophic bacteria, autotrophic cyanobacteria (also called blue-green algae), and the microflagellate predators that eat them. Organisms 1 to 3 micrometers in diameter account for much of the ocean's productivity.

Many of these organisms can live as either autotrophs or heterotrophs. Thus they are able to survive for long periods in the deep-ocean waters. Such organisms are often the first to bloom in newly upwelled waters. Indeed, the highly productive upwelling waters, dominated by diatoms, are isolated oases.

Minute organisms 3 to 10 micrometers in diameter preying on the tiny one-celled plants consume most of the organic matter in the ocean. There is roughly an equal quantity of organic matter in all size classes of organisms, ranging from bacteria to whales. Since microorganisms process matter and energy much faster than larger organisms, a large fraction of the organic matter in the ocean is processed by these abundant but little-known organisms.

BOX 12-1

Red Tides

Red tides are masses of discolored surface water formed by intense blooms of planktonic organisms, usually dinoflagellates. They occur in nearshore waters and have nothing to do with tides. They often give the water a tomato-soup or rusty color. Toxins in the cells cause problems. When filter-feeding organisms eat dinoflagellates, they can retain toxins in their tissues. These toxins may not immediately harm fish or clams, but they can poison people or marine mammals who eat them. Thus fisheries are often closed by health authorities when red tides occur. Toxins can also be released to surrounding waters and kill marine organisms. Toxins transported by spray cause lung irritation to people on nearby beaches. Economic damage to coastal communities can run into many millions of dollars.

Red tides are extreme cases of plankton blooms. They can occur when conditions are especially favorable. For example, a sudden increase in stability of surface waters (due to a heavy rain) keeps planktonic organisms in sunlit surface waters. Then they are no longer limited by scarcity of light, and a bloom can follow.

Availability of nutrients can also trigger red tides. Rainfall often contains nutrients derived from automobile exhaust. Street runoff contains animal wastes and lawn fertilizers. When nitrogen compounds are mixed into nutrient-limited waters, they can trigger plankton blooms. It is also possible for runoff to provide trace amounts of metals or possibly compounds that remove toxic substances (such as copper) from the water.

Dinoflagellates usually cause red tides. These one-celled organisms can photosynthesize in waters that contain few nutrients. Thus they are widely distributed, especially in nutrient-poor tropical ocean waters. Dinoflagellates can also swim, using a hairlike organ called a flagellum, toward the surface to concentrate in surface waters.

Sustained growth of dinoflagellates usually depletes the nutrients dissolved in the waters. Thus a red tide usually ends after a few days. But if there is a sustained source of nutrients, a red tide can persist for weeks or months. One source of nutrients is nitrogen-fixing blue-green algae (also called cyanobacteria). They take nitrogen gas from the water and produce ammonia, which is readily used by organisms. If blue-green algae and dinoflagellates occur together, they can support a red tide for many weeks. Eventually the tide is dispersed by currents and winds, or the weather changes, removing conditions favorable to plankton growth.

Dinoflagellates can form spores, a seedlike stage, when conditions are not favorable. The spores settle out on the bottom and may remain alive in sediments for years. When conditions are favorable, the spores develop into normal dinoflagellates and can seed future red tides. Thus once an area has a red tide, it is subject to future ones.

ZOOPLANKTON

Lower trophic-level consumers in marine food webs are mostly **zooplankters** (Fig. 12-9, p. 308). Some swim and can pursue prey. But most are suspension feeders, bearing tiny hairs or mucous surfaces to capture floating food particles. Because these animals usually depend on food particles of a particular size, their distributions depend largely on the availability of food and how the currents have moved them.

Another important factor limiting zooplankton distribution is the narrow temperature range (generally only a few degrees) in which they reproduce. Adult populations have greater temperature tolerances and so may be carried far out of their breeding range by currents.

Not all zooplankton remain free-floating throughout life. Those that do, known as **holoplankton,** are generally more important in marine food webs than **meroplankton,** which are the larval stages of bottom-dwelling (benthic) organisms. (About 80% of shallow-water benthic organisms in the tropics have planktonic larvae.) Holoplankton dominate in the open ocean, whereas meroplankton are more common in shallow coastal-ocean waters. Next, we discuss each type of plankton, giving examples of common organisms.

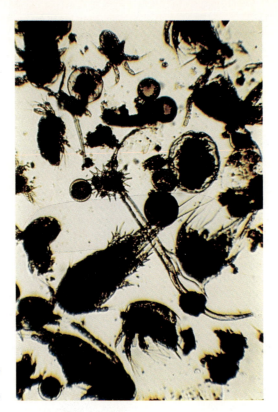

FIGURE 12-9
Common zooplankton from Chesapeake Bay.
(Photograph courtesy Michael J. Reber.)

HOLOPLANKTON

Many consumers of nannoplankton are single-celled **protozoans** including, for example, **Foraminifera** [Fig. 12-10(a)] and **Radiolaria** [Fig. 12-10(b)]. Foraminifera live nearly everywhere in the ocean—both in the water and on the bottom. Many have delicate, porous shells made of calcium carbonate. Thin protoplasm extrusions extend through holes in the shells to capture food particles.

Radiolaria, which are entirely pelagic, have siliceous (glasslike) or strontium sulfate skeletons. Protoplasmic strands project in all directions outside the shell as long, sticky filaments. These trap tiny particles, which are then borne by protoplasm toward the center of the body to be digested. Individuals range from 0.1 to more than 10 millimeters in diameter. Reproduction is by division into many small flagellated cells.

Crustacea are the most numerous. They constitute 70% or more of the zooplankton. **Copepods** (Fig. 12-11) and **euphausiids** (Fig. 12-12) are most important in marine food webs. Crustaceans have been called the "insects of the sea." In fact, both insects and crustaceans are arthropods ("jointed-feet") characterized by segmented bodies and appendages. They have stiff, chitinous outer shells which serve as skeletons. The appendages are specialized for various functions, such as feeding, moving, sensing, or reproducing. Copepods are approximately 0.3 to about 8 millimeters long and have feathery, curved bristles that form a filter chamber behind the mouth. Copepods occur throughout the ocean; they are among its most numerous animals, certainly the most numerous marine herbivores. Depending on temperature and availability of food, large copepods can double their numbers a few times in a year. Smaller ones reproduce even more frequently.

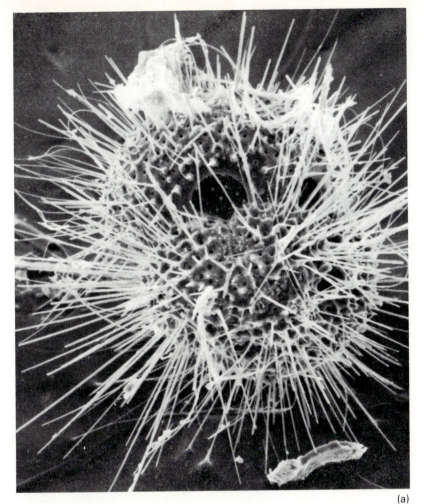

(a)

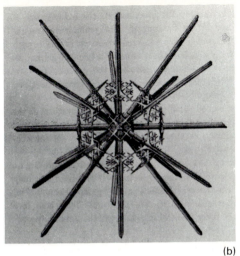

(b)

FIGURE 12-10
Some open-ocean zooplankton. (a) A foraminiferan. Two apertures, through which food is ingested, are visible just above center. Excluding spines, the organism is about 300 micrometers in diameter. (Photograph courtesy J. M. Sieburth and University Park Press, Baltimore, Md.) (b) This delicate shell made by an acantharian—a close relative of the radiolaria—is made of the soluble mineral strontium sulfate. The shell dissolves within a few hours after the organism dies (Drawn by Ernst Haekel. Report of the Scientific Results, H.M.S. *Challenger. Zoology* 18:1887.)

FIGURE 12-11
Copepods are the most abundant planktonic animals and dominate many oceanic food webs. The largest free-living species live at depths below 1000 meters and are about 20 mm long. (Photograph courtesy Institute of Oceanographic Sciences, U.K.)

FIGURE 12-12
Euphausiids (also known as krill) are large planktonic organisms that are especially abundant around Antarctica. They are eaten in great quantities by whales. (Photograph courtesy U.S. Antarctic Program, National Science Foundation.)

FIGURE 12-13
Pteropods—planktonic snails—are abundant herbivores. This pteropod is feeding on its delicate mucus web. (Photograph by W. Hamner, UCLA.)

Shrimplike euphausiids, also known as **krill,** are another important crustacean group. Dense swarms of these animals feed on diatoms. Krill, in turn, are the chief food of many fishes and of large whales. Euphausiids are larger than copepods, up to 5 centimeters long. They occur throughout the water column. In Antarctic and cold temperature waters, euphausiids mature slowly, living up to 2 years. A large baleen whale eats an average of 850 liters of euphausiids per day.

Besides crustaceans, other types of zooplankton can be extremely abundant. **Pteropoda** (''wing-footed'') are small pelagic snails that occur in dense swarms in all seas (Fig. 12-13). In this group the characteristic snail foot is modified into fins which permit them to swim vertically hundreds of meters each day. One group of pteropods is herbivorous and forms carbonate shells which are preserved in pteropod oozes. The other group is carnivorous and has no shell.

MEROPLANKTON

Planktonic larval forms of benthic animals (called **meroplankton**) are abundant in coastal waters. They are an important food for fishes. Most benthic animals have a free-swimming larval stage that lasts a few weeks. Eggs and sperm of benthic animals are discharged in great clouds to fertilize in the water. Tens of millions are produced per individual per year, but mortality due to predation and other hazards is high, and only a few reach adulthood.

The ability of maturing larvae to find suitable bottom material on which to settle is a limiting factor. Currents, for example, may carry worm larvae into rocky-bottomed areas where there is no sediment in which to burrow, and so the larvae die.

Fish larvae (Fig. 12-14) are also among the meroplankton. Some fishes, such as herring and sand eel, attach their eggs to rocks or vegetation; some lay them in ''nests'' near the shore; and some lay them in gelatinous masses. But most fish eggs are released and fertilized in coastal ocean waters. These eggs drift; when sufficiently developed, they hatch and begin to feed. Depending on temperature and species, it may be as soon as 1 or 2 days after release. They then live as plankton for weeks.

A

B

C

D

PACIFIC WHITING

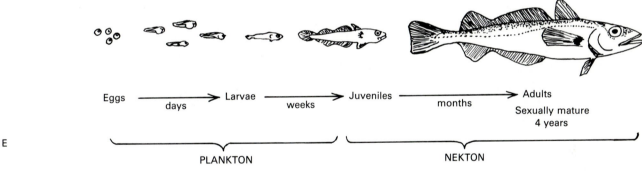

Eggs ——days——→ Larvae ——weeks——→ Juveniles ——months——→ Adults

Sexually mature
4 years

PLANKTON   NEKTON

E

PRINCIPAL PREY

Small copepods   Euphausiids   Euphausiids
                                                Shrimp
                                                Small fishes

PRINCIPAL PREDATORS

Suspension   Copepods   Fishes   Seals
Feeders   Sea Lions
                                                 Whales
                                                 Humans

FIGURE 12-14

Development of the white seabass (also called croaker) is shown in these photographs as they develop from eggs (a) about 1 mm diameter. Note that the fish are already recognizable in the eggs. When the fish is 3 days old (b), it still has the yolk sac attached which it loses by the time it is 7 days old (c). As the larval fish grows it acquires its characteristic coloration by the time it is 55 days old (d) and 27 mm long. (e) As Pacific whiting (a type of cod) grows, it changes from a planktonic existence (eggs and larvae) to become nekton (juveniles and adults). Its food also changes, and it too is eaten by different and larger organisms. (Photographs courtesy H. G. Moser, Southwest Fisheries Center, National Marine Fisheries Service.)

Success of the young fish in a particular **year class** (i.e., those spawned during a single season) is affected by environmental factors. The young fish must find enough food to survive before the yolk sac, shown in Fig. 12-14(b) is completely absorbed. Fish larvae depend on currents to carry them into waters where detrital particles or recently hatched larvae are the right size for the developing fish to eat. If the larvae are transported offshore where food is scarce, the entire year class may die.

As fish larvae grow to be juveniles and then adults, they are able to swim better. Thus they become part of the nekton, since they can swim against currents. Their food changes [Fig. 12-14(e)] as they grow, and they in turn are eaten by larger animals.

GELATINOUS PLANKTON

So far, we have discussed animals that are important members of food chains that lead to fishes exploited by humans. Other zooplankters are part of food webs not exploited—for example, jellyfish. They have a continuous, two-layered body wall surrounding a digestive cavity. The mouth is surrounded by tentacles bearing stinging cells. The tentacles capture particles and move them into the mouth, through which wastes are also eliminated. The animal moves by rhythmic pulsations of its bell [Fig. 12-15(a)].

FIGURE 12-15
(a) The adult jellyfish (*Chrysaora*) measures about 10 centimeters across the bell. It occurs worldwide (Photograph courtesy Michael J. Reber, Chesapeake Biological Laboratory, University of Maryland.) (b) The Portuguese man-of-war (*Physalia*) has a purple, air-filled bladder that floats at the sea surface. This is actually a colony of specialized individuals (called polyps), each performing various functions. Some entrap, paralyze, and engulf their prey, digesting and absorbing their juices. Other polyps have reproductive functions, producing sexual medusae. (Photograph courtesy Wometco Miami Seaquarium.)

(b)

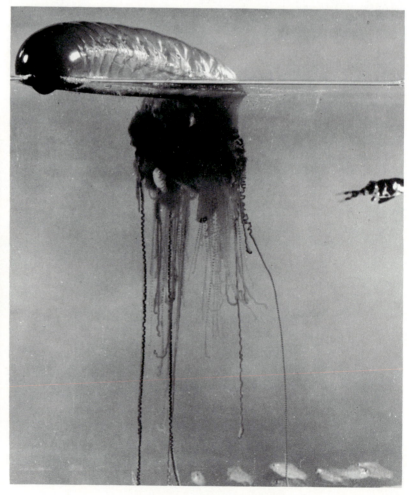

(a)

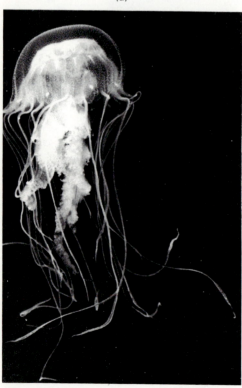

This group includes colonial **siphonophores** like the Portuguese man-of-war pictured in Fig. 12-15(b) and its deep-water relatives (Fig. 12-16). Jellyfish and siphonophores (Fig. 12-17) paralyze their prey with stinging cells that consist of barbs attached to poison sacs. Dangling tentacles entangle the food and sweep it toward the mouth, inside the bell (Fig. 12-18). Some forms swim upward and then sink with tentacles extended, trapping prey beneath. Siphonophores are colonies of individuals that live together and function as one animal.

Ctenophores (Plate 12-19) look something like jellyfish. Small and jellylike, they are sometimes known as sea walnuts or comb jellies. Some have trailing tentacles for capturing prey. Voraciously carnivorous, ctenophores often occur in great numbers and can substantially reduce populations of crustaceans and young fishes.

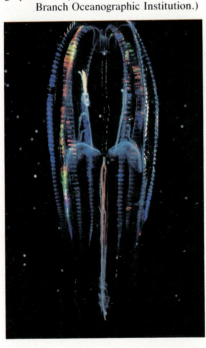

FIGURE 12-16
A deep-sea medusa, a relative of jelly-fish. The animal captures its prey with its tentacles and digests it in the central cavity, seen here in red. (Photograph courtesy M. J. Youngbluth, Harbor Branch Oceanographic Institution.)

FIGURE 12-17
Colonial siphonophores are active predators in the aphotic zone. (Photograph by M. J. Youngbluth, Harbor Branch Oceanographic Institution.)

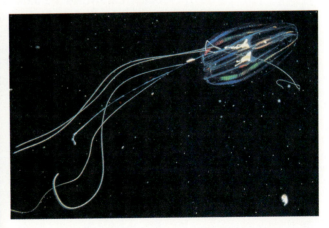

FIGURE 12-19
Ctenophores are major predators, often occurring in great abundance. (Photograph M. J. Youngbluth, Harbor Branch Oceanographic Institution.)

FIGURE 12-18
Sea wasps are poisonous planktonic medusa. (Photograph courtesy M. J. Youngbluth, Harbor Branch Oceanographic Institution.)

FIGURE 12-20
Planktonic tunicates form gelatinous tube-like structures in which the animals live. The animal pumps water through the tube which brings in food and propels the structure through the water. A filtering system captures tiny nannoplankton organisms, including bacteria, on mucus-covered structures, which the animal eats. (Photograph courtesy B. Robinson, MBARI.)

Tunicates (primitive creatures having backbones) form transparent, barrel-shaped gelatinous structures in which the animal lives (Fig. 12-20). They feed on tiny plankton by pumping water through their bodies. Mucus-covered structures capture food, which is eaten when the animal transfers the mucus to its gut. In surface waters, these organisms sometimes form large masses. They also occur in deep waters.

FEEDING STRATEGIES

Feeding strategies used by marine organisms vary with depth in the ocean (Fig. 12-21). The reason is obvious, since the abundances of different kinds of food are greatly different at the various depths.

Herbivores dominate near-surface waters where phytoplankton are abundant. They decrease markedly in abundance below the pycnocline. **Carnivores** are most abundant in the middepths, where they can feed on herbivores in the surface zone. Many of the vertically migrating organisms are carnivores.

Omnivores occur throughout the ocean, but they are especially common in deeper waters (Fig. 12-22). Their survival depends on being able to eat anything that sinks out of the surface zone.

FIGURE 12-21
Change in the abundance of different feeding strategies of planktonic organisms with depth.

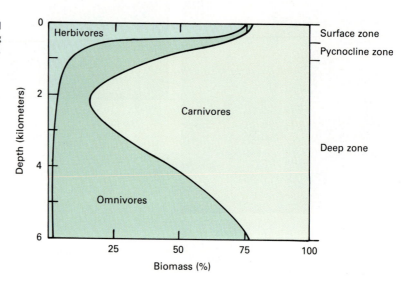

FIGURE 12-22
Amphipods are scavengers in the deep ocean. Bait lowered to the sea floor quickly attracts swarms of such organisms as well as fish. (Photograph courtesy P. I. Blades-Eckelbarger, Harbor Branch Oceanographic Institution.)

Many planktonic organisms are **suspension feeders,** that is, they feed by removing particles suspended in the water. This was long thought to involve simply moving a filter through the water to collect particles, much as one catches butterflies in a net. The net was thought to be the hairlike structures (called setae) on the organism's mouthparts. Particles thus captured were then moved into the mouth and ingested. This was called filter feeding.

Careful examination of the process using microscopes and cameras showed a markedly different picture. Because of the small size of the organisms involved, water's viscosity is important. The process is much like removing a particle from molasses.

Apparently, organisms can sense the presence of particles, probably by chemical means (much like our sense of smell). Then they use their appendages to capture individual particles. Many suspension feeders can pursue and capture relatively large prey by grasping their target with their feeding apparatus. Thus suspension feeders can switch from one type of feeding to another and can select their prey to obtain the quality of food they want as well as the quantity. Many organisms exhibit strong preferences for various foods.

Some fishes, such as sardines and herrings, swim through the water with their mouths open. In this way, they strain out both phytoplankton and zooplankton. In their case, viscosity is less of a problem because they are larger.

Another form of suspension feeding is the use of *mucous nets.* Many organisms, especially gelatinous plankton, secrete mucous structures that capture particles of all sizes from the water. The organism then eats the captured particles by consuming the entire net.

REPRODUCTIVE STRATEGIES

Two different reproductive strategies are used by organisms. One is suited for sparsely inhabited but resource-rich environments, where rapid population growth is advantageous. The other is appropriate for resource-poor environments already saturated by organisms, where slow population growth is more desirable.

The first strategy is used by opportunistic organisms that develop rapidly to take advantage of hazardous environments or variable resources which other organisms cannot exploit. These organisms mature early, have large numbers of small-sized offspring (eggs for example), and provide no parental protection. Much of the energy from food goes into producing eggs and sperm. Many fishes and invertebrates follow this strategy. Each female lays millions of eggs. Each male discharges sperm into the water where the eggs are fertilized and later develop. This so-called opportunistic or saturation strategy is used by most planktonic organisms, as well as fishes. (You can also think of it as the ''many-small'' reproductive strategy.)

The opportunistic strategy has some major drawbacks. The young are small and are therefore prey for many other organisms, including adults of their own kind. They must essentially grow their way through the food chain. At each stage, eggs and larvae are food for many predators. Such small organisms are also carried by currents. Thus they may well be carried out of areas necessary for survival and growth. This leads to large year-to-year differences in population sizes. Opportunistic reproduction is common among planktonic organisms.

The other reproductive strategy involves large, slow-growing organisms that take a long time to reach sexual maturity. They have few, relatively large offspring and protect them against predators. In this strategy, most of the energy from food goes into growth. Whales, sea birds, and mammals use this reproductive strategy. (You can think of this one as the ''few-large'' or the nurturing reproductive strategy.) This nurturing reproductive strategy is more common among larger animals, which we discuss in the next chapter.

PATCHINESS OF PLANKTONIC ORGANISMS

Planktonic organisms are not distributed uniformly in space. Instead, they tend to occur in patches of all sizes, from a few centimeters to many kilometers across. These patches are caused by physical processes, such as those we discussed in Chapter 7 and by the behavior patterns of the animals involved.

Let's deal first with the physical processes involved. We have already discussed upwelling and its effect on abundances and distributions of organisms. And we have seen how the major current systems affect where organisms occur.

Langmuir cells (discussed in Chapter 7) cause organisms to aggregate in the convergence zones between cells. The cells are a few meters across, so the patches have similar dimensions.

At a much larger scale, rings spun off by western boundary currents also aggregate organisms. These rings are several hundred kilometers across. Within a ring, the plankton are often concentrated in relatively small areas.

Much of the patchiness of marine organisms arises from their behavior. One of the most common causes is schooling, which we have already discussed. Other patches arise from reproductive behavior. Eggs and larvae are often patchy because that is how they were released by their parents. Some patches may arise from reproduction itself, since many organisms aggregate to spawn.

As we shall see later, patches of organisms on the ocean bottom are often due to changes in sediment characteristics. We will have more to say about that when we discuss bottom-dwelling organisms.

BIOLOGICAL PROVINCES

The major surface currents (which we discussed in Chapter 7) separate the open ocean into biological provinces. Within each province, temperature, light, and nutrient concentrations—the chief factors control-

SUBPOLAR GYRE
 Large Seasonal changes, especially
  in light intensity
 High nutrients, supplied by upward-
  moving subsurface waters

SUBTROPICAL GYRE
 High salinity (Evaporation)
 Low nutrients, resupplied by
  vertical diffusion

EQUATORIAL ZONE
 High rainfall
 High nutrients, supplied by
  upward-moving subsurface
  waters, upwelling on equator

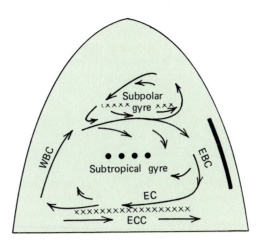

WBC Western Boundary Current
EBC Eastern Boundary Current
EC Equatorial Current
ECC Equatorial Counter Current

×××× Divergence
● ● Convergence
▬ Upwelling

SUBPOLAR
 Relatively few species
 Large biomass

SUBTROPICAL
 Relatively large number of
  species
 Small Biomass

EQUATORIAL
 Large number of species
 Large biomass
 Substantial east-west
  variability

FIGURE 12-23
Schematic representation of open-ocean surface currents in the Northern Hemisphere. Physical processes active in each gyre control the biological characteristics of the corresponding biological province. (Adapted from J. L. Reid and others. 1978. Ocean circulation and marine life. In *Advances in Oceanography*, ed. H. Charnock and G. Deacon. New York: Plenum Press.

ling distribution of planktonic organisms—are relatively constant. Each current gyre supports distinct assemblages of organisms. The major open-ocean provinces are the subpolar and subtropical gyres and the equatorial zone (shown in Fig. 12-23).

The *subpolar gyres* are highly variable in light intensity, rich in nutrients, and have relatively few species but contain high biomasses. Divergence within the subpolar gyres brings nutrients into the photic zone, a form of upwelling.

Subtropical gyres are relatively constant in water temperature and light intensity, low in nutrients, and have large numbers of species but a low biomass. Nutrients are only slowly replenished from below and by inward transport of nutrient-rich waters on the margins of the gyre.

The *equatorial zone* is characterized by high variability in abundance of organisms, large numbers of species, and intermediate-to-high biomass, partly because of upwelling along the equator.

Distributions of biomass are mostly controlled by water movements and availability of light and nutrients. Areas high in nutrients generally maintain high biomasses whenever light is sufficient. Those areas low in nutrients do not support high biomasses regardless of light intensity.

Phytoplankton tend to be widespread, and few species are restricted to a single gyre. Apparently phytoplankton are easily spread by currents from one gyre to the next. A single "seed cell" is sufficient to establish a new phytoplankton population. Also, phytoplankton tend to be able to survive unfavorable conditions, thus facilitating their dispersal. In fact, decreasing light intensities with depth in the open ocean are more of a barrier to phytoplankton growth than lateral differences. In other words, phytoplankton cells can more readily survive moving hundreds of kilometers horizontally than sinking a few hundred meters below the photic zone.

Distributions of zooplankton species are much more restricted. Many open-ocean zooplankton species occur in a single gyre. The subtropical gyres are the most extensive of the major oceanic habitats. They share a large fraction of their zooplankton forms. A few species are restricted to particular current systems, such as the equatorial zone. Mixing of zooplankton from different systems occurs in the western boundary currents, which have the largest number of different kinds of organisms.

SUMMARY

The open ocean is a three-dimensional world. The principal boundaries for organisms are marked by subtle changes in temperature, salinity, and light intensity. These changes are most pronounced in the vertical dimension.

There are three different oceanic life styles. Drifting organisms, called plankton, move with the currents. Swimming organisms can move independently of currents. Attached organisms remained fixed to their substrate, and waters move past them.

Planktonic organisms are small and lack skeletons. Most live in chosen levels in the ocean. Since they cannot swim, they may sink out of their preferred level. Most are coldblooded.

Open-ocean organisms have several defense strategies. They can be small or transparent. Some form schools. Others migrate vertically to feed in near-surface waters at night and spend daylight hours in dark, subsurface waters.

Phytoplankton are one-celled plants that produce most of the food in the ocean. The larger netplankton include diatoms, which are common in nutrient-rich waters of upwelling, coastal, and estuarine systems. Dinoflagellates are almost as common but less well known. Certain types of dinoflagellates sometimes grow in profusion (called red tides). They release toxins that poison marine organisms and any humans that eat them.

Minute organisms (less than 10 micrometers in diameter) produce 50 to 80% of the total organic matter in the ocean. They dominate the open ocean. Some marine bacteria are able to produce their own food. Flagellates feed on these tiny organisms.

Zooplankton are planktonic animals that feed on phytoplankton and other zooplankton. Some are planktonic throughout their life (called holoplankton). These include foraminifera, radiolarians, and other organisms. Gelatinous plankton are also important holoplanktors. These include jellyfish and other less well known organisms. Some gelatinous plankton organisms use mucous nets to capture nannoplankton.

Meroplankton are planktonic for part of their life cycle. Examples are fish larvae or the larvae of oysters. As fish grow, they become strong swimmers, becoming nekton. Many benthic organisms, such as oysters, use planktonic larvae to disperse.

Feeding strategies are different in the various parts of the ocean. Herbivores dominate near the surface. Carnivores are most abundant beneath the herbivores. Omnivores dominate the deep ocean.

Reproductive strategies also differ, depending on the amount of resources available. Where resources are abundant, opportunistic forms reproduce rapidly to take advantage of the resources. Where resources are limited, slow-growing forms can dominate. Plankton often appear in patches in part due to physical processes and in part due to behavior, such as reproduction.

Major currents in the ocean outline biological provinces. Their abundances are determined by availability of light and essential nutrients. Certain types of zooplankton are confined to a single gyre.

STUDY QUESTIONS

1. Discuss the different oceanic life styles.
2. Discuss the advantages and disadvantages of small size to a phytoplankton cell.
3. Show how diatom reproduction influences its shell size.
4. What is the difference between meroplankton and holoplankton? Give examples of each.
5. Discuss defensive strategies used by open-ocean planktonic organisms.
6. Discuss the factors that make coastal and estuarine areas prolific producers of fishes and invertebrates.
7. Discuss the factors causing red tides.
8. Discuss how plankton provinces correspond to the major current gyres.

FRASER, J. 1962. *Nature Adrift: The Story of Marine Plankton*. London: G. T. Foulis. 178 pp. Concise survey of plankton.

HARDY, A. 1965. *The Open Sea: Its Natural History*. Boston: Houghton Mifflin. 355 pp. A classic, beautifully illustrated.

Blacktip reef shark feeding on mackerel. (© Tom McHugh, Steinhart Aquarium, Photo Researchers.)

CHAPTER 13

Nekton

OBJECTIVES

1. To explain the roles of large marine organisms;

2. To understand the relationships between oceanic processes and the abundances of marine organisms;

3. To describe variability in marine communities.

*T*he larger animals in the ocean that can swim well enough to move independently of currents are called **nekton.** This includes fishes, squids, marine mammals, and marine reptiles. In this chapter, we consider some of the most common nektonic animals, including seabirds.

In this chapter, we discuss:

Adaptations to life in the open ocean;
The role of the nekton in ecosystems;
Life histories of typical nektonic animals; and
Processes causing variations in their abundances and distributions.

SWIMMING

The principal characteristic of nekton is their swimming ability. They have body shapes adapted to moving rapidly through the water. Any object moving through water experiences **drag,** partly because of the viscosity of water. Many nekton animals secrete slimes to reduce drag; others have special ribbing of the skin to reduce drag. (Ribbing of this sort has been used in designing yachts to reduce their drag and thereby increase their speed through the water.)

Another aspect is the shape of the animal's body. A cone moving through the water experiences resistance because of the turbulence behind the cone when the point goes first (Fig. 13-1). There is less turbulence when the blunt end is first. The least drag comes from a *fusiform* body, which is cylindrical, with a blunt end first and a tapered end following (Fig. 13-1). The fastest-swimming fishes have this body shape. (We discuss this more when we discuss fishes.)

There are other adaptations for speed. Tuna are good examples of fishes adapted for fast swimming (Fig. 13-2). They have characteristic lunate-shaped tails and a thin connection to the body (called a caudal peduncle). This increases their swimming efficiency.

In swimming, the fish's body is moved in a series of waves that begin at the head of the animal and progress toward the tail (Fig. 13-3). The head moves relatively little, while the tail moves the most. These motions accelerate the water around the fish, thereby propelling the fish forward.

FIGURE 13-1
A cone pointed into a flow (a) experiences more drag than a cone with its blunt end (b) facing the direction of flow. The fusiform shape (c) typical of many fishes has the least drag.

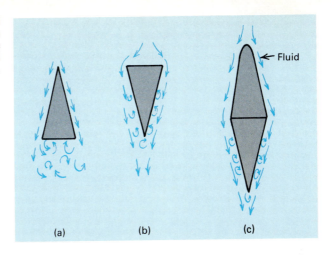

FIGURE 13-2
A bluefin tuna has a streamlined body for speed in capturing prey. An important sport fish, it grows to be 5 meters long. (Drawing by G. T. Sundstrom, NOAA.)

FIGURE 13-3
Swimming motions in an eel (a) and a cod (b). In both cases, a wave passes along the body of the fish. (After J. Gray. 1933. Directional control of fish movements. *Proc. Roy. Soc. B* 113.)

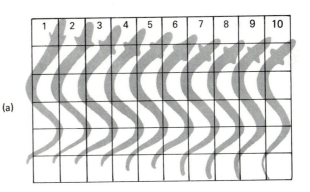

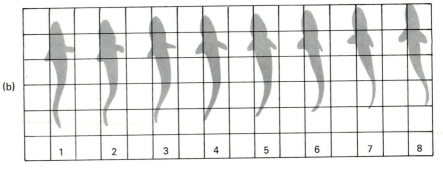

BUOYANCY

Since living matter has a density between 1.02 and 1.06, it is usually heavier than seawater (density 1.03). Thus actively swimming animals must be able to adjust their density to maintain themselves at desired depths. Bottom-dwelling fishes have densities of 1.06 to 1.09, since they rest on the bottom.

Fishes, sharks, and rays lacking swim bladders must swim continuously. Fish can also partially control their density by adjusting the fat content of their body to compensate for the weight of bones. (More about this when we discuss Antarctic fishes.) Continuous swimming results in higher energy expenditures by the animals.

To assist in maintaining their depths with minimum energy costs, many fishes have **swim bladders,** organs that contain gas, to adjust their density. The gas in swim bladders can be obtained by gulping air at the surface. Most often the gas in the swim bladder comes from the bloodstream of the fish. Typically it consists of oxygen and carbon dioxide, rarely much nitrogen.

Gas can be released from swim bladders through external openings. This is well known in herrings. In fact, fishermen in the Mediterranean can tell that a school of herring is coming to the surface by the appearance of gas bubbles expelled from their swim bladders. Other fishes must absorb excess gas back into their bloodstream. This is a slow process, so these fishes cannot readily change their depths. Thus the fishes that vertically migrate each day do not have swim bladders.

Swim bladders serve other purposes as well. In some fishes they are adapted to assist in the detection and production of sound. They serve somewhat as an ear to detect other animals. Most fishes have a connection between their swim bladder and their inner ear.

Many fishes have muscles that insert into their swim bladders, thus permitting them to make sounds. When the muscles are contracted, the swim bladder vibrates, producing a variety of sounds that have been described as grunts, thuds, barks, or growls. In some species, the sounds are made during reproductive activities; in others they are made during feeding.

VISION UNDERWATER

Seeing underwater is difficult for humans. Conditions there are radically different from those in the atmosphere. One writer characterized seeing underwater as attempting to see through a fog, wearing prescription sunglasses made for a very nearsighted person.

The fog in our analogy comes from the **scattering** of light by particles in seawater. In all but the clearest ocean water, this severely limits visibility. Our closest experience is indeed with fog, which is a mass of tiny water droplets in the air.

Light is absorbed as it passes through water, as we have already discussed (Fig. 13-4). The clearest open-ocean water is most transparent to (absorbs least) blue-green light. As the amount of dissolved matter and suspended particles increases, the color of least absorption shifts to yellowish green in nearshore waters to red in the most turbid estuarine waters. Thus as the light gets dimmer with increasing depth, its color also changes. In general, this is not a problem for humans because our eyes are sensitive to all the color ranges involved.

Focus underwater is a problem for humans. Since water is denser than air, our eyes do not function in water as well as they do in air.

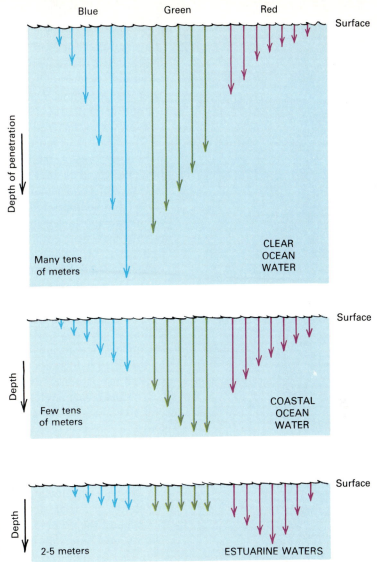

FIGURE 13-4
Penetration of different colors of light below the surface in open-ocean, coastal-ocean, and estuarine waters.

Some animals have extremely flexible eyeballs whose shape they can change to compensate for the reduced ability of the lens to focus the light. Humans do not have this ability. Scuba divers compensate by having air in their face mask so that their eyes can function effectively underwater.

To compensate for the reduced intensity of light, many animals have evolved large eyes to gather more light. This is especially obvious in fishes from the dimly lit middepths. Some fishes have eyes that can detect bioluminescence, so they are sensitive to the blue-green light given off. Also, the eyes point either forward or upward, depending on the behavior of the animal involved.

There is too little light in the deep sea to be useful to animals. Animals living at great depths in the ocean have either no eyes or vestigial ones.

FISHES

The many different shapes of fishes reflects the varied ways in which they capture their food. The typical streamlined shape of many pelagic fishes shows the value of speed for capturing prey or avoiding enemies. Most can swim rapidly for short periods, about 10 times their body length in a second. During migrations, medium-sized fishes, such as salmon or cod, travel hundreds of kilometers in a few days. Large oceanic fishes that feed at high trophic levels, such as **bluefin tuna** (Fig. 13-2) swim 100 or more kilometers per day for weeks at a time while hunting schools of smaller fish. Estuarine and bottom-dwelling species, however, tend to hide near rocks or in soft sand. Their bodies are adapted for concealment rather than speed (Fig. 13-5).

Of the **demersal fishes,** those that live on or near the ocean floor, the most important to humans are the **cods** and their relatives. They are mostly first-level carnivores, feeding on invertebrates, small fishes, and larvae. Another group, the **flatfishes** (Fig. 13-5) are modified to lie on the ocean floor. As flatfishes mature, their eyes migrate to one side of the head so that as adults they can conceal themselves on the bottom with only their eyes exposed. They eat animals in the sediments.

The upper sides of many fish are darker than the lower. This is a common protective adaptation in pelagic as well as demersal fishes; organisms viewing a fish from below are looking toward the light, so a light-bellied fish is less visible. But when viewed from above, a dark-backed fish is more difficult to see. This is called countershading.

Mackerel and related species are well adapted for strong, continuous swimming in open waters, both near the surface and at depth. Open-ocean species feed at high trophic levels and are voracious carnivores. They, in turn, are heavily fished by humans.

Mackerel are migratory fish. They leave surface waters about October and aggregate near the ocean floor. During this period they eat crustaceans and small fishes. In January mackerel move to the surface in schools and migrate (April) to spawning grounds. They spawn near the edge of the continental shelf, gradually moving closer to the land. During this time they feed on plankton, especially copepods. From June to July they form smaller schools and move close to the shore, changing their diet from plankton to small fishes in inshore bays. In the fall they again seek deeper waters.

FIGURE 13-5
A flatfish (turbot) lies on the bottom and takes on the color of the underlying sediment as camouflage. Both of the fish's eyes are on the same side of the body. (Photograph C. Arneson.)

FIGURE 13-6
Chinook salmon range from Southern California and are prized as sport fish. They grow to be 1.5 meters long. (Photograph courtesy Washington Department of Fisheries.)

Another group of valuable food fishes are **salmon** and **trout.** Both are high-level carnivores, generally confined to northern, fairly cold waters. They typically spend part of all of their lives in freshwater. Salmon spawn in rivers but attain most of their growth in ocean waters (Fig. 13-6).

HERRING

Herrings are abundant fish in coastal ocean waters. These fast-growing fish support some of the world's largest fisheries, including fisheries for sardines off California and for anchovies off Peru. These fish have been overexploited in many areas, leading to drastic declines in their abundance. Overfishing combined with the effects of El Niños in the Pacific have resulted in spectacular collapses of these fisheries—the California sardine in the 1950s and the Peruvian anchovy in the early 1970s.

Herrings (Fig. 13-7) spawn throughout the year in coastal waters. Most spawning occurs in the summer and fall in Mid-Atlantic waters. A single female can spawn 50,000 to 700,000 eggs, depending on her size and age. Older and larger fishes produce the most eggs.

FIGURE 13-7
Herrings are important food and commercial fish in many parts of the world. (a) Anchovies are abundant in warm seas, near shore, and in the open ocean. (b) Pilchard occur in coastal waters between 12° and 20°C. (c) Sardines, (d) herring, and (e) menhaden occur from Brazil to Nova Scotia.

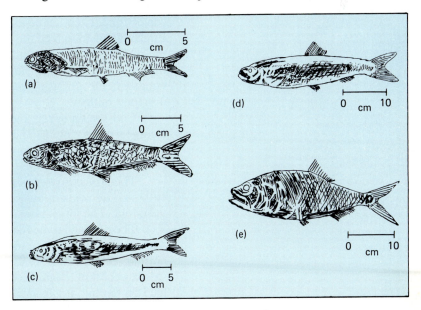

Newly fertilized eggs float near the surface and hatch in about 2 days. Larval fish feed on planktonic organisms. The larval fish enter estuaries when they are about 1 inch long. There they spend about 8 months before returning to offshore waters.

Herrings grow rapidly during their first 3 years. Adult herrings swim with their mouth open, filtering everything from the water—both plants and animals. One-year-old fish typically weigh about 250 grams and about 500 grams as 2-year-olds. Then their growth slows, so they weigh only about 1 kilogram as 6- or 7-year-old adults.

Spawning begins around 2 years of age. Older, larger fish are the most successful. Thus the abundance of the larger fish is critical to spawning success. Environmental factors also affect spawning success. If eggs and larvae are carried offshore by currents, the larvae cannot reach the estuaries and wetlands, and they starve to death. On the other hand, if surface currents carry the young toward shore, they have a much better chance of reaching suitable nursery areas. Those years when large numbers of young survive produce large year-classes that can sustain fisheries for years.

Heavy fishing eliminates the larger fish, which are a large fraction of the population in an unexploited fish stock. As fishing pressure increases, only younger fish remain. Some heavily exploited fisheries take the young fish spawned that year. Fishery management schemes seek to protect young fish so that some will reach spawning age.

Herrings school when frightened, forming large aggregations. These schools are spotted by light airplanes. Small boats then set their nets around the schools. The schools are also marked by seabirds feeding on the fish.

Herring and their relatives are oily. Some are consumed directly as food. Most are processed to make meal to feed to chickens and livestock or to make fish oil for paints, soaps, and lubricants. Fish oils are also used as diet supplements. Proteins extracted from the fish can be used to make fish products (called *surimi*) such as artificial crab legs or lobster meat.

COD

Cod are an extremely abundant fish in the mid- and high latitudes of the Northern Hemisphere. They were especially important in Early American history. Their abundance in the waters off Massachusetts and New England helped to feed the New England colonies and thereby ensured their success. Another indication of their abundance is the fact that they are still the third largest in world fish-catch statistics. Cod is familiar to most of us in frozen fish packages or fish fillet sandwiches at fast-food outlets.

Cod spawn in relatively shallow areas of the continental shelf—Georges Bank, for instance, off the Gulf of Maine. A single female produces up to 10 million eggs. The larger females are the more successful spawners.

The larvae grow to 4 centimeters in 4 months. Individuals apparently mature sexually in about 2 years but may live to be more than 15 years old. The largest ones grow to be 90 kilograms in weight.

The pelagic larvae feed entirely on plankton while they are small. After they grow to be about 10 centimeters, they begin to feed on the bottom, eating clams, crabs, mussels, and mollusks. When they grow still larger, they begin feeding on other fishes, although they always feed on benthic organisms when available.

ANTARCTIC FISHES

Ocean conditions around Antarctica are harsh for fishes. The average water temperature at McMurdo Sound (the principal U.S. research station) is −1.87°C, ranging from −1.4 to −2.15°C. During winter there is no sunlight for 4 months. Even during summer, the ice cover cuts light intensities at depth to 1% of the incoming solar radiation.

Ice is the principal threat to fish. It can form and penetrate a fish's gills and skin. But ice will not form in a fish until it is supercooled by more than 1°. If ice is present, blood will freeze with supercooling of only 0.1°C. The most abundant fishes, about 90% of those around Antarctica, have special adaptations to avoid freezing. They can withstand freezing until water temperatures drop below −2.2°C. Salts in the blood of fishes cause it to freeze at temperatures lower than the freezing point of pure water.

Successful Antarctic fishes have special antifreeze compounds that keep their blood and tissues from freezing. These compounds apparently coat ice crystal surfaces as they form. This prevents further growth of the ice crystals until water temperatures drop further. In short, these antifreeze compounds lower the freezing point of the fish's blood. It appears that antifreeze compounds bond to oxygen and hydrogen in ice.

Antarctic fishes have yet another problem: They must be able to survive the long winter when there is little to eat. One adaptation is to store energy as liquid fats (much like cooking oil) instead of the more familiar solid forms we know as fats. The abundant fats are stored in special sacs under the skin and between the muscles. The presence of these fats also reduces the fish's density. As previously mentioned, this permits the fish to float at midwater depths without having to swim to maintain its position.

These fishes also have unusual bone structures. They are made primarily of cartilage, which is less dense than normal bone. This thus reduces their weight and reduces the amount of energy that the fish must expend to maintain its position while waiting for food.

MIGRATIONS

A *migration* is an animal's movement between two places that occurs at a predictable time. Many nektonic animals have long migrations. The longest is the 20,000-kilometer annual migration of the California gray whale (discussed in Box 13-2).

Many fish migrate as part of their life cycle. Salmon are called **anadromous** fish. They spawn in fresh water and remain there for about a year. They live as adults in the open ocean before they return after 5 to 6 years to their home stream to spawn. These migrations cover many thousands of kilometers. Their return to their home stream occurs within a few days of the same time of year that their parents returned in their life cycle.

North Atlantic eels are **catadromous** fish (the opposite of anadromous). They spawn in the Sargasso Sea in the North Atlantic. As larvae they live planktonically for up to 3 years. When mature, the eels migrate to the coastal waters of either North America or Europe (Fig. 13-8). There they metamorphose into adult eels and return to fresh water for 4 to 7 years.

There are many unanswered questions about how eels migrate from the Sargasso Sea back to their native rivers. One hypothesis is that the two different types of eels (European and North American)

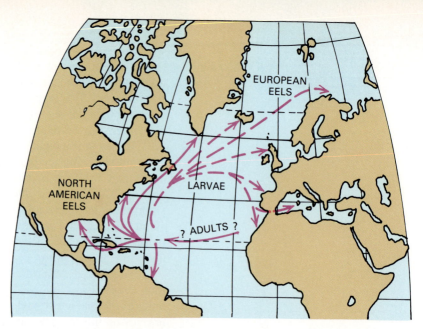

FIGURE 13-8
Migration routes of eel larvae from the Sargasso Sea to Europe and North America. Note the similarity between the migration paths of larvae and the Gulf Stream. Migration paths of adults are unknown but likely also follow currents as indicated here.

orient themselves at the fronts associated with the Subtropical Convergence. From there they are thought to navigate using the sun as a compass and the Earth's magnetic field. This permits them to return to the vicinity of their origin. Finally, they home in on their native river by the odors of the river waters.

FISHERIES

Fish provide about 3% of protein and about 10% of all animal protein in human diets. These are luxury items (lobster, shrimp, tuna) in diets in most developed countries. But in developing countries, fish are often major sources of protein—80% in Bangladesh, for example.

For years, many people thought that the ocean could provide an unlimited source of protein to feed the world's hungry people. The reason for this optimism was the rapid growth in fish catches in the 1950s and 1960s. Between 1950 and 1970, marine fish production increased from 20 to 50 million metric tons, due primarily to increased fishing efforts. With proper management, it was hoped the ocean's yield of protein could be virtually unlimited.

The first signs that this was not possible came with the collapse of the California sardine fishery in the 1950s. This was followed in the early 1970s by the collapse of the Peruvian anchovy fishery. At the time of its collapse in 1972, the Peruvian fishery was the world's largest, producing 12 million metric tons per year. In the late 1980s, Peru's fish production was 1 to 2 million tons annually.

In 1986, fish and shellfish production was about 76 million tons. About 60% of the catch was used directly for food. The remainder was used to make fish meal or oil, much of it going to feed chickens or livestock. Considered on the basis of value, the most important fisheries are salmon, tuna, and shrimp. These fish command premium prices worldwide.

The prognosis for major increases in fish catches is guarded. Reported fish catches have increased since the mid-1970s. This is probably as much a result of better reporting as of increases in the amount caught. The maximum amount that is likely to be produced from the ocean is thought to be about 100 million tons, which is not much higher than the late-1980 levels. Most of the increases will come from increased utilization of fishes or other organisms not now harvested.

SHARKS

Sharks occur throughout the ocean. The common image of a shark is a large, swift predator. Indeed, many sharks are able to kill large prey, including humans on rare occasions. Sharks are scavengers. Like large cats and wolves, sharks catch sick and diseased animals. They use their rows of serrated teeth to remove large bites of flesh or to tear off whole limbs.

Sharks also attack humans. To a shark a swimmer may appear to be a disabled animal, which the shark therefore attacks. Most shark attacks on humans occur in shallow, murky waters or at dawn or dusk, when visibility is limited.

Groups of sharks also engage in feeding frenzies, stimulated by blood or bits of food in the water. The reasons for these frenzies are not well understood.

Other sharks feed on plankton, including **whale sharks,** which grow to be more than 15 meters long. Another plankton feeder is the **basking shark,** which grows to be 12 meters long. Sharks have been hunted for their livers, which are rich in vitamin A. Sharks are also taken for food. The meat is quite tasty when properly prepared to remove the high levels of urea in the blood. In Asia, shark fins are highly prized as an ingredient for soups.

Sharks are a primitive animal, having cartilage rather than bone for their skeleton. They also lack scales but have small toothlike plates, called **denticles,** embedded in their skin, which makes it extremely abrasive. Shark teeth are specialized denticles; they occur in rows, up to seven deep, in the shark's mouth and are easily replaced if lost.

Reproductively, sharks are also different. They and a few bony fish produce eggs that are kept in the female's reproductive tract until hatched. The developing embryos obtain all their food from the egg yolks rather than from the mother, as mammals do. The young are delivered live into the ocean. Recall that most bony fishes lay eggs which are fertilized and hatch in the water.

SQUID

Squid (Fig. 13-9) are among the most common animals in the ocean. They are fast swimmers and effective predators. They have been compared with sharks in their ability as predators. They are taken commer-

FIGURE 13-9
Squid (about 25 cm long) are active, fast-swimming predators at all depths in the ocean. This transparent squid lives at depths around 500 m. (Photograph by P.I. Blades-Eckelbarger, Harbor Branch Oceanographic Institution.)

BOX 13-1

Whaling

Hunting of whales began many centuries ago with coastal dwellers using small boats operating together near shore. By the sixteenth century, European whalers using large vessels reached Arctic waters, opening new whaling grounds. One of the richest early whaling grounds was near Greenland. There the *right whale* was hunted until it became nearly extinct. (It was called the right whale because it was easily killed and floated after death for easier recovery. In other words, it was the "right whale" for relatively primitive hunting technology.) Whales provided meat and oil for lamps before the days of petroleum. Whalebone (from baleen) was used in dresses and corsets. **Sperm whales** provided spermaceti, which was used in cosmetics and candles. Whale oil still has many industrial uses.

As whale stocks declined, whalers had to go far-

ther to find new stocks. In the 1920s, whaling moved into the Southern Ocean around Antarctica. Fast, steam-driven vessels for hunting, factory vessels for processing whales at sea, and explosive harpoons greatly increased whale harvests (Fig. B13-1-1). Near extinction of several whales, including *blue whales,* led to their protection. The large blue whales (90 tons) were taken first until their abundances dropped. Then whalers moved to the next largest species, the fin whale (50 tons). The small *Minke whales* (6 tons) have not been taken in great numbers. Eventually, commercial whaling on the high seas was phased out during the late 1980s, although some whaling continued for research purposes. Since whales are long-lived animals, their recovery will take decades. Some whale populations may never recover.

FIGURE B13-1-1
World catch of principal whale species.
(Compiled from various sources.)

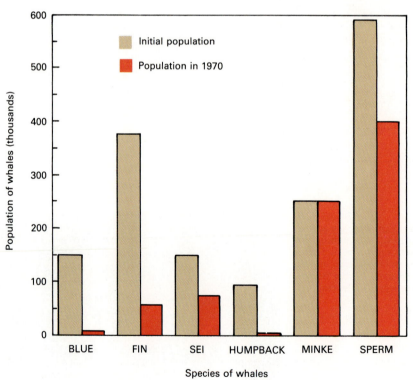

cially, and the annual production is around 1 to 2 million tons. Because of their swimming abilities, they are difficult to catch for study and are thus rather poorly known.

Most squid live at middepths in the ocean during the day and come to the surface at night as part of the vertically migrating animals.

The most common squid are fairly small, around 10 centimeters to a meter long. The largest squid is the rarely seen giant squid, which remains between 300 and 600 meters depth. It is eaten by sperm whales. Giant squid grow to be 18 meters in length. (Humans, for scale, are generally around 2 meters.) These giants are sometimes brought to the surface by feeding sperm whales, and a few dead ones have washed ashore.

Squid have a distinctive way of swimming. They have a muscular mantle cavity that they fill with water. They then contract the mantle and forcibly expel the water through a funnel. By moving the funnel,

---

BOX 13-2

California Gray Whales

Familiar sights along the Pacific coast of North America are the spring and fall migrations of gray whales. These animals can be seen from cliffs as they exhale, forming heart-shaped vapor plumes that can be seen for miles. These animals migrate 8000 to 9000 kilometers (one of the longest mammal migrations known) from coastal lagoons of Baja California (Mexico), where they breed, to feed in the highly productive waters between Alaska and Siberia. Half of the herds travel west to the Siberian coast. The other half go north into the Arctic Sea. Those entering the Arctic risk being trapped in the ice at the end of the summer. (This happened to two young whales in 1988 and required the cooperation of the native peoples, the oil industry, and a Soviet icebreaker to get them out.)

The trip takes the 20,000 whales about 2 months. During that time the animals do not eat. Nor do they eat during the 2- to 3-month stay in Baja California where they breed and where the young are born a year later. Thus these 20- to 40-ton animals migrate to feed for 6 months and then fast during the rest of the year. One problem is to account for the high level of secondary production in this region—in other words the number of whales sustained by diatom production.

The Bering Strait and Chuckchi Sea areas are highly productive. Primary production is more than 300 grams of carbon per square meter per year—equal to the most productive coastal areas anywhere in the world. Furthermore, this diatom production occurs during only a few months when the area is ice-free and well lighted during the long Arctic-summer days. The waters are shallow, so most of the diatoms

sink to the bottom in less than a day. Apparently most of the diatoms sink to the bottom uneaten.

The high levels of secondary production appear to be due to the short food chain, which involves only diatoms and amphipods. The diatoms are eaten by amphipods—benthic organisms related to the common pillbug often found under rocks. These large (3-centimeter long) amphipods live in burrows in the sandy bottom and eat diatoms after they sink there. The amphipods store fat in their tissues through the year, which makes them a good food for the whales. The whales suck up large volumes of sediment off the bottom and strain out the amphipods through their baleen, much as other whales feed by filtering plankton from the water as they swim. In short, the abundant food source can easily be taken by the whales.

Transfers of energy among diatoms, amphipods, and whales also appear to be extremely efficient. At least half of the annual amphipod production is needed to sustain the approximately 20,000 animals. There are questions about the number of whales that the normal phytoplankton and amphipod production can sustain, since the whale population is increasing at a rate of about 2.5% per year, or doubling in about 30 years.

Since commercial whaling ceased, these animals have reached levels comparable to those before whaling began. Their existence is threatened by possible oil and gas development in Alaskan coastal areas through which they migrate and where they feed. They are also threatened by human activities in the Mexican coastal lagoons where they breed and bear their young.

the animal controls its direction of movement. Squid can actually propel themselves out of the water like flying fish. They have been recovered from the decks of ships 3.5 meters above the water. Their flights have been measured at 50 meters in length.

Squid have well-developed eyes that function much like human eyes. Their eyes can focus over wide distances and adjust to large changes in light intensity. Clearly, vision is an important means of locating prey.

Squid also have long tentacles that they use to catch their prey. These tentacles have well-developed senses of touch and smell to assist in locating food. They feed primarily on small fish. Their typical way of feeding is to swim backward into a school of small fish and use their tentacles to catch the prey, which they then kill and eat.

MARINE MAMMALS

Marine mammals are warmblooded air breathers and include the largest animals on Earth. Some are herbivores, such as sea cows or *sirenans,* which graze on vegetation in shallow coastal waters. Some marine mammals spend most or all their lives in coastal waters.

Seals, walruses, and sea lions are more familiar, since they live in coastal areas. They belong to the *pinnipeds* or feather-footed animals, so named because of their distinctive swimming flippers. While spending much of their time in the water, they also spend considerable time ashore, where they breed and raise their young in large rookeries (Fig. 13-10). Because of their valuable furs, they were hunted almost to extinction in the late nineteenth century. They have recovered somewhat under the protection of international treaties.

Cetaceans, which include whales (Fig. 13-11), dolphins, and porpoises, are completely oceanic mammals. They spend all their life at sea. Like other mammals, the young are born alive and nursed by their mothers.

The largest whales feed on plankton. They swim slowly with their

FIGURE 13-10
Fur seals in a large rookery, St. Paul, Pribilof Islands, Alaska. (Photograph courtesy NOAA.)

mouth open to take in water and plankton. The tongue acts as a piston to push the water through plates of horny material, called **baleen,** that hang down from the roof of the whale's mouth. These plates filter plankton from the expelled water. The trapped plankton are licked off the baleen and swallowed. The blue whale, the largest animal that has ever lived, reaches lengths of 30 meters and weighs 150 metric tons.

Gray whales feed on bottom-living crustaceans. Feeding whales suck up large amounts of mud from the bottom and filter out the abundant crustaceans when they expel the mud and water through their baleen. Feeding by these animals leaves large scars on the shallow sea floor around the Arctic Sea, near Alaska.

Other smaller whales, such as porpoises, sperm whales, and killer whales, are rapid swimmers and capture their prey using their teeth and jaws. They feed primarily on fish and squid, but killer whales also prey on other whales.

Many toothed marine mammals use echo-location to find their prey. The common dolphin, in which echo-location has been most studied, emits clicks from its forehead, probably its blowhole, and detects its prey using the echoes. Low-frequency clicks are used to locate distant objects. High-frequency clicks are used to discriminate among objects by determining their size and shape. Dolphin clicks can be heard by human ears. Sperm whales can locate squid, their primary food, using low-frequency scanning clicks. Whales also produce a variety of other sounds, apparently for communication.

Whales are powerful swimmers. Their streamlined body shapes permit them to swim efficiently. They also have an unusual skin structure that deforms in response to water pressure, which reduces turbulence as they swim, thereby saving energy.

Whales commonly migrate many thousands of kilometers. Many breed and bear their young in the warm low-latitude waters. They migrate to exploit rich food sources, such as the intense blooms in the high latitudes.

FIGURE 13-11
A humpback whale, a medium-sized baleen whale, is typically about 16 meters long and weighs 35 tons. Note the extremely long flippers. It is found in all oceans between the Arctic and Antarctica. It feeds by swimming with its mouth open through schools of small organisms. (Drawing courtesy NOAA.)

MARINE REPTILES

Few reptiles live in the ocean. The best known are the sea *turtles* (Fig. 13-12), which live in the ocean but nest on land. Turtles graze on seagrasses and rooted aquatic plants growing in shallow waters around Caribbean islands and in south Florida. They, like cows, have bacteria in their intestines to digest cellulose in the grasses and provide needed nitrogen.

Each female turtle lays about a hundred eggs, which she buries in the beach, and then returns to the sea, leaving the eggs to hatch in the warm sand. As the newly hatched turtles cross the beach on their way to the ocean (Fig. 13-13), they are preyed on by birds and other animals, including humans. Once in the ocean, they must then escape predatory fishes. Mature turtles make long migrations between feeding and breeding areas. Turtles have been hunted nearly to extinction because of their valuable meat and shells, which are used for making combs and jewelry.

In the tropical Indian and Pacific oceans, there are sea *snakes*. These are truly oceanic creatures which reproduce at sea, bearing live young. Their bodies are flattened, which improves their swimming efficiency, and their nostrils can be closed while swimming. Sea snakes are extremely poisonous. They are quite timid and have a small mouth, so they pose relatively little threat to humans.

In the Galapagos Islands of the equatorial Pacific, there are also lizards, marine *iguanas*. They live on cliffs along the shore and dive into the water at low tide. Using a flattened tail to swim, they feed on seaweed.

FIGURE 13-12
A loggerhead turtle swimming. (Photograph courtesy NOAA.)

FIGURE 13-13
Young loggerhead turtles (8 hours old) cross the beach to reach the ocean. (Photograph courtesy Ralph Dresge, NOAA.)

SEABIRDS Among the most distinctive sights on the ocean are seabirds, adapted to living near or over the ocean. Although their food comes from the ocean, seabirds must go ashore to breed and to raise their young. No bird has evolved a truly marine life style, totally independent of land. Eggs are hatched at 40°C, and the chicks require feeding and protection until they can fly. This requires nesting ashore.

Wading birds have long legs and bills which permit them to obtain food by wading in wetlands (herons) or on beaches (shore birds). Pelicans (Fig. 13-14), ducks, cormorants, loons, grebes, puffins (Fig. 13-15), and most gulls and terns live along the shores and feed in shallow nearshore waters.

The truly *oceanic birds*—auks, albatrosses, petrels, penguins (Fig. 13-16), and gannets—have come closest to being completely independent of the land. For example, after learning to fly, an albatross may not set foot on land for 5 years. These open-ocean birds generally lay only one egg, and their chicks have a long adolescence. Adults live relatively long lives, tens of years.

Since most seabirds must return to land during breeding, most have retained their ability to fly. To do so they must remain relatively

FIGURE 13-14
Brown pelicans are fish-eating birds that live in coastal regions. (Photograph courtesy NOAA.)

FIGURE 13-15
Puffins are common seabirds of the North Atlantic. (Photograph by Burger, courtesy Newfoundland and Labrador Department of Development and Tourism.)

FIGURE 13-16
Emperor penguins travel across the ice at McMurdo Station, Antarctica, some walking, others sliding on their bellies (U.S. Navy photograph courtesy PH2 Jeff Hilton.)

small with lightweight bodies. Thus they cannot optimize their bodies or wings for maximum swimming efficiency. This limits their diving depths and the length of time they can stay underwater.

Penguins are well adapted for life at sea and have lost the ability to fly. The emperor penguin (see Fig. 13-16), the largest penguin, can swim underwater as well as seals. These birds can dive to depths of 250 meters, swim at speeds of 10 kilometers per hour, and remain submerged for nearly 20 minutes. Thus they can catch fast-swimming fish and squid.

Seabirds undertake long annual *migrations* in response to food availability and suitable weather for breeding. For instance, terns and red knots spend summers in the Canadian Arctic and winters off South America and in Patagonia (the southern tip of South America). Likewise, shearwaters travel from Tristan da Cunha (a volcanic island in the cool, temperate South Atlantic) to comparable climates off eastern North America. These migrations are tied to coincide with seasonal food availability. For example, many shore birds stop in Delaware Bay on their spring flight to South America. There they fatten up for the long flight by feeding on the abundant eggs of horseshoe crabs which are found on the beaches.

DEEP-OCEAN NEKTON

Conditions below the surface zone are markedly different for the organisms living there. Light intensities decrease, and the waters get progressively colder with increasing depth below the pycnocline. Both factors influence the types of animals living there and their life styles.

The dimly lit deep ocean has many bioluminescent animals. For example, angler fish (Fig. 13-17) are common in waters 100 to 500 meters deep. These fishes have distinctive patterns of light-emitting

FIGURE 13-17
A female angler fish (length about 10 cm) from around 1000 m depth. The organ in front of the fish's mouth is bioluminescent, to lure prey to where the fish can seize it. The fish has a large mouth so that it can swallow food of any size. Male angler fishes are smaller and in some species are parasitic upon the females. (Photograph B. Robison, MBARI.)

FIGURE 13-18
Deep-water squids have photophores (bioluminescent organs) on their tentacles. These organs may function as lures for prey or possibly to frighten predators. (Photograph courtesy M.J. Youngbluth, Harbor Branch Foundation.)

organs, which help them locate and identify potential mates. They also use lighted lures to attract prey. Most deep-ocean animals are black or red, both colors being invisible in the dim light.

Below 1000 meters, there is no light, and waters are uniformly cold, typically less than 4°C. Animals at these depths are widely dispersed, but they occur throughout the ocean. A few suspension-feeding organisms feed on detritus sinking out of surface waters, but predation is the norm. Many organisms can eat anything they can get into their mouth, including animals twice their size. Meals are infrequent in these depths—a fish may eat only a few times a year. On the other hand, food requirements are less because metabolism is slow in cold waters and at the high pressures of the deep ocean.

Squid (Fig. 13-18) are important predators at the middepths as well as in surface waters. They, like lantern fish, are luminescent.

VARIABILITY

Abundances and distributions of marine organisms also vary greatly from year to year. Every coastal area experiences years of unusual abundance of certain organisms—fish or crabs. A particularly good year for the growth of young fish may result in a large population that supports a fishery for many years. Much of this variability can be

BOX 13-3

Fisheries Management

Overexploitation combined with environmental changes has destroyed many fisheries. Depending on the growth rate of the organisms involved, recovery takes years to decades for a rapidly growing fish and centuries for slow-growing animals like whales. Let's look at how a fish stock reacts to increased harvesting to understand the effects of overfishing and environmental changes.

A fish stock is largest before fishing starts. Growth of young fish balances losses by death and predation. There is a large proportion of mature and older individuals. As a fishery is exploited, older animals are caught first. (They are the most valuable.) Then the number of young fish reaching maturity actually increases, because there is less competition for food and living space. In a young fishery, therefore, the number of young fish increases rapidly.

When the harvest reaches about 70% of the maximum potential yield, the rate of increased yield per unit of fishing effort begins to drop rapidly. Experience shows that the maximum sustainable yield is the amount the fishery can produce while continuing to be productive. Additional fishing effort may increase yield for a time, but eventually the population shrinks. Then its rate of natural increase approaches zero.

Overfishing, or harvesting beyond the maximum sustainable yield, causes future yields to decline. It also damages the stock's capacity to recovery, even if fishing is later restricted. The stock is also vulnerable to collapse if there are major environmental changes, such as droughts or El Niños. Protection of fisheries is especially important as many developing countries are turning to the ocean to provide protein for growing populations. At the same time, they are often destroying wetlands to make more agricultural land, and polluting rivers as they industrialize. In some parts of south China, fishermen have been reduced to catching plankton and harvesting sea cucumbers. The fish have been so reduced in numbers by overfishing and by destruction of the nursery grounds for young fish that the coastal fish stocks have been destroyed.

explained as the result of environmental changes, many of them very subtle. El Niños are an example of such environmental changes.

El Niños (discussed in Chapter 7) are caused by large-scale changes in atmospheric conditions in the Pacific and Indian oceans. Effects on marine organisms are most clearly observed along the equator and the coast of Ecuador and northern Peru.

During an El Niño, the thick lens of warm surface waters that collects near Indonesia and northern Australia produces pulses of warm water that move eastward along the equator. This thickens the surface layer above the thermocline. Although upwelling continues, it now involves warm surface waters and thus cuts off the nutrient supply required to support the growth of phytoplankton. Fish and other organisms that feed on the phytoplankton cannot obtain enough food to sustain themselves. They either starve or go elsewhere.

Seabirds nesting on the equatorial islands depend on catching fish and squid for themselves and their chicks. When the food supply is cut off, the birds either lay no eggs, or if the chicks have already hatched, they leave the chicks to starve when the adults leave to find food. Since seabirds are long lived, they can return to breed another year. But the loss of a year-class results in smaller bird populations.

Along the coast of South America, the warm waters from the equator also cause catastrophic changes in fish and bird populations. First, the productivity of nearshore waters is greatly reduced. During the 1982–1983 event, phytoplankton production off Peru was only 5% of normal. Even though local winds continue to cause upwelling, the upwelled waters come from the nutrient-poor surface waters because the thermocline is much deeper. The nutrient-rich waters are much deeper, below the depth of upwelling and do not reach the surface zone.

Furthermore, anchovies prefer cold waters. So they seek out small pools of cool water, where they are trapped by the surrounding warmer waters. In these pools they either starve due to lack of food or are easily caught by fishermen and birds, thus further reducing the breeding stocks. Either way, the year-class of anchovies is much smaller than normal, and the fishery takes several years to recover.

For many fish, this irregular (every 3 to 7 years) feast and famine would be intolerable. But it appears that fast-growing, rapidly maturing anchovies are well adapted to cope with these changing conditions. As we know from the collapse of the fishery in the 1972 El Niño, an unusually severe event combined with heavy overfishing can decimate the stocks, which take many years to recover. The fishery that produced 12 million metric tons in 1970 produced only about one hundred thousand tons in the mid-1980s and a few million tons per year in the late 1980s.

Effects of El Niños are felt along the Pacific coast of North and South America far from the equator. Migration of salmon in the North Pacific was disturbed by the 1982–1983 El Niño. Growth of kelp and breeding of marine mammals along the California coast were also disturbed.

SUMMARY

Nekton are adapted for swimming against currents. Most fish swim rapidly and are active predators. Many have a swim bladder to permit them to remain at a selected depth without continuously swimming. Some fishes are modified so that they can lie camouflaged on the bottom to wait for their prey. Fish are adapted to see in dim light.

Herrings are common fish in many parts of the ocean. They mature and breed rapidly, so they are able to take advantage of rapidly changing environmental conditions. They school when frightened. Cod are also abundant. When young they feed on plankton, and on benthos when older. Antarctic fishes are specially adapted to cope with freezing temperatures and scarcity of food. Many fish migrate as part of their life cycle, including salmon and eels.

Fisheries provide about 3% of the protein consumed by humans. Present food production cannot be greatly increased without utilizing many other types of marine organisms not now taken.

Sharks occur throughout the ocean. The largest sharks are slow-swimming plankton feeders. Others are predators that sometimes attack humans. Squid are also common nektonic animals and effective predators.

Marine mammals are warmblooded air breathers. The whales include the largest animals on Earth. They feed on plankton. Others take small benthic animals. Still others capture squid and other active nekton. Whales were extensively hunted until their populations were greatly reduced. Most commercial whaling has been prohibited, except for research.

Marine reptiles include turtles, snakes, and one marine lizard.

Seabirds are common over the ocean, feeding on fishes and other animals in near-surface waters. They must return to land to lay eggs and raise their young.

Deep-ocean benthos are markedly different from the surface zone. Light is dim or absent, and temperatures are low. Many organisms utilize bioluminescence either to signal for mates or to attract prey.

Abundances and distributions of nekton vary greatly from year to year due to changed environmental conditions. Changes associated with El Niños are especially noticeable in the eastern Pacific.

STUDY QUESTIONS

1. Describe how fishes swim.
2. How do swim bladders work?
3. Why is vision relatively unimportant to nekton in locating prey?
4. Why do deep-ocean nekton have no eyes or only vestigial ones?
5. List some of the adaptations of fish for rapid swimming.
6. Describe how herrings feed.

7. What are some of the adaptations of Antarctic fishes for dealing with low temperatures?

8. Why do fish migrate?

9. Describe some of the factors limiting further growth of fish catches.

10. List some of the adaptations necessary for nekton to live in the deep ocean.

11. What factors control the decline in the abundances in whales?

12. Discuss the reproductive strategy used by seabirds.

13. How do El Niños control abundances of organisms?

SELECTED REFERENCES

IDYLL, C. P. 1964. *The Abyss: The Deep Sea and the Creatures that Live in It*. New York: Thomas Y. Crowell. 396 pp. Elementary.

MARSHALL, N. B. 1958. *Aspects of Deep-Sea Biology*. London: Hutchinson. 380 pp. Ecology of the deep ocean.

MARSHALL, N. B. 1971. *Exploration in the Life of Fishes*. Cambridge, Mass.: Harvard University Press. 204 pp.

MINASIAN, S. M., K. C. BALCOMB III, AND L. FOSTER. 1984. *The World's Whales: The Complete Illustrated Guide*. Washington, D.C.: Smithsonian Books. 224 pp.

VOGEL, S. 1981. *Life in Moving Fluids: The Physical Biology of Flow*. Princeton, N.J.: Princeton University Press. 352 pp.

Large tube worms are conspicuous members of the benthic community that lives near hydrothermal vents. (Courtesy Jack Donnelly, Woods Hole Oceanographic Institution, Photo Researchers.)

Benthos

OBJECTIVES

1. To describe the life environment for bottom-dwelling organisms;

2. To understand oyster and coral reefs as communities of organisms;

3. To understand the trophic relationships among organisms on the deep-ocean bottom and in vent communities;

4. To understand the effects of water disposal and oil spills on organisms.

*T*he benthic world at the bottom of the ocean and along its shores is home to a great variety of organisms. Their world is basically two-dimensional—like ours—living in and on the bottom. In this chapter we focus on how organisms and communities of organisms adapt to that world. In particular, we are concerned with:

General features of the ocean bottom and shores;

Strategies for competing for scarce space and food resources; and

Adaptations to unusual sources of energy, such as hydrothermal vents.

BENTHIC LIFE

Benthic (bottom-dwelling) organisms live in a two-dimensional world, quite unlike the three-dimensional world of the planktonic and nektonic organisms we discussed in Chapters 12 and 13. The three *benthic life strategies* are (1) *attachment to firm surfaces,* (2) *free movement on the bottom,* or (3) *burrowing in sediments.* These life styles correspond to the principal *ways of obtaining food* among benthic organisms: *filtering from seawater, predation,* or *swallowing and digesting sediment.* Benthic organisms must compete for living space as well as food. In many ways, benthic life is easier for us to understand than planktonic or nektonic life.

A single organism can exhibit more than one life style. Crabs or worms, for instance, may take shelter in a sand or rock burrow but emerge to hunt their prey or to scavenge for detritus. Slow-moving animals with heavy shells, such as snails or sea urchins, feed on attached organisms or on detrital particles. Some attached animals, such as sea anemones, are also predators. They capture organisms that swim or float past them in the water.

Factors controlling distributions and diversity of benthic life include:

Light levels (if plants are involved);

Availability of food;

Temperature;

Salinity; and

Nature and stability of the bottom.

Stability favors highly diverse communities where many kinds of plants and animals live together. Where waves or currents frequently disturb the bottom, benthic organisms are usually scarce. In shallow polar seas, a short growing season for plants and marked salinity changes (due to melting and freezing of sea ice) create a rigorous climate where few species thrive. Frequently there are many individuals of the few species involved. On the other hand, quiet waters in nearshore tropical areas normally have highly diverse bottom communities—in other words, many different kinds of organisms. Availability of food and of suitable substrates controls the abundance of benthic organisms. Food is most abundant near land and below upwelling areas. Benthic organisms are generally less abundant with increasing water depths and distance from land.

SUCCESSION

The process by which organisms invade uninhabited surfaces is known as **succession.** It begins with an accumulation of bacterial slime [Fig. 14-1(a)]. Benthic diatoms and protozoans appear next. They multiply rapidly, utilizing absorbed organic compounds and products of bacterial decomposition. Hydroids and multicellular algae come next, followed by planktonic larvae of tunicates [Fig. 14-1(b)], barnacles, mussels, and snails. Eventually an ecosystem reaches a balanced state or climax community in which no further colonization occurs. Ecological succession ceases unless a disturbance of the system causes the process to start afresh. This process of colonization is familiar to boat owners and is known as **fouling.** It is the first stage in the development of the communities we discuss next.

ROCKY-SHORE COMMUNITIES

Intertidal areas provide a familiar sequence of quite different living conditions for attached organisms. Conditions range from nearly always dry (at the highest high tide level) to nearly always submerged (at the lowest low tide level). Associated population zones are often sharply divided. Each zone or horizontal band is occupied by a particular, well-adapted assemblage of plants and animals (Fig. 14-2), which are especially obvious on rocky shorelines.

This zonation reflects intense competition for living space. At a line where two populations meet, neither enjoys a marked advantage. Above or below that line, assemblages of organisms are determined by differences in an organism's ability to endure exposure to air and to large changes in water temperatures and salinities. In addition they must survive diseases and avoid being eaten by predators. Zonation among barnacles on the Maine coast can be seen in Fig. 14-3.

Where winds, sun, or waves create unfavorable environments or where the rock face is steep, attached seaweeds are sparse or absent. In more protected areas, many different kinds of plants grow. Above the high tide mark, seawater covers the rocks only at higher spring tides and during storms. But spray from breaking waves usually wets the rocks. Resistance to drying is a prime requirement for plants and animals in this region.

Between high and low tide levels, especially where there is heavy surf or scouring by strong currents, firm attachment is necessary. Barnacles (Fig. 14-3) dominate the upper, more exposed regions, where rocks are covered by water for less than half of each tidal day. In

(a)

(b)

FIGURE 14-1

(c)

Bacteria colonize submerged surfaces, creating a slimy coating that supports benthic organisms. (a) Rod-shaped, ring-shaped, and filamentous bacteria attached to the surface of a seaweed. The longest filament here is about 50 micrometers long. (b) The tunicate *Molgula,* a common fouling organism in Chesapeake Bay, feeds on organic detritus from the side of an aquarium. (Photograph courtesy Michael J. Reber.)

FIGURE 14-2
Rockweed, barnacles and brown algae grow on rocks in the intertidal zone on the Maine coast. (© Mary M. Thacher, Photo Researchers.)

FIGURE 14-3
Zonation among barnacles on the Maine coast. A small species of barnacles dominates the upper layer, separated by a sharply defined boundary from larger barnacles below.

Small species
of barnacles

Large species
of barnacles

shady, protected areas, barnacle zones extend farther up rock faces than on dry, sunny surfaces. These animals feed when submerged at high tide, filtering small particles from the water. Barnacle shells have a four-part lid that shuts tightly, protecting the animals from drying when exposed at low tide.

Requirements for light govern distributions of benthic plants. *Green algae* live somewhat above the low tide mark down to depths of about 10 meters. Varieties include bright green sea lettuce, up to 1 meter in length, and delicate, mossy plants only a few centimeters long (Fig. 14-4).

Red algae occur worldwide. They are most abundant in temperate and tropical waters and prefer dim light in deep waters or shaded pools. Brown algae flourish in colder waters, although some kinds occur on rocky coasts around the world.

Grazing by snails and other animals affects the presence of benthic plants. Areas actively grazed may have little attached algae left. If the animals are removed (or killed by an oil spill), the plants can grow luxuriantly. Normally they are kept grazed down.

Storms also affect plants and animals on rocky coasts. Storm waves can scour rocks and shallow bottoms, leaving nearly bare surfaces for later recolonization. This then leads to the growth of a succession of plants and animals, comparable to the events observed in fouling of newly cleaned boat bottoms.

Tide pools contain specialized plants and animals that can cope with highly variable environments. Protected environments permit more delicate organisms to live in such pools. A large variety of plants and animals can live in a small area (Fig. 14-4). Tiny shrimplike crustaceans (amphipods), swimming worms, and many kinds of snails are

FIGURE 14-4
A rocky intertidal area on the Maine coast shows zonation of attached animals and plants: barnacles above, brown algae below. At the bottom, which is underwater even at low tide, green and red algae grow in a shady pool. Snails, worms, sea stars, and tiny crustaceans live among the plants.

Barnacles

Brown algae

Tide pool

common in tide pools, especially where deep crevices or beds of sea-weed retain water during low tides (Fig. 14-4). Evaporation, overheating, and oxygen depletion occur on warm, sunny days. On the other hand, heavy rains can markedly lower salinity within a few minutes. Survival for tide-pool organisms requires the ability to tolerate sudden changes in temperature, salinity, and dissolved oxygen levels in the waters around them.

The most populous zone on a rocky beach is around and below low tide level. Starfish and crabs are common, usually hidden in crevices. Small scavenging snails inhabit protected niches containing stagnant water and decaying debris. Sea anemones [Fig. 14-5(a)], sea ur-

FIGURE 14-5
(a) Warm-water sea anemones. The animals at left center and far right are open and feeding. The mouth opening is at the center of the open "flower." When startled, they close instantly and then look like a wrinkled stump (right center). (Photographed at the New York Aquarium.) (b) Barnacles and mussels compete for space on a rocky intertidal zone at La Jolla California. Both are filter feeders but seal themselves in their shells during low tide to avoid drying out. (Photograph C. Arneson.)

(a)

(b)

FIGURE 14-6

The coelenterate *Chrysaora quinquecirrha*, a native of Chesapeake Bay. (a) Hydroid polyps grown on an oyster shell. (b) Detail of a typical feeding polyp, about 1 millimeter tall. The mouth, surrounded by tentacles, is at the top of the polyp. (c) Reproductive polyp is budding and may produce a new stack of free-swimming medusae in 2 to 4 weeks. Each polyp can produce a new stack two to four times in a single summer. (d) Full-grown medusa, about 10 centimeters across the bell. Its unpleasant sting sometimes prevents swimming when the organism is abundant. (Photographs courtesy Michael J. Reber.)

chins, sea cucumbers, and mussels [Fig. 14-5(b)] are locally abundant. Hydroids (Fig. 14-6) grow in quiet but not stagnant waters as well as the nudibranchs (Fig. 14-7) that feed on them.

Lobsters and crayfish (spiny lobsters, Fig. 14-8) scavenge on subtidal hard bottoms, both near shore and far out on the sandy and rocky continental shelf. They walk about at night, searching for worms, mollusks, and organic debris. They usually seek shelter under rocks or seaweed during daylight. In autumn, Maine lobsters move offshore to breed, probably to avoid cold shallow waters in winter.

The octopus (Fig. 14-9) is a common resident of rocky sea floors. It prefers to hide in crevices and feeds at night.

KELP

Large brown benthic algae, known as *kelp,* grow in shallow subtidal areas in subtropical to subpolar coastal oceans. In waters cooler than 20°C, kelp grows abundantly where it finds suitable substrates. (Kelp distributions are almost the opposite of coral reefs, which grow in waters warmer than 18°C.) Kelp often forms dense beds or underwater forests, often many kilometers long. In clear waters, kelp grows down to depths of 40 meters.

Kelp plants alternate sexual and asexual generations. One form is the conspicuous plants that reach lengths of tens of meters. These large plants are asexual and produce microscopic spores that reproduce sexually, forming tiny filamentous plants. These plants settle on a suitable

(b)

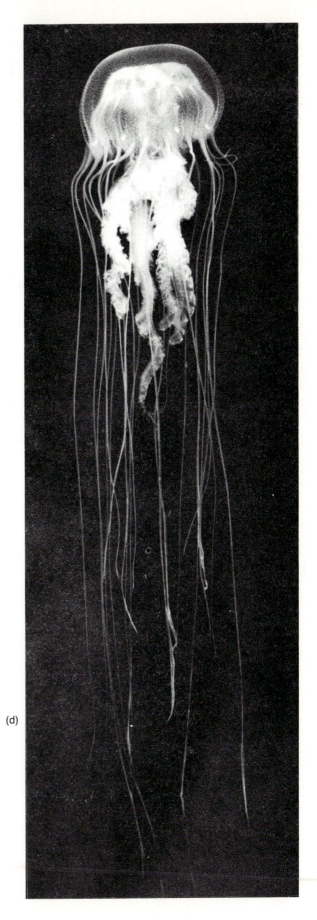

(c)

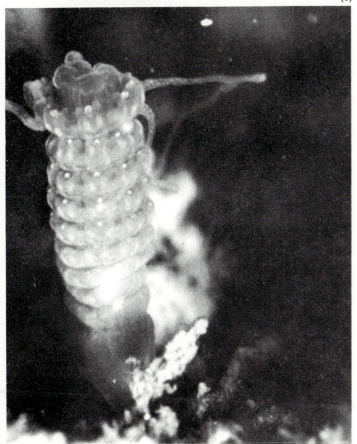

(d)

FIGURE 14-7
Nudibranchs, about 20 millimeters long, feed on hydroid polyps. This shell-less snail breathes through gill-like projections along the sides. Because of their unpleasant taste, they have few predators. (Photograph courtesy Michael J. Reber.)

FIGURE 14-8
Spiny lobster lacks the large claws of the true lobster. It occurs in warm waters of the Pacific and Gulf coasts. (Photograph courtesy Marineland of Florida.)

FIGURE 14-9
Octopus. The eyes are on the top of the head. Below is a siphon through which water is ejected after passing over the gills. The animal has eight tapered tentacles or arms, each equipped with double rows of suckers. Maximum size is around 50 kilograms with a 9-meter arm spread. (Photograph courtesy Marineland of Florida.)

rock or hard sand area. There they grow into the familiar large plants that live for several years in areas where they are not destroyed by waves.

The microscopic sexual generation is poorly known. They are apparently dispersed by currents, where they persist for months, perhaps up to a year, thus acting as seeds. In this way they disperse the kelp plants and replenish beds after winter dieoffs or after destruction of plants by storm waves.

Kelp plants have *holdfasts* that grip rocks or the hard bottom. They anchor the plants, much as roots anchor plants on land. Unlike roots, holdfasts do not take up nutrients from the bottom. Instead, nutrients are taken up by the leaves, called *fronds*. The fronds are kept in the photic zone by gas-filled bladders.

Kelp grows primarily in nutrient-rich waters. It is highly productive. Primary production in kelp forests ranges from 500 to 1500 grams of carbon per square meter per year. This is two to five times more than the average production by phytoplankton in coastal-ocean waters. It rivals the production of the best agriculture on land.

Kelp plants are easily damaged by waves. Often, ends of fronds are worn away by abrasion, forming fragments. As they are worn away, the fronds are renewed by continued growth. Thus they are like a conveyor belt of algal tissue. Much of the organic matter produced enters detritus food webs.

Kelp forests offer many different environments to animals in the water or on the bottom. Tops of the tallest plants float near the surface, forming a shaded canopy where animals can hide from predators. Many plants and animals also grow on the bottom among the holdfasts. Such plants must tolerate low light levels. Various animals live in and around the edges of kelp forests, using them as refuges.

Kelp plants are also eaten by fish, sea urchins, and snails (Fig. 14-10). Grazing by *sea urchins* is important because where they are especially abundant, they destroy the forests, leaving so-called urchin barrens.

Sea urchins are eaten by *sea otters* (Fig. 14-11) and lobsters. In the eighteenth and nineteenth centuries, when sea otters were hunted nearly to extinction, sea urchin populations were able to grow. Then

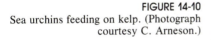

FIGURE 14-10
Sea urchins feeding on kelp. (Photograph courtesy C. Arneson.)

FIGURE 14-11
Sea otters feed on sea urchins. When sea otters nearly became extinct due to being hunted for their furs, the urchins became far more abundant and grazed back the kelp forests. Since sea otters have been protected against hunting and have become more abundant they have reduced the numbers of sea urchins and the kelp forests have recovered. (Photograph courtesy Monterey Bay Aquarium, Kathleen E. Olson, photographer.)

they were able to greatly reduce the abundance of kelp beds. Sea otters have been protected from hunting since 1911, and their populations are now large enough to control sea urchin populations. Consequently, kelp beds have recovered much of their original extent along the Pacific coast of North America.

MUDDY- AND SANDY-BOTTOM COMMUNITIES

Marshes, beaches, and estuarine shorelines are inhabited by benthic plants and animals. Seagrasses in marshes contribute plant debris to sediments and protect deposits from erosion by currents and waves. On open coastlines, waves or currents can inhibit growth of rooted plants. These waves and currents can also carry away fine particles and organic debris, leaving behind hard sandy bottoms with little organic matter. These environments support many animals adapted to survive on sandy or muddy bottoms.

Infauna—animals that live buried in sediments—are much less conspicuous than surface-dwelling animals, called **epifauna.** Many infaunal animals are selective deposit feeders Fig. 14-12. They select food particles from among sediment grains. Others pump large quantities of water through filtering devices to remove edible materials. Still others are unselective feeders. They eat their way through sediments and digest the sediment, extracting their food from it. These include nudibranchs, sea cucumbers, and many worms.

Distributions of infaunal animals are governed, in part, by sediment grain size. Animals that feed on suspended plankton and detritus from seawater dominate stable, relatively coarse-grained sandy deposits below low tide. They require relatively clear, nonturbid waters that will not clog delicate, mucus-coated filtration devices.

Fine muds, easily eroded by currents, are generally not suitable for *suspension-feeding* organisms (Fig. 14-13). Muds are well suited to those organisms that feed unselectively by ingesting sediments. Smaller particles normally contain abundant organic matter. They also support bacteria and many other microbes that are food for deposit feeders.

Mud-burrowing animals have specialized breathing structures that are not clogged by sediment grains. Breathing through the skin—

FIGURE 14-12
Polychaete worms, relatives of the earthworm, are common among the infauna in soft sediments. Larvae of these worms are planktonic until they metamorphose. (Photograph courtesy P.I. Blades-Eckelbarger, Harbor Branch Oceanographic Institution.)

FIGURE 14-13
Benthic bivalves. (a) The cockle (right) is a filter feeder, while the clam *Telina* (left) selects edible particles from the sediment surface. (After Hedgpeth. 1957.) (b) Sipunculid worms can retract the tentacle-bearing head into the body. Many live in soft sediments and feed by swallowing mud and sand from which they digest organic matter. (Photograph of model in U.S. National Museum.)

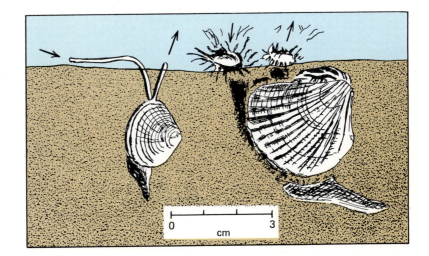

(b)

as sea stars do, for instance—is not suited to a muddy environment or turbid waters. One widely used method of feeding and breathing uses water pumped through burrows (Fig. 14-14) by the animals living there. Dissolved oxygen is supplied for respiration, and sediment particles are swept off the animal's gills. At the same time, food particles are captured.

In intertidal areas, deposit feeders generally dominate because unstable sediment surfaces prevent accumulation of phytoplankton or other particulate food utilized by filter feeders. Many deposit feeders excrete durable *fecal pellets*. These bind the sediment particles and thus reduce the turbidity of the sediment-water interface. In deep quiet

FIGURE 14-14

The pink echiuroid worm *Urechis caupo,* the so-called inn-keeper worm (a) is common on Pacific mud flats at or below the low tide level. The worm constructs a deep U-shaped burrow with narrow openings by scraping away mud with bristles at each end of its body. Inside, it secretes a funnel-shaped, fine-meshed mucous net that fits over its head like a collar (b). As the worm pumps water through the burrow, food particles are trapped in the net. When the funnel becomes clogged, the worm eats it and constructs a new one. Wastes and debris are ejected by blasts of water. A worm (c) lives behind the host along with tiny pea crabs and sometimes small fishes called gobies. (After Hedgpeth. 1957.)

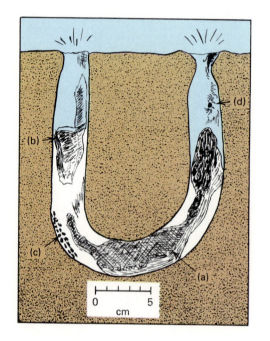

FIGURE 14-15

Yoldia limatula feeding. Half-buried in the mud, the clam uses its feeding palps to collect sediment and bring it into the mantle cavity. Inedible materials are ejected as a cloud of loose sediment, creating mounts of laminated, reworked sediment. When resting, *Yoldia* commonly burrows a few centimeters below the sediment surface (After Rhoads. 1963.)

waters, the bottom may be nearly covered by fecal pellets. This makes the deposits more resistant to erosion and keeps waters clearer.

Below the low tide level, in muddy sediment on continental shelves, there are usually many deposit-feeding bivalves such as clams (Fig. 14-15). Where waters are too deep for algae and eelgrass to grow, sediments are often soft and easily eroded, due in part to constant reworking by burrowing polychaetes and mollusks. A few tens of clams per square meter can rework all sediment to a depth of 2 centimeters once or twice a year. Attached animals are unable to colonize the semiliquid substrate, and filter feeders are choked by the turbid waters. Thus the clams are able to exclude their competitors.

Where deposit feeders are abundant, suspension feeders are scarce. This situation occurs when deposit feeders make the area less easily habitable by filter feeders. In this case one organism forces others out of the area. This is competition by creating an undesirable living environment rather than competing directly for food or living space. On the other hand, oxygen-bearing waters mixed into sediments by burrowing organisms (a process called *bioturbation*) permit aerobic bacteria, protozoans, and other small benthic animals to live much deeper below the sediment-water interface than is possible in oxygen-deficient muds.

SALT MARSHES AND SEA GRASSES

Salt marshes and sea grasses occur in shallow waters along most of the world's coastlines except in polar regions. Salt marshes are intertidal, whereas sea grasses are permanently submerged. These benthic communities have been well studied because of their importance and the convenience of studying them. These grasses are extremely productive. Primary production of 1500 grams of carbon per square meter per year (comparable to a corn field) is common.

In the absence of strong waves or currents, intertidal sediment deposits, rich in organic detritus, are trapped by marsh grasses, forming wetlands, which we discussed in Chapter 10. Walking through such an area at low tide, one sees few signs of animal life, but the ground is riddled with animal burrows.

Around high tide levels, *fiddler crabs* (Fig. 14-16) dig burrows. These burrows may be up to 1 meter long. They are dug obliquely and are thus shallow enough for the crabs to remain dry. Fiddler crabs need keep only their gills moist, although they can survive for weeks immersed in seawater.

Dense stands of eelgrass, a seed-bearing marine plant, grow in shallow-water sediments, as shown in Fig. 14-17(a). Eelgrass has underground stems and ribbonlike leaves, as much as 1 meter long. Specialized communities of plants and animals find food and shelter in eelgrass beds. The leaves are coated with epiphytic diatoms, cyanobacteria (also called blue-green algae), protozoans, organic detritus, and small hydroids [Fig. 14-17(b)]. Tube-building crustaceans and worms attach themselves to the leaves. Small grazing animals, such as shrimp and snails, scrape off detritus. This cleaning of eelgrass leaves is essential to its continued growth. If the leaves become too thickly coated with detritus, plants cannot photosynthesize and thus die.

A large fraction (up to half) of the organic matter formed in many North American marshes is carried out of the marsh by tidal currents. Part of this organic matter is consumed on the continental shelf, and part is deposited in recent mud deposits either on the continental shelf

FIGURE 14-16
Fiddler crab and its burrow. (a) Crab poised on the sand. (b) Crab ready to dart into its burrow, fighting claw held protectively over its head. (a)

(b)

(c)

FIGURE 14-16

(c) A freshly made burrow surrounded by lumps of compacted sediment excavated by the crab and deposited outside the burrow.

Disease

Marine organisms are affected by diseases. Entire populations have been destroyed by diseases. But little is known about diseases in the ocean. Still less is known about the organisms causing these diseases. The reason is easily understood: There are no domesticated marine organisms; thus marine organisms are not as well known as humans or domestic plants and animals. We have only limited observations of the effects of diseases on commercially harvested organisms or those that live in shallow water, where they can easily be observed.

The best-known examples of diseases in marine animals are those affecting oysters. Oysters are harvested in many coastal regions and cultivated in some regions. Oysters are known to be affected by several diseases. One of these is known as MSX or "Multinucleate Sphere X"—a fancy way of saying that no one knows what the disease organism is. What is known is that the organism is a round, single-celled protozoan that infects oysters in waters with salinities greater than 15‰. The organism is a parasite that weakens the host oyster. The sick oyster is then more subject to other predators or other diseases. Droughts cause MSX to spread, since estuarine waters are saltier than normal. Up to 95% of the oyster crop in Delaware Bay was destroyed by the disease when it first appeared in 1957. Since then, MSX has spread to Chesapeake Bay, where it devastated the oyster industry in the late 1980s.

Scientists are working to determine how healthy oysters normally protect themselves against the disease. One theory is that higher salinities weaken oysters, so that they are less able to fight off the parasites. One way of fighting the disease is to select animals that have lived through an episode of the disease. Presumably they are more resistant than those killed. Selectively growing resistant organisms might permit cultured oyster crops to withstand the disease.

(a)

FIGURE 14-17

(a) Model of an eelgrass community, Woods Hole Massachusetts. Scallops, shrimps, and crabs are shown in the eelgrass. Many kinds of worms live in the soft bottom. The burrow at left contains a polychaete worm, which maintains a current through its burrow by flapping leaflike parapods (side feet). The burrow lining is a tough parchmentlike material, shown in the W-shaped burrow at center. At right, the tube-dwelling worm with its gills is emerging from the tube. (Drawing courtesy American Museum of Natural History.) (b) Bacteria and diatoms grow as a crust on a blade of sea grass (magnified 200 times). (Photograph courtesy J. M. Sieburth and University Park Press, Baltimore Md.)

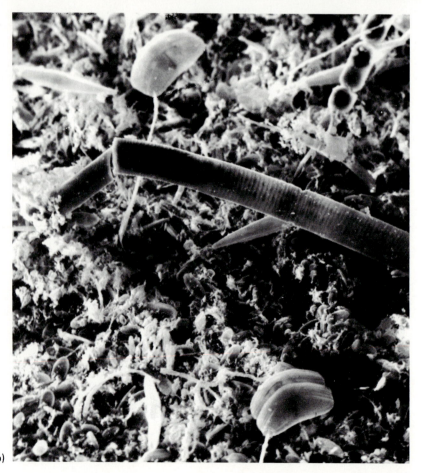

(b)

or on the upper continental slope. Whether a marsh exports organic matter is highly dependent on its connections with the adjacent coastal ocean.

Sea grasses are more thoroughly marine and live at greater depths, where they form underwater meadows. They are limited by the availability of light for photosynthesis. They too are highly productive, comparable to the intertidal salt marshes. Some live as deep as 90 meters.

Both salt marshes and sea grass beds are highly productive and provide refuge for young organisms. Thus they are important nursery beds for many organisms. In addition, they are grazed by sea turtles. Finally, they export large amounts of organic matter to surrounding continental shelf areas.

OYSTER REEFS

An *oyster reef* is a rigid, wave-resistant structure built by oysters. Oysters are sessile bivalved mollusks that are abundant in coastal oceans and estuaries worldwide. Each mature female produces millions of eggs yearly. Most are fertilized in the water (Fig. 14-18). Only a small fraction of the planktonic larvae survive (this stage lasts only a few weeks) and settle to the bottom to develop as adults. Larvae select suitable settling areas, preferring oyster shells to any other substrate and apparently favoring live oysters over dead shells. An oyster needs 1 to 5 years growth to mature, during which time most of those that settled together as larvae are killed by crowding, by competition for food, or by being eaten by predators.

Oyster reefs (Fig. 14-19) contain enormous numbers of individuals whose shells are cemented to rocks and to one another. Tidal flats of partially enclosed bays and river mouths provide advantageous con-

FIGURE 14-18
Oysters (male on left, female on right) expelling sperm and eggs into water. (Photograph courtesy Michael J. Reber.)

ditions, especially where moving water with little silt brings fresh supplies of plankton and oxygen and where few predators exist.

Each oyster pumps many gallons of water each hour. Plankton and food particles in the water are caught on a mucous net that is moved steadily toward the mouth by ciliary action. Some particles concentrated in this way are swept out of the shell, before the oyster can consume it. Nearby filter-feeding animals take advantage of this food supply.

FIGURE 14-19
Oysters form wave resistant reefs by cementing their shells together. Space between the shells provide shelter for many small organisms. (Photograph courtesy Michael J. Reber.)

BOX 14-2

Charles Darwin

Charles Darwin, the great nineteenth century scientist, contributed significantly to improved understanding of the ocean and its life. His career began as an unpaid naturalist on a 5-year surveying voyage.

Darwin graduated from Edinburgh and Cambridge universities, where he first intended to become a physician. But the sight of blood sickened him, so he decided to become a minister instead. He later decided that botany was far more appealing to him. Thus he jumped at the chance to join a British-navy-sponsored surveying expedition.

Darwin's ship—the *Beagle,* a 10-gun brig—was only 90 feet long and carried 74 people in cramped quarters. The ship sailed on December 27, 1831, and immediately ran into a vicious storm. Darwin was terribly seasick. Indeed, the many difficulties he encountered on this voyage may have contributed to his ill health in later life.

During the voyage, Darwin visited many parts of the world, especially South America, which was little known at the time. He collected, described, and sent back to England an enormous number of plants, animals, and fossils. Darwin was especially influenced

by the great variety of plants and animals in the Galapagos Islands, off Ecuador. The many different species of birds had evolved from a few ancestors to live on the isolated islands. These unusual animals (including a giant tortoise, many different kinds of birds found only in these islands, and a marine iguana) and the many fossils he collected provided the foundation for his theory on *Origin of Species.* This book shocked the Victorian world and is still controversial in some quarters.

Darwin also theorized that the many atolls in the Pacific formed on subsiding volcanic islands. This theory was finally proved by drilling on several atolls in the 1950s and 1960s and finding volcanic rocks underlying atolls at Midway Island and at Eniwetok in the Pacific.

Darwin married the daughter of Josiah Wedgwood, founder of the famous pottery firm, and fathered 10 children. Much of his later life was devoted to working out his theory of the origin of the species and to studying the materials he collected while on the *Beagle.*

(a)

(b)

FIGURE 14-20
Sabellarian polychaete worms (a) secrete a mucous tube that hardens into a parchmentlike substance. Feathery gills, often brightly colored, are the only part of the animal that emerges from the end of the tube. Sabellarian worms can build reefs in surf zones where the shoreline is rocky enough to support their tubes. Currents supply food particles and sand grains which they cement with their mucous secretion to make wave-resistant structures that may extend for several meters along a beach or breakwater. (b) Calcareous tube worms commonly encrust shells of oysters or other mollusks. (Photograph courtesy Wometco Miami Seaquarium.)

Attached organisms, such as barnacles, mussels, and tube worms (Fig. 14-20), add to the bonding, thus increasing the reef's stability. Crevices between shells shelter small filter feeders.

Oysters can live in low-salinity waters. In Chesapeake Bay, for instance, large natural beds occur in intermediate-salinity waters (7 to 18‰). Oysters survive there, but their principal enemies and most diseases cannot. In the absence of predation, oysters grow faster and are more prolific in higher salinity waters. If the salinity increases, as a result of drought, the oyster may then be attacked by both predators and diseases.

CORAL REEFS

Coral reefs occur in tropical and subtropical waters warmer than 18°C. Like oysters, they build their environment. Reef-building corals are colonial animals that build calcareous skeletons (Fig. 14-21). New individuals bud from their parent's side, an example of asexual reproduction. This is the primary means of enlarging coral colonies. Other corals build flexible, fan-shaped structures (Fig. 14-22) attached to rocks.

FIGURE 14-21
Corals grow in many different forms on reefs—a branching form in front, a solid form in the rear. (Photograph courtesy C. Arneson.)

FIGURE 14-22
A red sea fan, a close relative of the reef-forming corals, grows on a coral reef. The minute animals that form the structure capture food from waters moving past the fan. Note the rough surface of the reef and the brilliant colors of the corals and encrusting algae. (Photograph courtesy C. Arneson.)

As individuals die, others build over and around the dead skeletons, eventually forming massive structures. Some corals build large branching structures; others build compact forms such as brain coral. Still others grow by encrusting reef surfaces. Corals also reproduce sexually. They apparently take 7 to 10 years to reach sexual maturity. Sexual reproduction results in free-swimming larvae. When larvae settle and attach, they start new colonies. These larvae are also produced asexually. Apparently, sexual reproduction accounts for about one-quarter of the new coral colonies in Jamaican reefs.

Unicellular algae called *zooxanthellae* (a type of dinoflagellate) live in coral tissues. This is an example of a mutually beneficial relationship, called **symbiosis.** The algae make photosynthetic products directly available to the host coral. In return, the algae receive nutrients and carbon dioxide. Photosynthesis by the algae alters the carbon dioxide concentrations in the coral tissues, greatly increasing the animal's ability to secrete calcium carbonate to make its skeleton.

Corals are the most conspicuous contributors to the reef framework. Encrusting **red algae** are important cement depositors. In addition, red and green calcareous algae are important cement depositors. They produce much of the carbonate sediment that collects and is later cemented into the reef mass.

Coral reefs are complex, shallow-water benthic environments that are among the most productive communities in the ocean. Certain unicellular plants supply nitrogen compounds to the reef waters. Nutrients are readily recycled in coral reefs. Currents in and around reefs tend to retain nutrients in the overlying waters. Reefs are also highly diversified in the many kinds of living spaces available that host many different kinds of organisms. Competition for living space rather than lack of food apparently limits the abundance of suspension-feeding organisms on coral reefs.

Tides and waves periodically expose parts of reef tops. There hardy, soft-bodied algae, barnacles, and coralline algae form rimmed pools. These pools are a few centimeters above the level of the average sea surface. Water splashed into the pools by waves drains slowly, keeping plants and animals in the pools moist. (This is similar to tide pools seen on rocky shorelines.) Coralline algae form a purplish red ridge, called the *algal* or *lithothamnion ridge,* at sea level.

In the photic zone, reef-building corals are most conspicuous. But attached plants greatly exceed the mass of living animal matter. Besides the coralline algae that add calcareous material to the reef framework, filamentous green algae embedded all over its surface manufacture food during the day. They use nutrients released by animals and bacteria. At night, coral polyps extend their feeding apparatus (Fig. 14-21) to capture plankton and detritus from the water.

Many kinds of invertebrates and fish (Fig. 14-23) live in coral reefs. But there are relatively few individuals of any one kind. Some,

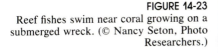

FIGURE 14-23
Reef fishes swim near coral growing on a submerged wreck. (© Nancy Seton, Photo Researchers.)

FIGURE 14-24
A large crown-of-thorns eats corals, leaving behind bare coral skeletons. Under certain conditions, not well understood, these animals become extremely abundant and decimate large areas of reefs. (Photograph courtesy C. Arneson.)

like *sea stars* (Fig. 14-24) consume coral polyps and algae. Others feed on detritus. Still others—for example, moray eels and sea anemones—prey on animals living on or near the reef. Parrot fish and other browsers graze on the reef surfaces to eat algae and animals living in it.

Due to continual browsing of their surfaces, coral reefs are constantly changing. One organism with a competitive advantage gains space or position over another, while perhaps losing space to a third. Some coral species even practice chemical warfare. When their tissues touch, one form can release substances that destroy the tissues of the other.

A *disturbance*—a severe storm, an infestation of starfish, or an oil spill—can set this competition for space in motion by leaving a barren surface [Fig. 14-19(b)]. Succession on coral reefs occurs much like the development of fouling communities that we have already discussed. The most abundant larvae will initially colonize the newly bare surface. Then as the organisms compete for space with other forms, the composition of that part of the reef may change.

VENT COMMUNITIES

Dense populations of benthic animals live near hydrothermal vents (Fig. 14-25), which discharge sulfides (midocean ridges) or oil and gas (continental margins). These communities contain 10,000 to 100,000 more living matter than normal deep-ocean benthic communities. They are indeed oases of benthic life. Vent communities are also unusual in that the organisms depend on chemosynthetic bacteria for food instead of photosynthesis.

Vents and their associated organisms are best known on rapidly spreading midocean ridges, such as the East Pacific Rise. Vent communities have been found around hot-water discharges on the Mid-Atlantic Ridge and on active submarine volcanoes. Similar communities of organisms occur around cold-water discharges on continental margins and in subduction zones. A few active vents have been found with no organisms living on them. One example is the young volcanic center, Loihi, southeast of the island of Hawaii.

The most conspicuous organisms in these communities are giant worms (Fig. 14-26), close relatives to much smaller worms that live in

FIGURE 14-25
A black smoker discharging superheated waters and sulfides (which give it the black color) which supply energy for chemosynthetic bacteria. The bacteria are food for communities of abundant, benthic organisms which live on the adjacent rocks. Many organisms filter the many particles seen here in the water. (Photograph courtesy Woods Hole Oceanographic Institution.)

FIGURE 14-26
Near active hydrothermal vents there are dense growths of organisms that depend on chemosynthetic bacteria. The bacteria are supported by the sulfide-bearing waters discharged from the crevices between the large rocks. Note the many different kinds of organisms, including crabs, fishes, and the large worms with their blood-red gills. (Photograph, J.M. Edmund, MIT, courtesy WHOI.)

sulfide-rich wetland sediments. Vent worms are unusual in that they grow to be nearly 3 meters long (Fig. 14-27) and 2 to 3 centimeters in diameter. These worms have red gill-like organs that they retract into a white plasticlike tube when disturbed. These worms have no digestive tract. Instead, they obtain food from sulfide-oxidizing bacteria that live in specialized organs. Wastes from the animals provide nutrients needed by the bacteria. This is another example of symbiosis, in which two organisms live together to their mutual benefit.

FIGURE 14-27
The worms growing on the hydrothermal vents are exceptionally large. Here scientists examine them on the deck of a research vessel. (Photograph by J. Donnelly, WHOI.)

FIGURE 14-28
Clams growing near the vents grow very large in a few years because of the abundance of food which they filter from the water. These organisms have red flesh because of the hemoglobin in their blood used to transport oxygen, just as it does in mammals. (Photograph courtesy Woods Hole Oceanographic Institution.)

Other organisms inhabiting vents (Fig. 14-28) also depend on chemosynthetic bacteria. Clams and mussels have such bacteria on their gills. Clams grow to dinner-plate size in less than 2 years. (Recall that most deep-ocean organisms are very slow growing.) These organisms have hemoglobin in their blood (like land animals) to transport oxygen. Consequently, their tissues look like raw liver. They also have special enzyme systems to protect them from hydrogen sulfide poison-

CHAPTER FOURTEEN   BENTHOS

ing. Without such protection, hydrogen sulfide in the vent discharges would poison the hemoglobin and kill the animals.

Dense communities of benthic animals also grow on the continental slope off Florida near discharges of cold, sulfide-bearing, ammonia-rich waters. Similar communities have been observed off the Oregon-Washington coast, growing on sediments in subduction zones where cold, sulfide-bearing waters are expelled during subduction. Offshore oil seeps in the Gulf of Mexico also have comparable communities.

At first glance, hot hydrothermal vents seem unlikely places to support such rich growths of organisms. In addition to poisonous hydrogen sulfide discharges, the vent waters are extremely hot; temperatures up to 400°C have been recorded. But most vent organisms do not experience such high temperatures, due to rapid mixing with cold bottom waters. Some bacteria living in the vents may, however, require temperatures of 100°C or more to grow. Most organisms at surface pressures are killed by such temperatures.

One unsolved question raised by the discovery of these isolated vent communities is how the organisms disperse. Hydrothermal vents on midocean ridges are short-lived, each one probably lasting only a few years to a few tens of years. When vents no longer discharge sulfides, nearby organisms die. Even the most active sites may be quiescent for decades or centuries.

How do organisms spread from one vent to another? Transport of long-lived larvae by near-bottom currents is one possible explanation. Another is that the larvae rise to the surface, where they are dispersed by currents, and then later return to the bottom. The question remains unanswered.

DEEP-OCEAN BENTHOS

Soft sediments and **deposit-feeding** animals (Fig. 14-29) dominate much of the deep-ocean bottom. Large, suspension-feeding animals are rare but conspicuous. Crinoids (Fig. 14-30), for example, stand on long stalks above the bottom to filter particles from near-bottom currents. Predatory forms, such as *brittle stars* move about on long legs that support their bodies well above the soft-sediment surface.

Suspension-feeding sponges (as in Fig. 14-31), coelentrates (Fig. 14-32) worms, bivalved mollusks, and crustaceans—in fact, all the major groups of animals in the shallow-water benthos—occur on the deep-ocean bottom. Uniform coloration (gray or black among fishes, often reddish among crustaceans) and delicacy of structures are typical among organisms living in these dark waters. Deep-sea animals are generally smaller than their shallow-water relatives. They live much longer and reproduce less frequently. All these differences are adaptations to scarcity of food.

Most of the detritus that falls from the surface zone is consumed (usually many times) or decomposed before reaching the deep-ocean floor. Thus relatively little usable food reaches the deep-ocean floor. Because of the scarcity of food, the biomass of animals living on the bottom is much lower than in shallow coastal-ocean waters. Eggs of some benthic forms float to the surface, where food is more plentiful for larvae. The eggs hatch there and then descend to the bottom. Most deep-water benthic animals produce "young adults" that can feed themselves. Alternatively, others produce eggs with large yolks. In either case, the larvae do not have to go into the photic zone to feed—or to be eaten.

FIGURE 14-29

(a) Deep-ocean sea cucumbers. (b) Drawn from below as if crawling on a pane of glass. As these echinoderms crawl along the bottom, the mouth scoops up sediment, debris, and small animals.
(c) A brittle star (lower left) moves along the bottom and an octopus (lower right) swims near the bottom in an area of abundant manganese nodules, about 4500 meters deep in the tropical North Pacific. The brittle star feeds on detritus while the function of the octopus is not known. (Photograph courtesy Deepsea Ventures, Inc.)

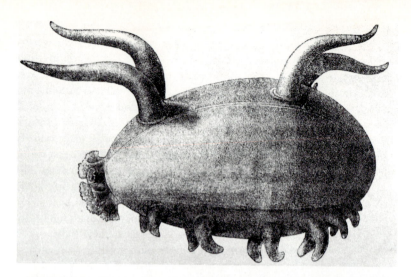

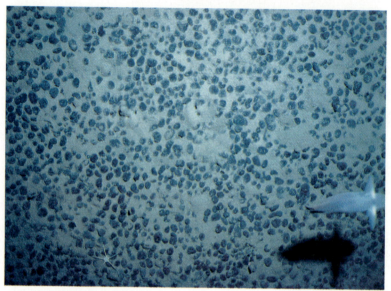

FIGURE 14-30
Crinoids (sea lilies) have cup-shaped bodies and long arms. The skeleton is made of calcareous plates, which are absent near the mouth. (Model photographed in the U.S. National Museum.)

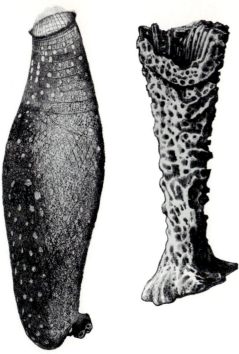

FIGURE 14-31
Deep-ocean sponges are filter-feeding organisms.
(From John Murray and A.F. Renard, 1891.
Report on Deep-sea deposits based on the
specimens collected during the voyage of *H.M.S.
Challenger* in the years 1872–1876. *Report of the
Scientific results H.M.S. Challenger*, vol. 5.)

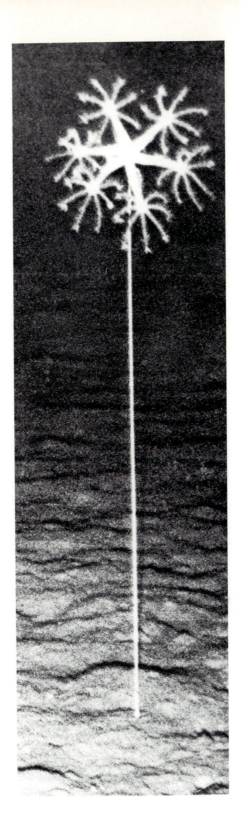

FIGURE 14-32
A filter-feeding stalked coelenterate, whose
pencil-thick stem is about 1 meter long, lives at a
depth of 5000 meters off the Atlantic coast of
Africa. (Official U.S. Navy photograph.)

In the midst of this scarcity of food, there are infrequent falls of
large amounts of food—bodies of sharks, whales. These carcasses
attract swimming scavengers, but such events are rare. Thus the ability
to sense food—probably by smell—from afar and to move quickly is an
important adaptation in this barren environment.

Benthic organisms live in a two-dimensional world. The three benthic life strategies are (1) attaching to a surface, (2) freely moving on the bottom, or (3) burrowing in sediments. Ways of obtaining food are (1) filtering from seawater, (2) preying, or (3) swallowing and digesting sediments.

Stable environments favor diverse communities, containing many different kinds of organisms. Availability of food controls their abundance. Abundance of food usually decreases with increasing water depth and greater distance from land. Thus benthic animals are usually most abundant on continental margins and scarce on the deep-ocean floor far from land.

Competition for living space controls which organisms live where. Requirements for light control distributions of benthic plants. Many organisms must cope with widely changing conditions, such as variable temperatures, salinities, and dissolved oxygen concentrations.

Succession on uninhabited hard surface begins with the growth of bacteria, which form a slime that attracts other organisms and benthic larvae. Other organisms attach, and eventually a community of organisms forms. When a climax community develops, there is no further succession. This process is also known as fouling.

Rocky shores provide living space for attached organisms. But these organisms must cope with changeable living conditions. Kelp are large algae that grow on hard surfaces. They form thick forests that provide shelter for many other organisms. Sea urchins feed on the kelp. They in turn are eaten by sea otters.

Soft sediment deposits support many different kinds of organisms. Infauna live buried in sediment and usually live on organic matter in the sediments. Epifauna live on the sediment surface. Some organisms are selective feeders, selecting food particles to be ingested. Unselective feeders simply ingest sediment and digest the usable organic matter. Sediment particle size controls distribution of infauna. Finer-grained sediments usually contain more organic matter and are better suited for unselective feeders. Fine-grained deposits are easily eroded. Resulting high sediment concentrations are often unfavorable for filter-feeding organisms.

In wetlands, grasses growing on sediment surfaces stabilize the deposits. Many organisms can live in the deposits among the plants. Their position in the marshes depends on their requirements for submergence and tolerance for exposure at low tides.

Reefs are wave-resistant structures built by organisms such as oysters or coral-algal communities. Oysters bind together, forming large structures in which other types of organisms live in the interstices between shells. Oysters thrive in low-salinity waters, which kills their predators (boring snails, starfish).

Coral reefs grow in warm ($>18°C$), clear, tropical or subtropical waters of normal salinity. Corals are relatively fast growing because their tissues contain symbiotic zooxanthellae, photosynthetic organisms that live in the coral tissues and supply food to the corals. This association also helps the corals secrete their carbonate skeletons. Encrusting calcareous algae bind the coral skeletons together, forming the reef structure. The cavernous reef structure shelters many other kinds of organisms, and their skeletons contribute to the reef mass.

Dense growths of specialized benthic organisms grow near the hydrothermal vents on midocean ridges. These include large gutless worms that depend on bacteria living in special organs to provide their food. This is an example of symbiosis. Other organisms are resistant to the hydrogen sulfide given off by the vents, which kills most organisms.

Competition for space is a dominant control on the abundance and distribution of benthic organisms. Growths of abundant soft-bodied algae cover exposed reef surfaces.

Soft sediments cover most of the deep-ocean bottom. Sediment-feeding organisms dominate. Due to scarcity of food, these organisms are usually smaller and slower growing than their shallow-water relatives. The deep-ocean floor has low biomass. Many deep-ocean organisms are opportunistic feeders. They must quickly find and feed on the falls of large organisms from near-surface waters. Often, deep-ocean benthic animals can reproduce only after they have had one of their rare meals.

STUDY QUESTIONS

1. Contrast conditions for life in benthic and pelagic environments.

2. Describe the major lifestyles for benthic organisms. Give an example of an organism employing each lifestyle.

3. Contrast epifauna and infauna. Give an example of each.

4. Describe the role of marsh grasses in wetlands.

5. Discuss how the deep-ocean benthic environment differs from the inshore continental shelf environment.

6. Describe some adaptations of deep-ocean benthic organisms to their environment.

7. Describe the unusual features of organisms living around deep-ocean hydrothermal vents.

8. Discuss the different techniques involved in culturing fish and in culturing benthic organisms.

9. Describe some of the changes that waste discharge cause in coastal waters.

10. What are the principal effects of oil spills on coastal regions.

SELECTED REFERENCES

BOADEN, P. J. S., AND R. SEEDS. 1985. *An Introduction to Coastal Ecology*. London: Blackie. 218 p.

JOHNSON, M. E., AND H. J. SNOOK. 1967. *Seashore Animals of the Pacific Coast*. New York: Dover. 659 pp. Nontechnical.

KAPLAN, E. H. 1982. *A Field Guide to the Coral Reefs of the Caribbean and Florida*. Boston: Houghton-Mifflin. 289 pp.

LEVINTON, J. S. 1982. *Marine Ecology*. Englewood Cliffs, N.J.: Prentice-Hall. 526 pp. Intermediate.

MANN, K. H. 1982. *Ecology of Coastal Waters*. Berkeley, CA.: University of California Press. 322 pp. Intermediate to advanced.

STODDART, D. R., AND M. YONGE. 1978. *The Northern Great Barrier Reef*. London: Royal Society. 366 pp.

THORSON, G. 1971. *Life in the Sea*. New York: World University Library, McGraw-Hill. Elementary, emphasizing marine environments.

A coccolithophorid is a microscopic plant that forms calcareous plates. The plant lives in the sunlit surface waters. When it dies, the plates disaggregate and sink to the bottom, forming carbonate-rich sediments. (Photograph courtesy WHOI, Photographer S. Honjo.)

CHAPTER 15

Sediments

OBJECTIVES _____

1. To understand how sediment deposits form and how they relate to other ocean processes;

2. To describe methods of deciphering ocean history recorded in sediment deposits;

3. To understand the history of ocean basins and currents.

*T*he ocean receives the debris washed or blown off the land. Rock fragments from the land mix with shells and bones of marine organisms as they sink to the ocean bottom. There the particles accumulate, forming sediment deposits. These deposits record the history of ocean basins and the life in their waters. In addition, sediment deposits contain oil and gas as well as valuable deposits of various minerals.

In this chapter we examine:

Origins and characteristics of sediment particles and deposits;
Processes controlling distributions of deep-ocean sediment deposits;
Sediment transport in the ocean and atmosphere;
Sediment deposits on continental margins; and
Deciphering ocean history.

SEDIMENT PARTICLES

Sediment deposits are accumulations of minerals and rock fragments from the land mixed with insoluble shells and bones of marine organisms and some particles formed through chemical processes occurring in seawater. Much of the information about Earth history comes from study of such deposits. Thus we need to know more about where and how sediment deposits form.

Particles in sediment deposits come from three primary sources:

Biogenous particles come from shells, bones, and teeth of marine organisms (Fig. 15-1). If a deposit has more than 30% (by volume) biogenous particles, it is called a **biogenous sediment** or an *ooze* (named by the *Challenger* Expedition scientists because of its fine-grained nature.)

Lithogenous particles are rock fragments and mineral grains from decomposition (weathering) of rocks on land and from volcanic eruptions (Fig. 15-2).

Hydrogenous particles form in seawater through chemical reactions. Manganese nodules (Fig. 15-3), also called iron-manganese nodules, are examples of hydrogenous sediments. In the central Pacific, metal-rich coatings on rock surfaces of extinct volcanoes are another type of hydrogenous deposit.

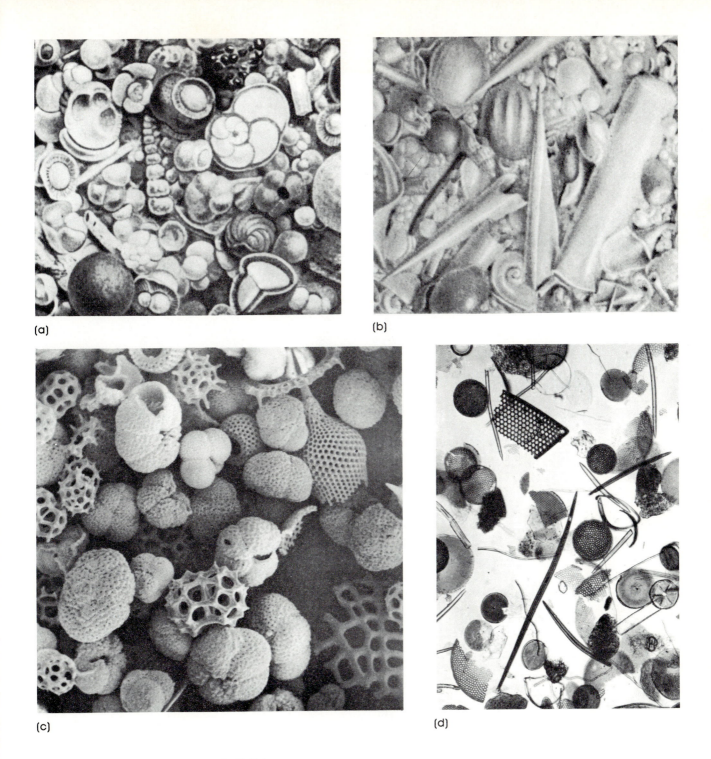

(a)

(b)

(c)

(d)

FIGURE 15-1
Constituents of biogenous sediments. (a) Foraminiferan (*Globigerina*) and (b) pteropod shells in deep-ocean sediments collected by the *Challenger* Expedition. (c) Skeletons of foraminifera and radiolaria (shells with large latticelike opening) from sediment in the western Pacific Ocean; (d) shells of diatoms. These particles typically have diameters of about 100 micrometers, the thickness of a human hair.

Sediment particles are also classified according to their grain size. (The names given to different sized particles are shown in Fig. 15-4.) Particle size is important because it determines how grains are transported and where they accumulate in the ocean. Various sources produce particles of different sizes, as indicated in Fig. 15-4. Sediments with a wide range of particle sizes are called *poorly sorted*. On the other hand, beaches usually consist of *well-sorted* sand because wave action has removed most of the fine particles. Deep-ocean deposits are usually poorly sorted mixtures, dominated by very fine grained particles. These we call *deep-sea muds*.

FIGURE 15-2
Ash cloud caused by an eruption of La Soufrière on April 17, 1979, on the island of St. Vincent in the West Indies. Such volcanic eruptions inject ash into the atmosphere. Winds transport ash for long distances, and much ash is deposited in the deep ocean.

FIGURE 15-3
Manganese nodules on the floor of the South Pacific. These nodules are fist-sized, 8 to 10 centimeters across. (Courtesy IDOE, National Science Foundation.)

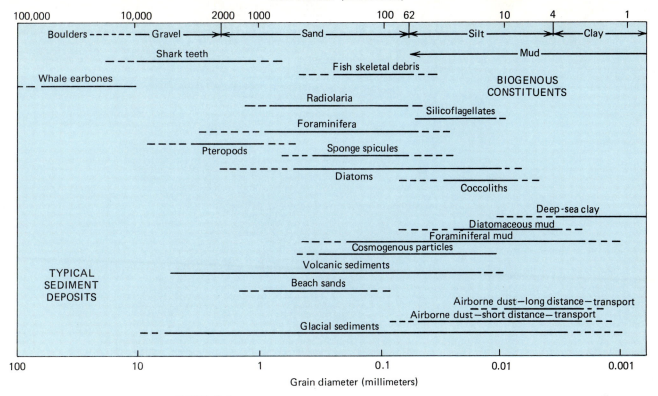

FIGURE 15-4
Grain-size distributions in common marine sediment and in various sediment sources.

Before we discuss sediment deposits, let us consider how sedimentary particles form.

BIOGENOUS SEDIMENTS

Biogenous sediments cover more than half the deep-ocean bottom (Fig. 15-5). Three processes control this distribution: *biological production* in overlying surface waters, *dilution* by other kinds of particles, and *destruction* of the shells, either while sinking through the water or on the sea floor. (Remember that biogenous sediments contain more than 30% shells, by volume.)

Siliceous sediments accumulate under highly productive waters. In the equatorial Pacific and Indian oceans, radiolarian-rich sediments accumulate, up to 1 kilometer thick. Diatomaceous sediments dominate the subpolar North Pacific and around Antarctica and under the major coastal-upwelling areas (Fig. 15-5).

Dilution of biogenous particles by mixing with materials from other sources is most obvious in the North Atlantic. Note in Fig. 15-5 that terrigenous sediments mask diatomaceous sediments in the subpolar North Atlantic. Also, rock fragments and fine-grained rock flour dominate sediment deposits near Antarctica.

Where a deposit is less than 30% by volume biogenous particles, it is called a *deep-sea mud*. In the Atlantic, iron-stained particles are common, forming *red clays,* which dominate the deep-ocean floor downwind from the North African deserts.

DESTRUCTION OF BIOGENOUS PARTICLES

Siliceous shells and phosphatic bones and teeth dissolve everywhere in the ocean. Thus deposits of such materials are common only below highly productive regions, such as upwelling areas. Otherwise, only

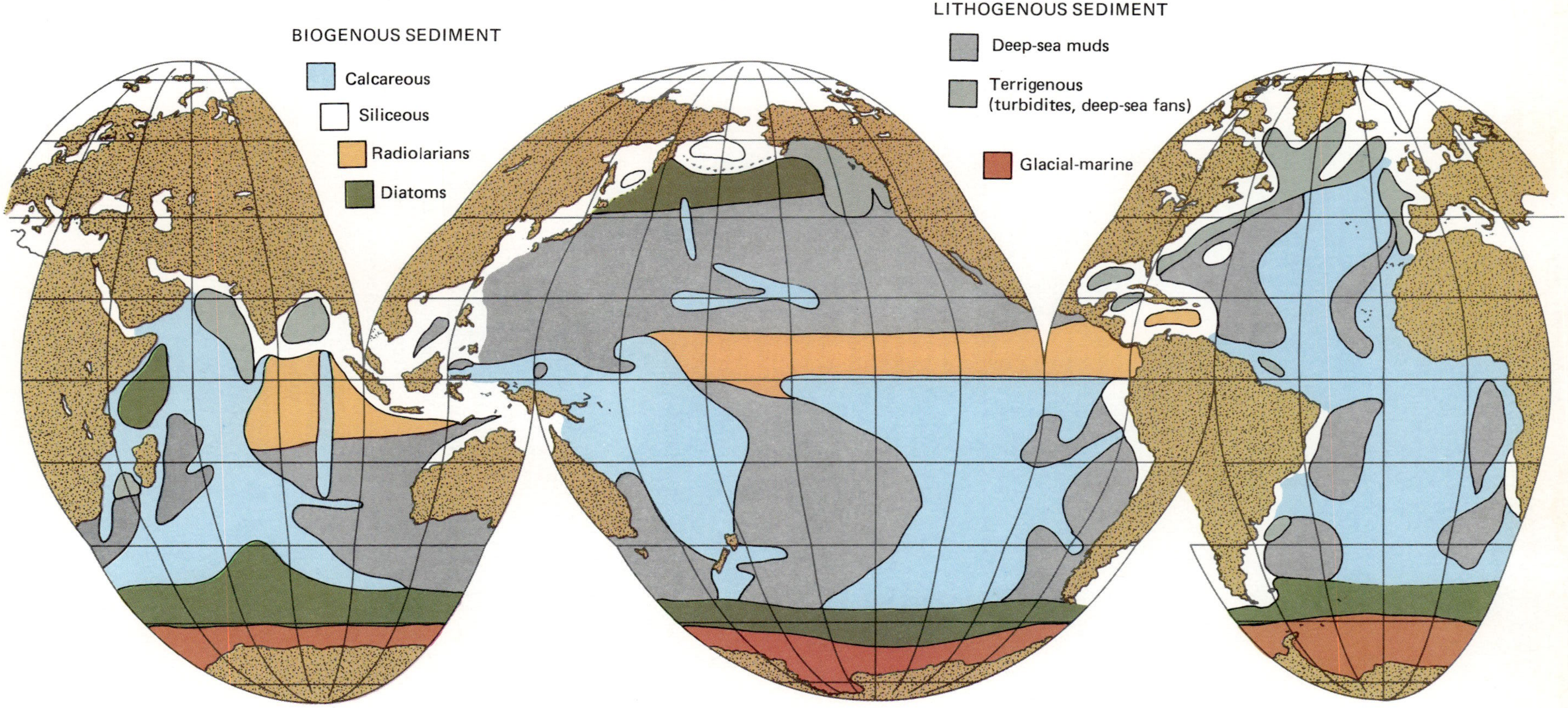

FIGURE 15-5

Distributions of deep-sea sediment deposits. Continental shelf and slope sediments are not shown.

BIOGENOUS SEDIMENT

Calcareous

Siliceous

Radiolarians

Diatoms

LITHOGENOUS SEDIMENT

Deep-sea muds

Terrigenous (turbidites, deep-sea fans)

Glacial-marine

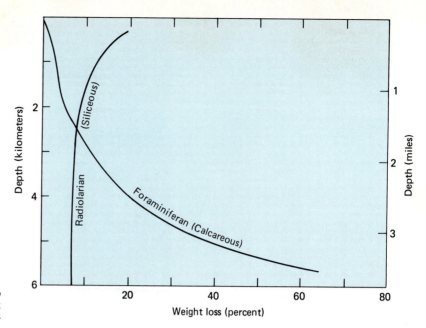

FIGURE 15-6
Loss of weight of particles due to dissolution at various depths in the central Pacific after 4 months of submergence.

fish-bone fragments and scales or the most robust siliceous shells occur in marine deposits. Resistant particles such as fish teeth or whale ear-bones can persist for a long time on the ocean floor. These resistant objects are often the nuclei of manganese nodules.

Destruction of carbonate particles by dissolution is more complex. Calcium carbonate minerals dissolve below the surface zone throughout the ocean. They dissolve more rapidly at low temperatures and high pressures in the deep ocean. Warm surface waters are supersaturated with calcium carbonate and therefore cannot dissolve carbonate shells. The *carbonate saturation level* (depth at which carbonate particles dissolve) occurs at about 4000 meters (Fig. 15-6). There rates of weight loss for foraminiferal shells increase markedly. Below the carbonate saturation level, carbonate particles dissolve and do not form calcareous sediments, except below highly productive waters.

Some organisms, such as *pteropods* (pelagic snails), make their shells from an especially soluble carbonate mineral. Hence, pteropod shells dissolve rapidly and form deposits only on shallow volcanic peaks in the Atlantic (Fig. 15-7).

FIGURE 15-7
Calcareous deposits are common on the shallow portions of seamounts and ridges. These deposits often contain easily destroyed constituents, such as pteropod shells.

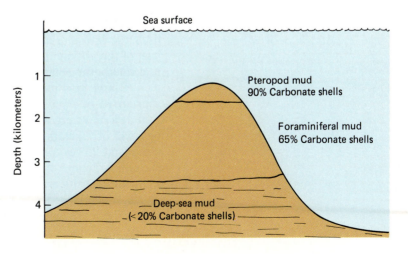

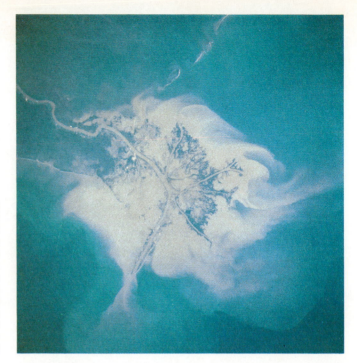

FIGURE 15-8
The delta of the Mississippi River during a flood in early April 1984. The turbid, sediment-laden waters from the river are discharged from all the river mouths. Note also that the areas between the river channels are flooded by turbid waters. Sediment in the waters caught between the distributaries are trapped by vegetation and are incorporated into sediment deposits. Turbid waters discharged through the river mouths are moved by currents; suspended sediments are deposited on the continental shelf. (Courtesy NASA.)

LITHOGENOUS SEDIMENTS

Lithogenous particles come from silicate rocks, primarily from the land. Most rocks form at high temperatures and pressures with no oxygen and with little free water. These rocks are broken down at the earth's surface by chemical and physical processes, known collectively as **weathering.** During weathering, some constituents are dissolved and carried in solution by river waters to the ocean. The remaining rocks are broken up by physical processes, such as freezing, into small particles. These are carried by running waters and by winds. Some minerals, quartz for example, resist chemical alteration and enter the ocean essentially unaltered. The slightly altered remains of mica minerals form clays, common in deep-sea muds.

Rivers transport most of the lithogenous sediment to the ocean—about 13 billion tons each year (Fig. 15-8). Most of it comes from Asia, where four rivers supply about one-quarter of the total sediment discharge from all the continents. Most large sediment-transporting rivers discharge into marginal seas. These seas, along with estuaries, trap most of the sediment load near the mouth of the river. Thus relatively little riverborne sediment directly enters the major ocean basins.

A large fraction of the sediment load coming to the ocean is transported by winds. Most of the wind-transported sediment comes from deserts where rainfall is inadequate to support an erosion-resisting cover of vegetation. Mountain building also results in rapid erosion of the land. Long-continued agriculture also causes high sediment discharges. Around Antarctica, the effects of glaciers are apparent. There sediment deposits are dominated by rock fragments released by melting ice. Even large rocks are rafted out and then released as the ice melts. These are called **glacial-marine** deposits.

FIGURE 15-9
Growth rings of a manganese nodule are seen in this cross section. The rings are usually formed by the remains of organisms that lived on the nodule surface as it formed.

HYDROGENOUS SEDIMENTS

Chemical and biological processes form **hydrogenous particles** or **coatings.** Potato-sized nodules of iron and manganese (Fig. 15-9) are conspicuous examples. There are also coatings on rock surfaces on long-extinct volcanoes. The processes forming these nodules and coatings are very slow.

Under the central gyres, far from continents, sediment particles accumulate very slowly. There nodules form around objects, such as volcanic rock fragments or resistant ear bones and teeth. Over millions of years, iron-manganese coatings collect on such objects, forming a nodule with roughly concentric rings (Fig. 15-9). Nodules and coatings contain metals such as copper, cobalt, and nickel, which make them potential sources for these metals.

Slow-growing manganese nodules are buried by rapidly accumulating sediments, so nodules are rare in the Atlantic. In the central Pacific, where sediments accumulate slowly, nodules cover an estimated 20 to 50% of the ocean bottom. Tunneling and churning by sediment-feeding organisms helps keep nodules at the surface, so they continue to grow.

COSMOGENOUS SEDIMENTS

Particles from space also occur in oceanic sediments and in glacial ice in Antarctica. An estimated 30,000 metric tons per year of dust falls into the ocean, mostly derived from meteorites that burn up in the atmosphere.

Cosmic spherules were first recognized in marine sediments by John Murray early in this century when he was studying the sediment samples collected by the *Challenger* Expedition. These particles are about 200 to 300 micrometers in diameter. Most are magnetic and consist primarily of iron or iron-rich minerals. Many have distinctive surface features that originate in their having melted, either during their formation or in passing through the earth's atmosphere. Others are composed of silicate minerals, which makes them harder to recognize and differentiate from other types of lithogenous grains.

SEDIMENT TRANSPORT

Particle transport in the ocean is controlled primarily by physical processes. Particle sizes and current speeds primarily control sediment transport. Sinking of individual particles is easiest to visualize. A particle's settling speed is controlled primarily by its size; large particles sink more rapidly than small ones. Particle settling times are given in the following table:

PARTICLE DIAMETER (micrometers)	SETTLING VELOCITY (centimeters per second)	TIME TO SETTLE 4 KILOMETERS
Sand (100 μm)	2.5	1.8 days
Silt (10 μm)	0.025	185 days
Clay (1 μm)	0.00025	50 years

Large grains settle out near the place where they enter the ocean. Smaller grains can be transported great distances during the time it takes a particle to settle 4 kilometers. Clay-sized particles can theoretically be transported throughout the ocean during the 50 years an individual particle takes to sink to the deep-ocean bottom. Settling times can also be increased by upward movements of water during upwelling and turbulence. Silts and clays are also transported hundreds or thousands of kilometers by winds. (More about this later.)

Organisms, however, remove particles as they feed by filtering water. Inedible particles removed along with the food are compacted into pellets and then excreted. These **fecal pellets** sink rapidly (within a few days) to the bottom. In nearshore waters, oysters, mussels, clams, and many other animals remove particles from suspension and bind them into pellets that accumulate nearby.

Most riverborne sediment particles are trapped by estuaries. For example, U.S. Atlantic coast rivers transport about 20 million metric tons of sediment each year to their estuaries, but virtually none reaches the outer continental shelf or slope. Thus the continental margin of eastern North America has almost no modern sediment deposits.

A large sediment load can quickly fill an estuary. Then sediments can be carried out into the coastal ocean. The Mississippi River is an example. Its large sediment load has filled the former estuary and is now building the Mississippi Delta far out into the Gulf of Mexico and onto the deep floor of the Gulf. Levees on the river to prevent flooding have contributed to the seaward transport of riverborne sediment.

ATMOSPHERIC TRANSPORT

About 100 million metric tons per year of sediment particles are carried to the ocean by winds, primarily from deserts and high mountains. Wind-transported sediment is especially important in open-ocean areas in the midlatitudes of the Northern Hemisphere, where there is a lot of land. It is less important in the Southern Hemisphere, where the ocean dominates. Volcanic eruptions also contribute immense quantities of ash to the ocean each year.

Particles smaller than 20 micrometers in diameter are carried great distances by winds. Volcanic rock fragments smaller than about 10 micrometers may be carried around the world if the ash is injected into the stratosphere.

Most particles eroded by winds remain in the lower atmosphere, however. This is particularly noticeable off Africa, where sand from the Sahara Desert is carried hundreds of kilometers to sea (Fig. 15-10). Rust-coated grains from North African deserts are common in Atlantic sediment deposits, which have a characteristic reddish color. Such deposits are called **red clays.**

Large volcanic eruptions produce very large quantities of ash. The 1815 eruption of the Indonesian volcano *Tambora* released 80 cubic kilometers of ash. This left a halo of volcanic ash deposits that are still visible on the deep-sea floor near the volcano. On August 26, 1883, *Krakatoa,* a volcano between Java and Sumatra in Indonesia, erupted and discharged about 16 cubic kilometers of ash into the atmosphere. Nearly 4 million square kilometers surrounding the volcano were covered by the ash. The 1980 eruptions of Mount St. Helens in

FIGURE 15-10
Dust from the Sahara Desert in North Africa is blown out over the North Atlantic, sometimes reaching all the way to Florida. (Photograph courtesy NASA.)

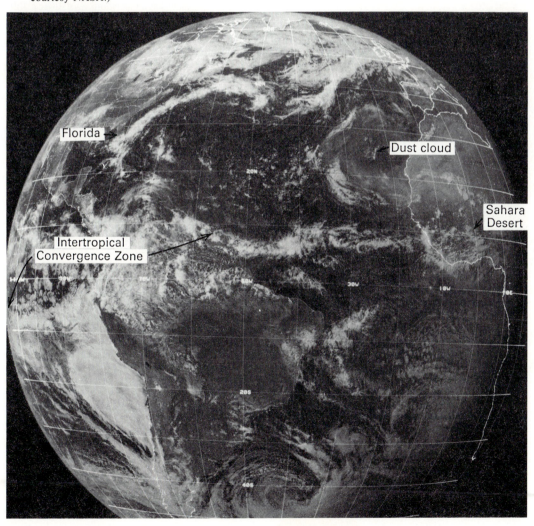

Washington were small by comparison, injecting only about 1 cubic kilometer of ash into the atmosphere.

Because of the enormous amount of energy released during volcanic explosions, the ash can be injected into the stratosphere, where it is not readily removed. Thus the ash can be transported around the world, having distinct climatic consequences. The haze from the Tambora eruption caused unusually cool summers in 1816 and 1817. Crops failed all over Europe and much of China. The location of the eruption near the equator made the dust cloud more effective, since it reduced the heat input to the earth.

DEEP-SEA SEDIMENTS

Much of the deep-ocean floor is covered by deposits called **pelagic sediments,** which accumulate slowly, particle by particle. Typical accumulation rates are between 0.1 and 1 centimeter per thousand years. These deposits blanket the bottom, preserving the original outlines much like a newly fallen snow.

Since deep-ocean sediments accumulate slowly (Table 15-1), particles may spend years suspended in seawater or exposed to the overlying waters while being slowly buried on the bottom. As a result, there is ample time for the particles to react chemically with seawater. For instance, dissolved oxygen in deep-ocean waters reacts with iron on the particles, forming an iron-rust coating. Also, exposure to oxygenated waters and consumption by benthic organisms result in very low carbon contents in deep-ocean sediments. (When we discuss the deposits formed during times of low dissolved oxygen in the bottom waters, we will see the consequences of the abundance of dissolved oxygen.)

The thinnest deposits generally occur on the young crust of midocean ridges. Strong currents there also tend to keep ridge crests free of sediment deposits, which usually accumulate only in protected pockets on midocean ridges. Sediment deposits thicken away from the ridges and are usually thickest over the oldest crust near the edges of the ocean basins. And as we have already noted, sediment deposits are often thick in the marginal ocean basins near the mouths of major sediment-transporting rivers.

TABLE 15-1
Typical Sediment Accumulation Rates

AREA	AVERAGE ACCUMULATION RATE (centimeters per thousand years)
Deep ocean	
Deep-sea muds	0.1
Coccolith muds	1
Marginal ocean basins	10–100
Continental margins	
Continental shelf	30
Continental slope	20
Estuary, fjord	400
Delta (Fraser River, Canada)	700,000
Wetlands	150

TURBIDITY CURRENTS

Turbidity currents are dense mixtures of water and sediment. Because of their density, they move along the bottom, transporting sediment [Fig. 15-11(a)] onto the deep-ocean floor. A turbidity current is best visualized as a huge avalanche of very muddy water.

Earthquakes can cause turbidity currents. In November 1929, the Grand Banks earthquake, near Newfoundland, triggered a turbidity current that hurtled down the continental slope breaking submarine telegraph cables in its path. The flow moved at speeds of up to 100 kilometers per hour. Sediments deposited by the currents covered a large part of the nearby North Atlantic ocean bottom. Sudden large discharges of riverborne sediment (during floods) can also trigger turbidity currents.

When a turbidity current first forms, it moves at high speeds and carries sand-sized particles, eroding the bottom as it passes. Later, as the flow moves more slowly, it transports primarily finer particles, which settle out on top of the earlier-deposited coarser-grained de-

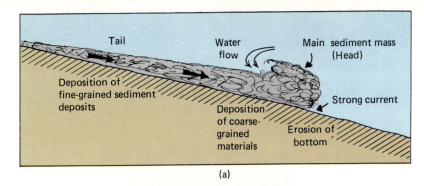

(a)

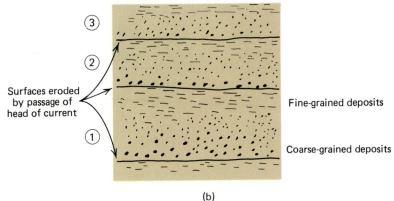

Surfaces eroded
by passage of
head of current

Fine-grained deposits

Coarse-grained deposits

(b)

FIGURE 15-11
(a) Schematic representation of a turbidity current moving down a slope. Note that the current first erodes as it passes. As current speeds decrease when the main sediment mass passes, coarse-grained sediments are deposited first; fine-grained sediments are deposited later. (b) Deposits formed by turbidity currents.

posits. The result is a deposit that forms rapidly and shows distinctive gradations in grain size—coarsest at the bottom and finest at the top. This **graded bedding** [Fig. 15-11(b)] is characteristic of **turbidites** (turbidity current deposits). Turbidity currents carry shallow-water organisms and plant fragments onto the deep-ocean floor. Plant materials—some of them still green—have been recovered from the ocean bottom during repairs of cables broken by turbidity currents.

Turbidity currents are most common near narrow continental shelves. They are least common off wide shelves. In the Atlantic, turbidity-current deposits near the margins have buried the older sea-floor topography. Now turbidity currents can flow all the way to the Mid-Atlantic Ridge. Similarly, sediment brought into the northern Indian Ocean can flow unimpeded over great distances of deep-ocean bottom. In the Pacific, on the other hand, trenches and ridges around the basin trap turbidity currents before they reach the deep-ocean floor.

Besides carrying sediment and burying ocean-bottom topography, turbidity currents also erode submarine canyons that indent continental margins. In many areas, turbidity currents remove sediment from canyons, keeping them open. Evidence for this comes from frequent breaks in submarine cables off the mouths of large sediment-transporting rivers, such as the Congo in Africa or the Ganges-Brahmaputra in Asia.

CONTINENTAL-MARGIN DEPOSITS

Locations of particle sources and the transport processes involved control sediment distributions around basin margins and in shallow waters. In general, river-transported sediment is restricted to continental margins. Most riverborne sediment remains on continental shelves

BOX 15-1

Mineral Resources

Sand, gravel, and shell are mined in large quantities from the shallow ocean bottom. Such materials are used to pave roads, construct buildings, and fill low-lying areas. Sand is used to restore beaches damaged by storms and rising sea level.

Heavy minerals—gold, tin, chromium, and titanium—are also recovered from the ocean bottom. These minerals are much denser than normal sand. They are concentrated and left behind when other grains are removed by river flows or waves. Channels formed by rivers that cut across the continental shelf when sea level was lower sometimes contain heavy minerals, especially tin near Indonesia and Malaysia. Other deposits formed on beaches, when waves and longshore currents separated heavier particles from the lighter sand grains, concentrating the heavy minerals.

Phosphorite nodules, containing phosphorus (used in fertilizers) occur on continental shelves, when they are easily dredged. Phosphate content of the nodules tends to be relatively low, but many billions of tons are available.

Manganese nodules are potentially valuable for their copper, nickel, and cobalt contents. Potentially commercial nodule deposits occur in regions far from major sediment sources. For instance, the most interesting region for nodule production is in the central Pacific, south of the Hawaiian Islands.

Metal-rich muds occur on the bottom of the Red Sea. These muds contain 2% zinc and 0.7% copper, rich enough to be potentially exploitable. Other metal-sulfide deposits containing high concentrations of zinc, copper, and iron have been recovered from active hydrothermal vents in the Atlantic and Pacific oceans. Similar ancient marine deposits are mined for copper on the island of Cyprus in the Mediterranean.

---

or rises except where turbidity currents can carry materials into the deep ocean. Thus most ocean margins have thick sediment deposits.

Deposits that formed under conditions no longer existing are known as **relict sediments.** Such sediments have distinctive features, such as remains of organisms that no longer live there. For example, oyster shells are found on the U.S. continental shelf, a relict of lower sea levels. Other characteristics of relict sediments include iron stains or coatings on grains that could not have been formed under marine conditions. About 70% of the world's continental shelves are covered by relict sediments (Fig. 15-12). Many of these are unburied landscapes, now submerged. This is a consequence of the relatively recent rise of sea level.

Where rivers bring sediment to the coastal ocean, relict deposits and surfaces are buried by modern sediments. Alternatively, relict materials are reworked by waves as the shoreline moves across the area when sea level rises and falls. In the process, the relict deposits are destroyed.

Sand deposited in nearshore waters—near a river mouth, for example—is moved by longshore currents. The close association of sand beaches and river mouths is especially obvious on the U.S. Pacific coast. There the large area of beaches and dunes on the Washington-Oregon coast occur near the Columbia River mouth.

Most sediments deposited on continental shelves accumulate so rapidly that particles have too little time to react chemically with seawater. Coastal-ocean sediments therefore retain many characteristics acquired during weathering. These rapidly accumulating deposits also contain organic matter, causing them to be rather dark colored—gray, greenish, or sometimes brownish.

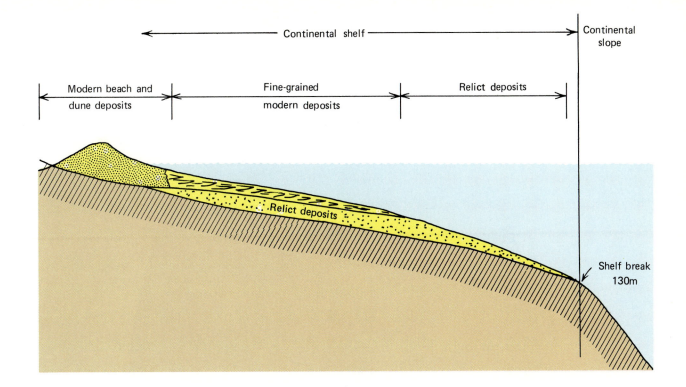

Continental shelf

Continental slope

Modern beach and dune deposits

Fine-grained modern deposits

Relict deposits

Relict deposits

Shelf break 130m

FIGURE 15-12
Continental shelves in midlatitudes frequently have an active sand beach and dune complex along the shore that merges with deposits of finer-grained modern sediments at middepths on the shelf. At the outer edge of the shelf are much older deposits.

In polar regions, **glacial-marine sediments** are common. These deposits contain particles of all sizes—from boulders to silt (Fig. 15-4)—derived from melting icebergs. Sea ice normally contains sediment only if it has gone aground and incorporated material by freezing to the bottom. This happens often in the Arctic Sea. When such sea ice melts, any sediment load is deposited on the bottom.

Sediment deposits directly derived from continents cover about 25% of the ocean bottom. Accumulations are thickest in marginal ocean basins. Although accounting for only about 2% of the ocean area, these basins contain about one-sixth of all oceanic sediment and are major sources of oil and gas.

General relationships between ocean processes and sediment distributions are summarized in Fig. 15-13. Effects of high productivity can be seen on the deep-ocean bottom and on continental shelves. On the deep-ocean bottom, bands of diatomaceous muds in high latitudes and radiolarian muds in equatorial regions directly reflect the high biological productivity of surface waters in these regions. On the continental shelf, the abundance of recent carbonate sediment in tropical waters is again a result of the locally high productivity.

DECIPHERING OCEAN HISTORY

The history of the ocean basins and of the organisms that lived in them is preserved in sediment deposits. Oceanographers use these deposits much as archaeologists use broken pottery and refuse piles to study ancient civilizations. Sediments are sampled by various techniques, some as simple as lowering a weighted pipe into the bottom to core the sediments, like coring an apple. The sediment cores are then sampled for detailed study (Fig. 15-14).

Samples are analyzed by many different techniques. Relative

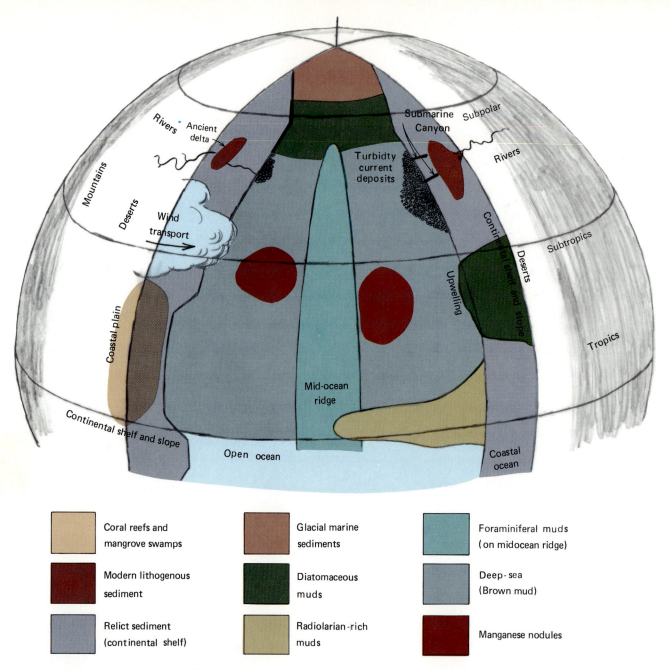

Coral reefs and mangrove swamps

Glacial marine sediments

Foraminiferal muds (on midocean ridge)

Modern lithogenous sediment

Diatomaceous muds

Deep-sea (Brown mud)

Relict sediment (continental shelf)

Radiolarian-rich muds

Manganese nodules

FIGURE 15-13
Schematic representation of the general distribution of sediment in the deep and coastal ocean.

ages of layers are determined by using the fossils in them (Fig. 15-15). The presence (or absence) of certain fossils can be interpreted to determine when the layer was deposited. This technique gives *relative ages*. That is, we can determine which layer is oldest and which is youngest. We can also correlate with other deposits that contain the same sequences of fossils. We cannot, however, determine absolute ages. In other words, we cannot say a particular deposit is so many million years old, using fossils alone.

Radioactive substances in particles are used as clocks to determine *absolute ages*. Each **radionuclide**—radioactive isotope of an element—decays at a fixed and known rate. The time elapsed since that

FIGURE 15-14
Scientists sample a sediment core for further study. (Courtesy David Ross, Woods Hole Oceanographic Institution, Photo Researchers.)

particle formed is indicated by the amount of the decay product in it compared with the amount of the original radionuclide present. The type of radionuclide used in dating depends on the expected age of the sediment.

For very young deposits (less than 30,000 years), radioactive carbon-14 is used. It is formed in the upper atmosphere when cosmic rays bombard the Earth. All living organisms contain carbon-14 mixed in the carbon of their tissues. When an organism dies, no more carbon-14 is taken up. Instead, the amount of carbon-14 begins to decrease at a steady rate through decay. One-half is gone in 5600 years (its **half-life**). After 11,200 years, only one-quarter of the original amount remains. Thus by comparing the amount of carbon-14 per unit of carbon in a fossil with comparable living organisms, it is possible to determine when that organism died. Assuming that the organism was incorporated into the deposit when it formed, we can find its absolute age.

Dating older minerals requires using radionuclides with longer half-lives. Most rock-forming minerals contain some potassium, including the radioactive potassium-40, with a half-life of 1.3 billion years. This is useful for deposits or rocks with ages of hundreds of millions of years. Uranium and thorium are also used for dating deposits of a few million years.

Other evidence is used to date parts of the ocean floor. We have discussed (in Chapter 3) how magnetic reversals can be used to determine when a part of the ocean floor formed. A comparable record of magnetic reversals is contained in cores, permitting them to be dated as well.

OXYGEN-DEFICIENT DEEP OCEAN

About 100 million years ago, the deep ocean was significantly different from present-day conditions. The Atlantic was much narrower, and the deep basins were more isolated. At that time, bottom waters in these basins were often devoid of dissolved oxygen, a condition called **anoxia.** With little or no dissolved oxygen, carbon-rich deposits formed in the North Atlantic and Arctic basins.

FIGURE 15-15
Microfossils (coccoliths and discoasters) from sediments are used to determine the relative ages of the deposits.

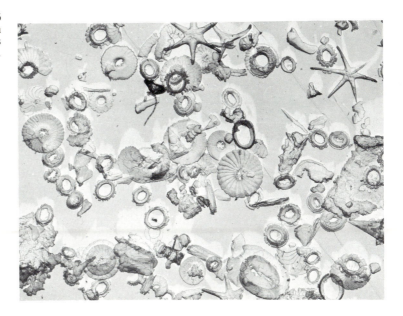

BOX 15-2

Oil and Gas

Oil and natural gas—the most valuable resources taken from the ocean—are the remains of organic matter buried in sediments and later altered by heat. In areas where productivity is exceptionally high and bottom-water circulation sluggish, dissolved oxygen is used up by decomposing marine plants. Growth of plant-eating organisms is thereby inhibited, permitting organic matter to accumulate in the sediment deposits. Higher-than-normal heat flow transforms organic matter to oil at temperatures of 100° to 150°C and at depths of around 5 kilometers. But if the heating continues too long or if temperatures are too high, all the organic matter may be transformed to natural gas or even destroyed.

Through time, sediments compact under the weight of younger overlying deposits. Fluids (water, oil, and gas) are expelled and move from the source sediments into nearby porous rocks, where they accumulate. Ancient reefs and buried beaches are especially favorable sites for accumulation of oil. Eventually these movements are stopped by impervious sediments through which fluids cannot readily move. Thus three factors are necessary for oil and gas deposits to form:

1. Deposition of sediments rich in organic matter (source beds);

2. Heating of source beds to form oil or gas from the organic matter; and

3. Permeable, porous (reservoir) rocks to hold oil and gas where it can be extracted.

Estimates of undiscovered petroleum resources are shown in Fig. B15-2-1. About 70% of the undiscovered oil and gas is expected to come from continental shelves and shallow marginal ocean basins. About 24% is expected to come from the continental slope, which cannot be exploited with present production techniques. Relatively little is expected to come from the deep-ocean floor, where little organic matter is now preserved.

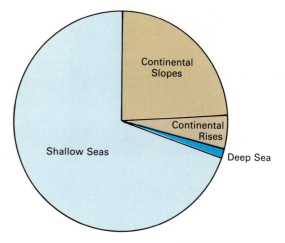

FIGURE B15-2-1
Estimated undiscovered total offshore petroleum resources (oil plus equivalent amount of natural gas). (From the National Academy of Sciences, 1975. *Mineral Resources and the Environment.* Washington, D.C., 348 pp.)

The primary requirement for anoxia to occur is a highly stable water column. Very dense bottom waters (resulting from either high salinities or low temperatures) promote stability. At that time, the earth's climate was much warmer than at present, so there was little probability of low bottom-water temperatures.

Partially isolated basins in the midlatitudes provided warm, saline bottom waters, which contained little dissolved oxygen (because of their high temperatures). Thus organic matter was protected against decomposition by bacteria or utilization by benthic organisms. The resulting carbon-rich deposits are the source beds for many of the world's largest oil and gas fields, such as the Prudhoe Bay field in Alaska.

Widening of the Atlantic basin and worldwide climatic cooling brought these conditions to an end. Now, anoxia occurs in only a few isolated basins, including estuaries. The present glacial climate leads to vigorous circulation of the deep ocean, preventing any accumulation of carbon-rich deposits.

Lithogenous sediments (derived from silicate rocks) and biogenous sediments (from skeletons of plants and animals) cover most of the ocean bottom. Sediment deposits are absent near the crest of midocean ridges, on newly formed crust, and where currents are too strong to permit sediment accumulation.

Biogenous sediment deposits cover more than half the ocean floor. They accumulate most rapidly in areas of high biological productivity, especially beneath upwelling areas, such as the equator. Destruction of soluble or fragile remains leaves only sturdy forms at great depth. Destruction of shells is especially important for carbonate deposits, which occur primarily in shallow ocean areas. At the greatest depths in the oceans, most of the biogenous particles have been dissolved, leaving only lithogenous materials.

Lithogenous sediment deposits are thickest near continents. They are also carried by winds and deposited in open-ocean areas. Rivers transport about 13 billion metric tons of sediment each year to the ocean. Much of the riverborne sediment is deposited in estuaries and on continental shelves near the river mouths. Most continental shelves are covered by deposits that formed under conditions no longer prevailing.

Turbidity currents—dense, sediment-laden waters—transport sediments out onto the deep-ocean floor. They form characteristic types of deposits, with the coarsest materials on the bottom and fine-grained materials on top. Many such cycles usually occur, each marking a single turbidity-current event. Turbidity currents are important transport mechanisms near major rivers, where they keep submarine canyons open.

Hydrogenous sediments (precipitated from seawater) occur in areas where they are no other major sediment sources. Manganese nodules are common in areas of slow sediment accumulation, such as the North Pacific.

Cosmogenous sediments (derived from space) also occur in marine sediments and in Antarctic ice. They are thought to be the remains of meteorites that survived passage through the atmosphere.

Ocean sediments contain a history of ocean-basin development, of changes in sea level and climate, and of life in the ocean. Ages of sediment layers are determined from the presence or absence of microfossils and from the amounts of selected radioactive constituents. Magnetic reversals recorded in cores also provide useful indications of sediment age.

STUDY QUESTIONS

1. What techniques are used to determine the ages of sediment deposits?

2. Describe the three major categories (by composition) of particles in oceanic sediment deposits. Where in the deep-ocean basins is each sediment type most likely to predominate? What is the primary source of each?

3. Describe where turbidity currents form, how they transport sediment, and the type of deposits they form. Where are turbidity currents most common?

4. Explain why calcareous biogenous deposits are rare or absent on the deepest parts of the ocean floor.

5. Define relict sediments. Explain why relict sediments are common on many continental shelves.

6. Draw a cross section of the ocean floor, showing the distribution of various types of sediments in relation to continental margins and midocean ridges.

7. Discuss how oil and gas deposits form.

8. List some of the minerals recovered from the ocean bottom.

9. What are the most valuable resources recovered from the ocean?

SELECTED REFERENCES

KENNETT, J. P. 1982. *Marine Geology.* Englewood Cliffs, N.J.: Prentice-Hall. 813 pp. Comprehensive treatment of ocean geology.

KENT, P. 1980. *Minerals from the Marine Environment.* London: Edward Arnold. 88 pp.

SIEVER, R. 1988. *Sand.* New York: Scientific American Library. 237 pp. Earth science from the perspective of a sand grain.

TUREKIAN, K. K. 1976. *Oceans,* 2d ed. Englewood Cliffs, N.J.: Prentice-Hall. 149 pp. Elementary.

Conversion Factors

EXPONENTIAL NOTATION

It is often necessary to use very large or very small numbers to describe the ocean or to make calculations about its processes. To simplify writing such numbers, scientists commonly indicate the number of zeros by **exponential notation,** indicating powers of ten, or the number of zeros. Some examples are given below with some common prefixes.

1,000,000,000 =	10^9	(one billion)	
1,000,000 =	10^6	(one million)	
1,000 =	10^3	(one thousand)	**kilo-**
100 =	10^2	(one hundred)	
10 =	10^1	(ten)	
1 =	10^0	(one)	
0.1 =	10^{-1}	(one tenth)	**deci-**
0.01 =	10^{-2}	(one hundredth)	**centi-**
0.001 =	10^{-3}	(one thousandth)	**milli-**
0.000 001 =	10^{-6}	(one millionth)	**micro-**
0.000 000 001 =	10^{-9}	(one billionth)	**nano-**

Multiplication: To multiply exponential numbers (powers of ten), add exponents. For example, $10 \times 100 = 1000$, which is written exponentially as $10^1 \times 10^2 = 10^3$

Division: To divide exponential numbers, subtract the exponent of the divisor from the exponent of the dividend. For example, $100/10 = 10$, or written exponentially as $10^2/10^1 = 10^1$

UNITS OF MEASURE

Length

1 **kilometer** (km) = 10^3 meters = 0.621 statute mile = 0.540 nautical mile

1 **meter** (m) = 10^2 centimeters = 39.4 inches = 3.28 feet = 1.09 yards = 0.547 fathom

1 **centimeter** (cm) = 10 millimeters = 0.394 inch = 10^4 micrometers

1 **micrometer** (μm) = 10^{-3} millimeter = 0.0000394 inch

Temperature

Conversion formulas

$$^\circ C = \frac{^\circ F - 32}{1.8}$$

$$^\circ F = (1.8 \times {}^\circ C) + 32$$

CONVERSION TABLE	
°C	°F
0	32
10	50
20	68
30	86
40	104
100	212

Area

1 **square centimeter** (cm^2) = 0.155 square inch
1 **square meter** (m^2) = 10.7 square feet
1 **square kilometer** (km^2) = 0.386 square statute mile = 0.292 square nautical mile

Volume

1 **cubic kilometer** (km^3) = 10^9 cubic meters = 10^{15} cubic centimeters = 0.24 cubic statute mile

1 **cubic meter** (m^3) = 10^6 cubic centimeters = 10^3 liters = 35.3 cubic feet = 264 U.S. gallons

1 **liter** (l) = 10^3 cubic centimeters = 1.06 quarts = 0.264 U.S. Gallon

1 **cubic centimeter** (cm^3) = 0.061 cubic inch

Mass

1 **metric ton** = 10^6 grams = 2205 pounds

1 **kilogram** (kg) = 10^3 grams = 2.205 pounds

1 **gram** (g) = 0.035 ounce

Time

1 **day** = 8.64×10^4 seconds (mean solar day)

1 **year** = 8765.8 hours = 3.156×10^7 seconds (mean solar year)

Speed

1 **knot** (nautical mile per hour) = 1.15 statute miles per hour = 0.51 meter per second

1 **meter per second** (m/s) = 2.24 statute miles per hour = 1.94 knots

1 **centimeter per second** (cm/s) = 1.97 feet per minute = 0.033 feet per second

Energy

1 **gram-calorie** (cal) = $\frac{1}{860}$ watt-hour = $\frac{1}{252}$ British thermal unit (Btu)

Useful Data About the Earth and Ocean

TABLE A2-1
Dimensions of the Earth

SIZE AND SHAPE OF THE EARTH		
DIMENSIONS	MILES	KILOMETERS
Equatorial radius	3963	6378
Polar radius	3950	6357
Average radius	3956	6371
Equatorial circumference	24,902	40,077

AREAS OF THE EARTH, LAND, AND OCEAN		
	MILLIONS OF	
PART OF EARTH	SQUARE MILES	SQUARE KILOMETERS
Land (29.22%)	57.5	149
Ice sheets and glaciers	6	15.6
Oceans and seas (70.78%)	139.4	361
Land plus continental shelf	68.5	177.4
Oceans and seas minus continental shelf	128.4	332.6
Total area of the Earth	196.9	510.0

DISTRIBUTION OF LAND AND WATER ON THE EARTH'S SURFACE*		
HEMISPHERE	LAND (percent)	OCEAN (percent)
Northern	39.3	60.7
Southern	19.1	80.9

* After Sverdrup, Johnson, and Fleming. 1942. p. 13.

TABLE A2-2
Heights and Depths of the Earth's Surface

LAND			OCEANS AND SEAS		
HEIGHT	FEET	METERS	DEPTH	FEET	METERS

Greatest height:
  Mount Everest
	29,028	8848	Greatest known depth:
			  Mariana Trench
				36,200	11,035

Average height	2757	840	Average depth	12,460	3800

TABLE A2-3
Volume, Density, and Mass of the Earth and Its Parts*

PART OF EARTH	AVERAGE THICKNESS OR RADIUS (km)	VOLUME (× 10⁶ km³)	MEAN DENSITY (g/cm³)	MASS (× 10²⁴ g)	RELATIVE ABUNDANCE (percent)
Atmosphere	—	—	—	0.005	0.00008
Oceans and seas	3.8	1370	1.03	1.41	0.023
Ice sheets and glaciers	1.6	25	0.90	0.023	0.0004
Continental crust†	35	6210	2.8	17.39	0.29
Oceanic crust‡	8	2660	2.9	7.71	0.13
Mantle	2881	898,000	4.53	4068	68.1
Core	3473	175,500	10.72	1881	31.5
Whole Earth	6371	1,083,230	5.517	5976	

* From Holmes. 1965.
† Including continental shelves.
‡ Excluding continental shelves.

TABLE A2-4
Ocean Provinces*

OCEAN†	SHELF AND SLOPE (percent)	CONTINENTAL RISE (percent)	DEEP-OCEAN FLOOR (percent)	VOLCANOES AND VOLCANIC RIDGES (percent)	RISE AND RIDGE (percent)	TRENCHES (percent)
Pacific	13.1	2.7	43.0	2.5	35.9	2.9
Atlantic	19.4	8.5	38.0	2.1	31.2	0.7
Indian	9.1	5.7	49.2	5.4	30.2	0.3
World ocean	15.3	5.3	41.8	3.1	32.7	1.7
Earth's surface	10.8	3.7	29.5	2.2	23.1	1.2

* After H. W. Menard and S. M. Smith. 1966. Hypsometry of ocean basin provinces. *J. Geophys. Res.* 71:4305.
† Includes adjacent seas—for example, Arctic Sea included in Atlantic Ocean.

TABLE A2-5
Surface and Drainage Areas of Ocean Basins and Their Average Depths*

OCEAN†	OCEAN AREA (millions of square kilometers)	LAND AREA DRAINED‡ (millions of square kilometers)	RATIO OF OCEAN AREA TO DRAINAGE AREA	AVERAGE DEPTH† (meters)
Pacific	180	19	11	3940
Atlantic	107	69	1.5	3310
Indian	74	13	5.7	3840

* From H. W. Menard and S. M. Smith. 1966. Hypsometry of ocean basin provinces. *J. Geophys. Res.* 71:4305.
† Includes adjacent seas. Arctic, Mediterranean, and Black seas included in the Atlantic Ocean.
‡ Excludes Antarctica and continental areas with no exterior drainage.

TABLE A2-6
Average Temperatures and Salinity of the Oceans, Excluding Adjacent Seas*

	TEMPERATURE (°C)	SALINITY (parts per thousand)
Pacific (total)	3.14	34.60
North Pacific	3.13	34.57
South Pacific	3.50	34.63
Indian (total)	3.88	34.78
Atlantic (total)	3.99	34.92
North Atlantic	5.08	35.09
South Atlantic	3.81	34.84
Southern Ocean†	0.71	34.65
World ocean (total)	3.51	34.72

* After L. V. Worthington. 1981. The water masses of the world ocean: some results of a fine-scale census, pp. 42–69. In *Evolution of Physical Oceanography,* ed. B. A. Warren and C. Wunsch Cambridge, Mass.: MIT Press. 623 pp.
† Ocean area surrounding Antarctica, south of 55°S.

TABLE A2-7
Water Sources for the Major Ocean Basins (centimeters per year)*

OCEAN	PRECIPITATION	RUNOFF FROM ADJOINING LAND AREAS	EVAPORATION	WATER EXCHANGE WITH OTHER OCEANS
Atlantic	78	20	104	6
Arctic	24	23	12	35
Indian	101	7	138	30
Pacific	121	6	114	13

* From M. I. Budyko. 1958. *The Heat Balance of the Earth's Surface,* trans. N. A. Stepanova. Office of Technical Services. Department of Commerce, Washington, D.C.

TABLE A2-8
Characteristics of Trenches*

TRENCH	DEPTH (kilometers)	LENGTH (kilometers)	AVERAGE WIDTH (kilometers)
Pacific Ocean			
Kurile-Kamchatka Trench	10.5	2200	120
Japan Trench	8.4	800	100
Bonin Trench	9.8	800	90
Mariana Trench	11.0	2550	70
Philippine Trench	10.5	1400	60
Tonga Trench	10.8	1400	55
Kermadec Trench	10.0	1500	40
Aleutian Trench	7.7	3700	50
Middle America Trench	6.7	2800	40
Peru-Chile Trench	8.1	5900	100
Indian Ocean			
Java Trench	7.5	4500	80
Atlantic Ocean			
Puerto Rico Trench	8.4	1550	120
South Sandwich Trench	8.4	1450	90

* After R. W. Fairbridge. 1966. Trenches and related deep sea troughs. In *The Encyclopedia of Oceanography,* ed. R. W. Fairbridge, pp. 929–38. New York: Reinhold Publishing Corporation.

Some Elements in Seawater

	ELEMENT	CHEMICAL FORM	AVERAGE CONCENTRATION AT 35‰	DEPTH DISTRIBUTION
Ag	silver	$AgCl_2^-$	25 pmol/kg‡	N
Al	aluminum	$Al(OH)_4^-$, $Al(OH)_3^\circ$	20 nmol/kg	M
Ar	argon	Ar (gas)		
As	arsenic	AsO_4H^{2-}	23 nmol/kg	N
Au	gold	$AuCl_4^-$	?	?
B	boron	$B(OH)_3$	0.42 mmol/kg	C
Ba	barium	Ba^{2+}	100 nmol/kg	N
Be	beryllium	$BeOH^+$, $Be(OH)_2^\circ$	20 pmol/kg	NS
Bi	bismuth	BiO^+, $Bi(OH)_2^+$	0.01–0.2 pmol/kg	ND
Br	bromine	Br^-	0.84 mmol/kg	C
C	carbon	CO_3H^-, organic C	2.3 mmol/kg	N
Ca	calcium	Ca^{2+}	10.3 mmol/kg	S
Cd	cadmium	$CdCl_2^\circ$	0.7 nmol/kg	N
Ce	cerium	$CeCO_3^+$, Ce^{3+}, $CeCl^{2+}$	20 pmol/kg	S
Cl	chlorine	Cl^-	0.546 mol/kg	C
Co	cobalt	Co^{2+}, $CoCO_3^\circ$, $CoCl^+$	0.02 nmol/kg	S, ND
Cr	chromium	CrO_4^{2-}, $NaCrO_4$	4 nmol/kg	N
Cs	cesium	Cs^+	2.2 nmol/kg	C
Cu	copper	$CuCo_3^\circ$, $CuOH^+$, Cu^{2+}	4 nmol/kg	N, NS
F	fluorine	F^-, MgF^+	68 μmol/kg	C
Fe	iron	$Fe(OH)_3^\circ$	1 nmol/kg	S, ND
Ga	gallium	$Ga(OH)_4^-$	?	?
Ge	germanium	$Ge(OH)_4$, $H_3GeO_4^-$	70 pmol/kg	N
H	hydrogen	H_2O		
He	helium	He (gas)		
Hf	hafnium			

ELEMENT		CHEMICAL FORM	AVERAGE CONCENTRATION AT 35‰	DEPTH DISTRIBUTION
Hg	mercury	$HgCl_4^{2-}$	5 pmol/kg	?
I	iodine	IO_3^-	0.4 μmol/kg	N
In	indium	$In(OH)_3$	?	?
K	potassium	K^+	0.2 mmol/kg	C
Kr	krypton	Kr (gas)		
La	lanthanum	La^{3+}, $LaCO_3^+$, $LaCl^{2+}$	30 pmol/kg	S
Li	lithium	Li^+	2.5 μmol/kg	C
Mg	magnesium	Mg^{2+}	53.2 mmol/kg	C
Mn	manganese	Mn^{2+}, $MnCl^+$	0.5 nmol/kg	ND
Mo	molybdenum	MoO_4^{2-}	0.11 μmol/kg	C
N	nitrogen	organic N, NO_3^-, NH_4^+, N_2 (gas)	30 μmol/kg	N
Na	sodium	Na^+	0.468 mol/kg	C
Nd	neodymium	$NdCO_3^+$, $NdSO_4^+$	20 pmol/kg	S
Ne	neon	Ne (gas)		
Ni	nickel	Ni^{2+}, $NiCO_3^o$, $NiCl^+$	8 nmol/kg	N
O	oxygen	OH_2, SO_4^{2-}, O_2 (gas)	0–300 μmol/kg	†
P	phosphorus	HPO_4^{2-}, $NaHPO_4^-$, $MgHPO_4^o$	2.3 μmol/kg	N
Pa	protoactinium			
Pb	lead	$PbCO_3^o$, $Pb(CO_3)_2^{2-}$, $PbCl^+$	10 pmol/kg	ND
Ra	radium			
Rb	rubidium	Rb^+	1.4 μmol/kg	C
Rn	radon	Rn (gas)		
S	sulfur	SO_4^{2-}, $NaSO_4^-$, $MgSO_4^o$	28.2 mmol/kg	C
Sb	antimony	$Sb(OH)_6^-$	?	?
Sc	scandium	$Sc(OH)_3^o$	15 pmol/kg	S
Se	selenium	SeO_4^{2-}, $HSeO_3^-$	1.7 nmol/kg	N
Si	silicon	$Si(OH)_4$	100 μmol/kg	N
Sn	tin	$SnO(OH)_3^-$	4 pmol/kg	††
Sr	strontium	Sr^{2+}	90 μmol/kg	S
Ta	tantalum			
Th	thorium			
Ti	titanium	$Ti(OH)_4^o$	<20 nmol/kg	?
Tl	thallium	Tl^+		
U	uranium	$UO_2(CO_3)_3^{4-}$		
V	vanadium	HVO_4^{2-}, $H_2VO_4^-$	30 nmol/kg	S
W	tungsten	WO_4^{2-}		
Xe	xenon	Xe (gas)		
Y	yttrium	YCO_3^+, YOH^+, Y^{3+}	?	?
Zn	zinc	Zn^{2+}, $ZnOH^+$, $ZnCO_3^o$, $ZnCl^+$	6 nmol/kg	N
Zr	zirconium	$Zr(OH)_4^o$, $Zr(OH)_5^-$	?	?

* After K. W. Bruland. 1983. *Trace Elements in Seawater. Chemical Oceanography*, ed. J. P. Riley and R. Chester, 8:157–221. New York: Academic Press.
C—conservative; S—surface depletion; M—middepth minima
N—nonconservative, depleted in surface waters, enriched at depth (nutrient-type)
ND—nonconservative, depleted at depth
NS—nonconservative, scavenging
†—mirror image of nutrient-type concentration
††—high in surface waters
‡—mol—molecular weight, in grams
    mmol—10^{-3} mol; μmol—10^{-6} mol
    nmol—10^{-9} mol, pmol—10^{-12} mol

TABLE A3-1
Major Types of Oceanic Phytoplankton

TYPE AND CHARACTERISTICS	LOCATION	COLOR AND APPEARANCE	METHOD OF REPRODUCTION
Diatoms: silica and pectin "pillbox" cell wall, sculptured designs; of major importance for coastal ocean productivity; has floating and attached forms	Everywhere in surface ocean, especially in colder waters, upwelling areas, even in polar ice; some heterotrophic below photic zone; some form "resting spore" under adverse conditions	Size: 0.01–0.2 mm Yellow-green or brownish; single cells or chains of cells; radial or bilateral symmetry; many have spines or other flotation devices	Division, splitting of nuclear material; average reduction of one cell-wall thickness at each division when limiting size is reached, cell contents escape, form new cell
Dinoflagellates: next to diatoms in productivity; many heterotrophic, ingest particulate food; some have cellulose "armor"; very small open-ocean species are naked	In all seas, and below photic zone; some parasitic; warm-water species very diverse; some have resting stage for protection; sometimes abundant in coastal areas as "red tide"	Size: 0.005–0.1 mm Usually brownish, one-celled; have two whiplike flagellae for locomotion; many are luminescent	Simple, longitudinal, or oblique divisions; daughter cells achieve size of parent before dividing
Coccolithophores: covered with calcareous plates, embedded in gelatinous sheath; important source of food for filter-feeding animals	Mainly in open seas, tropical and semitropical; sometimes proliferate near coasts; some heterotrophic forms at depths to 3000 meters	Size: 0.005–0.05 mm Many flagellated; often round or oval single cells; when present in great numbers, they give the water a milky appearance	Some individuals form cysts from which spores arise to develop into new individuals
Silicoflagellates: very small, have silica skeleton; some heterotrophic forms	Widespread in colder seas worldwide, especially in upwelling areas	Size: about 0.05 mm Single-celled, one or two flagellae; starlike or meshlike skeleton	Simple cell division
Cyanobacteria (also called Blue-Green Algae): small, relatively simple cell structure; cell wall of chitin	Mainly inshore, warmer surface waters, tropics	Size: filaments to 0.1 mm or more Blue-green or red rafts of mottled filaments; can cause a colored "bloom" in water	Simple division of each cell into two

Graphs, Charts, and Maps

In any scientific discipline, data are compiled and presented in graphic form at some point. Oceanography is no exception, although the wide variety of material obtained from studying the ocean poses some special problems. In this section we review some common means of graphically presenting data—graphs, profiles, maps, and diagrams—to show the uses as well as the limitations of each technique.

GRAPHS

Of the various techniques used to portray scientific data, **graphs** are perhaps the most widely used. A graph permits general aspects of a relationship between two properties to be understood at a glance. Furthermore, values of various properties can be measured from a graph.

In constructing a graph, data are often first organized into tables. For example, if we are studying the increase in the boiling point of seawater with salinity changes, we may arrange our data as follows:

SALINITY (parts per thousand)	BOILING POINT INCREASE (°C)
4	0.06
12	0.19
20	0.31
28	0.44
36	0.57

From this table it is apparent that the boiling point of seawater increases with increased salinity. To see this more clearly, we can draw a graph, as in Fig. A4-1. Note that salinity is plotted on the **x** (horizontal) axis, with values increasing to the right. The boiling point increase, in degree

Celsium (°C), is plotted on the **y** (vertical) axis, with values increasing upward. We plot our experimental points and then draw a line through these points.

Generally a straight line is the best first estimate. In this case, it is a reasonably good approximation. For many other graphs, we may need to use more complicated curves; but even for complicated curves, a straight line is a reasonable estimate for small portions of the curve.

Note that the graph shows that the boiling point is raised as salinity increases. Also, we can estimate boiling

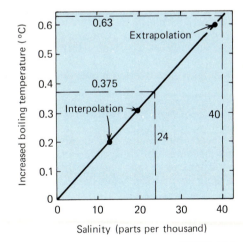

FIGURE A4-1
Graph of the increase in temperature of boiling and water salinity. The dots represent experimental data.

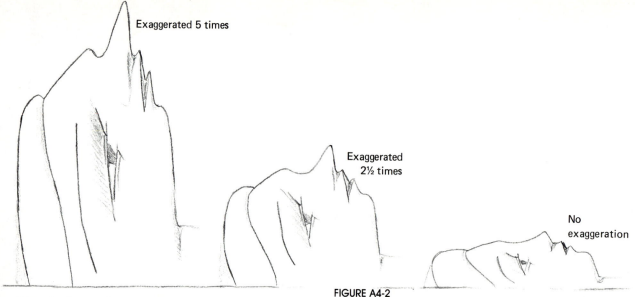

Exaggerated 5 times

Exaggerated
2½ times

No
exaggeration

FIGURE A4-2

Distortion in a profile of a human face caused by different vertical exaggerations.

point increases for salinities not studied here. For instance, for seawater with a salinity of 24 parts per thousand, we can interpolate a boiling point increase of 0.375°C; for a salinity of 40 parts per thousand, we extrapolate (or extend) the curve to indicate a boiling point increase of 0.63°C.

In reading a graph, always determine which property is plotted on each axis. Also, check both the scale intervals and the values at the origins. The appearance of the graph can be changed drastically by changing either. Advertisers, for instance, often make use of graphs where the scales and origins are chosen to present their data in the most favorable light.

PROFILES

Profiles are used to show topography, either of ocean bottom or of land. An ocean-bottom profile can be considered as a vertical slice through the earth's surface. Such a profile can be drawn with no distortion—distances are equal vertically and horizontally. But imagine the problems involved in drawing a profile 10 centimeters long of the Atlantic Ocean between New York and London, a distance of 5500 kilometers but only 3.4 kilometers deep at the deepest point. A pencil line would be too thick to portray accurately the maximum relief; thus such a profile conveys no useful information.

To get around this problem, profiles—including those in this book—are usually distorted. Profiles showing oceanic features are typically distorted by factors of several hundred or several thousand. Consequently, even gently rolling hills look like impossibly rugged mountains. The effect of profile distortion can be seen rather dramatically when it is applied to the human profile, as in Fig. A4-2.

CONTOURS AND CONTOUR MAPS

Various means have been used to portray land forms or topography; the most useful employ contours in contour maps. **Contour lines** connect points that are at equal elevations or at equal depths. Obviously not all elevations (or depths) can be connected by contour lines. Only certain

ones at selected intervals are shown; otherwise the map would be solid black. The vertical interval represented by successive contour lines is called the **contour interval.**

To interpret a contour map, imagine the shoreline of a lake. The still-water surface is a horizontal plane, touching points of equal elevation along the shore. The shoreline is thus a contour line. If the water surface were controlled to fall by regular intervals, it would trace a series of contour lines, forming a contour map on the lake bottom (or hillside). Note that the shorelines formed at different lake levels do not cross one another; neither do contours on maps.

Contours reveal topography. For example, contours that are closed on a map (do not intersect at a boundary) indicate either a hill or a depression. To find out which, look to see if the elevation increases toward the closed contour(s). If so, you are looking at a hill. If the elevation decreases toward the closed contour, it is a depression. Contours around depressions are often marked by **ha-chures**—short lines on the contour pointing toward the depression. When a contour line crosses a valley or canyon, the contour line forms a V, pointing upstream.

Contour lines also indicate the steepness of a slope. Closely spaced contour lines indicate a steep slope. A gentle slope is indicated by widely spaced contour lines.

Contours can also be used to depict properties other than elevation (or depth). For example, several maps in this book use contours to show distribution of properties, such as ocean surface temperatures and salinities. We would consider them to be temperature (or salinity) hills and valleys. High temperature corresponds to a hill, low temperature to a valley. Contours may also be used to show the distribution of temperature or salinity in a vertical section of the ocean.

COORDINATES—LOCATIONS ON A MAP

A **coordinate** is an address—a means of designating location. The most familiar coordinate system, found in many cities, is the network of regularly spaced, lettered or numbered streets crossing one another, usually at right angles, which enables us to locate a given address. This is an example of a **grid.** As soon as we determine how the streets and avenues are arranged, we can find Fifth Avenue and 42nd Street in New York or 16th and K Streets in Washington, D.C., even though we may never have been in those cities before.

A printed grid is used on large-scale maps to designate locations of features. In making maps of relatively small areas, it is simplest to assume that the world is flat. This works well for areas extending up to 100 miles from a starting point. For larger areas, the earth's curvature must be considered.

Since the earth is round, we must use **spherical coordinates**—a grid fitted to a sphere. A small town or even a state has rather definite starting points for a grid—the edges or the center. But a sphere has no edges or corners; so we must designate those points where our numbering system is to begin.

Distances north or south are easiest to deal with. We can easily identify the earth's geographic poles, where the axis of rotation intersects the earth's surface. Using these points, it is easy to draw a line circling the earth and equally distant from the North and South Poles. This is the **equator,** which serves as our starting point to measure distances north and south. Going from the equator, the distance to the pole is divided into 90 equal parts **(degrees).** The series of grid lines that circle the earth and connect the points that are the same distance from the nearest pole are known as **parallels of latitude.**

To see how this works, imagine the earth with a section cut out, as in Fig. A4-3. Now look at the angle formed by the line connecting any point of interest with the earth's center and the line from the earth's center to a point directly south of that point, on the equator. This angle is a measure of the distance between the chosen point and the equator. The North Pole has a latitude of 90°N, Seattle is approximately 47°N, and Rio de Janeiro approximately 23°S.

Latitude was easily measured by early mariners. The angle between Polaris (the pole star) and the horizon provides a reasonably accurate measure of latitude. At the equator, the pole star is on the horizon (latitude 0°N). Midway to the North Pole (latitude 45°N), the pole star is 45° above the horizon. At the North Pole, Polaris is directly overhead. Although there is no star directly above the South Pole, the same principle holds except that a correction is necessary to allow for the displacement from the South Pole of the star used.

Measuring east-west distances on the earth poses the problem; Where do we start? The answer has been to establish an arbitrary starting point—the **prime meridian**—and to indicate distances as east or west of that meridian. Several prime meridians have been used by different nations, but today the Greenwich prime meridian is most commonly used. It passes through the famous observatory at Greenwich (a suburb southeast of London).

Longitude—distance east or west of the prime meridian—is indicated on a map by north-south lines, connecting points with equal angular separation from the prime meridian. They are called **meridians of longitude** and converge at the North and South poles. Longitude in degrees is measured by the size of the angle between the prime meridian and the meridian of longitude passing through the given point, as in Fig. A4-3. Going eastward from the prime meridian, longitude increases until we reach the middle of the Pacific Ocean, when we come to the 180° meridian. Going westward from Greenwich, longitude also increases until we reach the 180° meridian, which represents the juncture between the Eastern and Western hemispheres. Through much of the Pacific Ocean, the 180° meridian is also the location of the **international dateline.** This designation of the 180th meridian as the international dateline—where the "new day" begins—is no accident. Its position in midst of the Pacific avoids the problem of adjacent cities being one day apart in time. This also explains, in part, the choice of the Greenwich meridian as the prime meridian.

Longitude and time are intimately related. The earth turns on its axis once every 24 hours. Since it takes the earth

FIGURE A4-3
Latitude and longitude for Cape Hatteras, North Carolina.

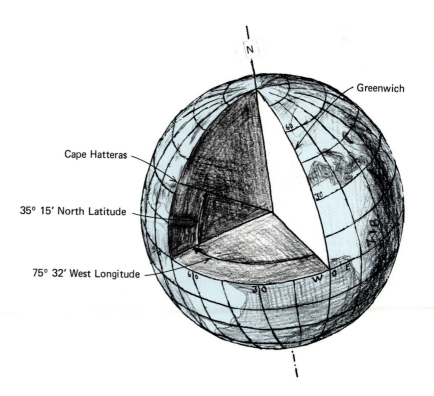

Cape Hatteras

35° 15′ North Latitude

75° 32′ West Longitude

Greenwich

24 hours to make one complete turn (360°), we calculate that the earth turns 15° per hour. We use this relationship to find our relative position east or west of the prime meridian.

Each meridian of longitude is a **great circle.** If we sliced through the earth along one of the meridians of longitude, our cut would go through the center of the earth. Of the parallels of latitude, only the equator is a great circle. All the other parallels are **small circles,** for a plane (or slice) passing through them would not go through the center of the earth. Great circles are favored routes for ships or aircraft because a **great circle** route is the shortest distance between two points on a globe.

To study time and longitude, let us begin at local noon on the prime meridian, when the sun is directly overhead. One hour later the sun is directly over the meridian of a point 15° west of the prime meridian; 2 hours later it is over a meridian 30° west of the prime meridian. And 12 hours later (midnight) it is over the 180th meridian and the new day begins.

If we have an accurate clock keeping "Greenwich time" (the time on the prime meridian), we can determine our approximate longitude from the time of local noon, when the sun is highest in the sky. Assume that our clock reads 2:00 P.M. Greenwich time at local noon. The 2-hour difference indicates that our position is 30° from the prime meridian. Since local noon is later than Greenwich, we know that we are west of Greenwich and our longitude is therefore 30°W. Another example: If our local noon occurs at 9:30 A.M. Greenwich time, we are 2.5 hours × 15° per hour = 37.5° east of the prime meridian; our longitude is thus 37.5°E.

Degrees—like hours—are divided into 60 parts known as **minutes.** Each minute is further divided into 60 **seconds.** Consequently, in the last example we would give our position as 37°30′E.

To determine longitude, therefore, a ship need only have accurate time. With modern electronic communications, this poses no problem. For centuries, however, seafarers had no means of keeping accurate time at sea. Not until the 1760s, when the first practical **chronometers**—accurate clocks for use aboard ship—were designed, was it possible for most ships' pilots to determine longitude. Even the Greek astronomer Ptolemy (A.D. 90–168) made maps with relatively accurate positions north and south. But he overestimated the length of the Mediterranean by 50%, an error that was not corrected until 1700.

Each map in this book shows latitude and longitude (usually at 20° intervals) to indicate the positions of the map features. The parallels of latitude may also be used to determine approximate distance on a map. Each degree of lati-

tude equals approximately 60 nautical miles (69 statute miles or 111 kilometers). Each minute of latitude is approximately 1 nautical mile, or 1.85 kilometers. At the equator each minute of longitude is 1 nautical mile but decreases so that 1 minute of longitude is only 0.5 nautical mile at 60° north or south latitude and vanishes at the poles.

MAPS AND MAP PROJECTIONS

A **map** is a flat representation of the earth's surface. Symbols are used to depict surface features. Because the earth is a sphere, making a flat map distorts the shape or size of surface features. The only distortion-free map is a globe, but a globe is not practical for the study of relatively small areas and so maps are used almost exclusively in science.

In making a map, we would like to make the final product as useful as possible. In general, we would like a map to preserve the following properties of the earth's surface:

Equal area: each area on the map should be proportional to the area of the earth's surface it represents.

Shape: the general outlines of a large area shown on a map should approximate as nearly as possible the shape of the region portrayed. A map that preserves shape is said to be **conformal.**

Distance: a perfect map would permit distance to be measured accurately between any two points anywhere on the map. Many common maps, such as the Mercator projection, do not accurately portray distances in a simple way.

Direction: ideally, it would be possible to measure directions accurately anywhere on a map.

No map has all these properties; only a globe preserves size, shape, direction, and distance simultaneously.

A **map projection** takes the grid of latitude and longitude lines from a sphere and converts them into a grid on a flat surface. Sometimes the resulting grid is a simple rectangular one where longitude and latitude lines intersect at right angles. In other projections, latitude and longitude lines are complex curves that intersect at various angles.

With the network formed from the grid lines, the map is drawn by plotting points in the appropriate spot on the new projection. In this way, the various types of maps are prepared. We shall consider only a few of the many map projections that have been developed to serve specific functions (listed in Table A4-1).

The **Mercator projection** is the most familiar. Parallels

TABLE A4-1
Characteristics of Various Map Projections

NAME OF PROJECTION (type)	DISTINCTIVE FEATURES	DESIRABLE FEATURES (UNDESIRABLE FEATURES)	USES
Mercator (cylindrical)	Horizontal parallels Vertical meridians	Compass directions are straight lines (Extreme distortion at high latitudes)	Navigation
Goode homolosine projection	Horizontal parallels Characteristic interruptions of outline	Equal area Little distortion of shape (Interruption of either continents or oceans)	Data presentation
Hoelzel's planisphere (modified)	Horizontal parallels Characteristic nearly oval outline	Shapes easily recognizable (Moderate scale distortion at high latitudes)	Index map Data presentation

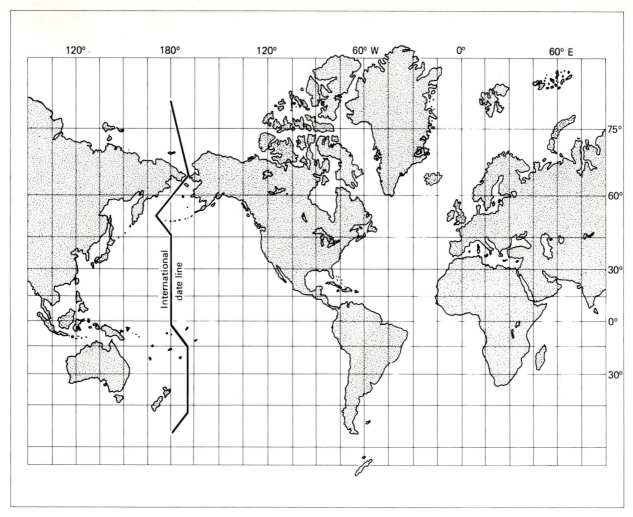

International date line

120°   180°   120°   60° W   0°   60° E

75°
60°
30°
0°
30°

FIGURE A4-4

Mercator projection. Compare the shape and size of Greenland in this projection with that in Figures A4-5 and A4-6.

of latitude and meridians of longitude are straight lines and cross at right angles. The outline shape is a square or rectangle, as shown in Fig. A4-4.

In its simplest form, the Mercator projection can be visualized as being made by a light inside a translucent globe projecting latitude and longitude onto a cylinder surrounding the globe. Even though the cylinder used for the projection is curved, it is easily made into the flat map desired.

The common Mercator projection in our example is most accurate within 15° of the equator and least accurate at the poles. Although shapes are well preserved by this projection, area is distorted, especially near the poles. For example, South America is in reality nine times the size of Greenland, but this fact is not obvious from common Mercator projections. The scale of a Mercator projection changes going away from the equator. The reader can use the length of a degree of latitude as the scale to avoid serious error.

Another property of the Mercator projection useful to mariners is that a course of constant compass direction (a **rhumb line**) is a straight line on this projection. Although a rhumb course is not a great circle and thus not the shortest distance between any two points on the earth's surface, it is

useful for navigation because a great circle course requires a constant changing of direction. A rhumb line is slightly longer but easier to navigate.

For world maps the Mercator projection has distinct limitations; but for relatively small areas, such as navigation charts, the Mercator projection is without equal. Nearly all navigation charts used at sea are Mercator projections.

For use in this book, a **Hoelzel planisphere** (shown in Fig. A4-5) is used. This modified cylindrical projection shows the continents well, permitting most of the world's coastal regions to be easily recognized. Like the Mercator, this projection distorts areas near the poles. Instead of converging to a point at the poles, the meridians converge to a line that is only a fraction of the length of the equator. Because the projection used in this book splits the Pacific Ocean in the middle, it is not overly convenient for showing properties in the ocean.

A special projection (the **interrupted homolosine,** shown in Fig. A4-6) was developed by J. P. Goode to show the ocean basins without interruptions. In addition, the projection shows area equally. In this projection the continents are interrupted to show the ocean basins intact.

FIGURE A4-5

Hoelzel projection. Note that the meridians (north-south lines) of longitude converge
to a line shorter than the equator but still not a point.

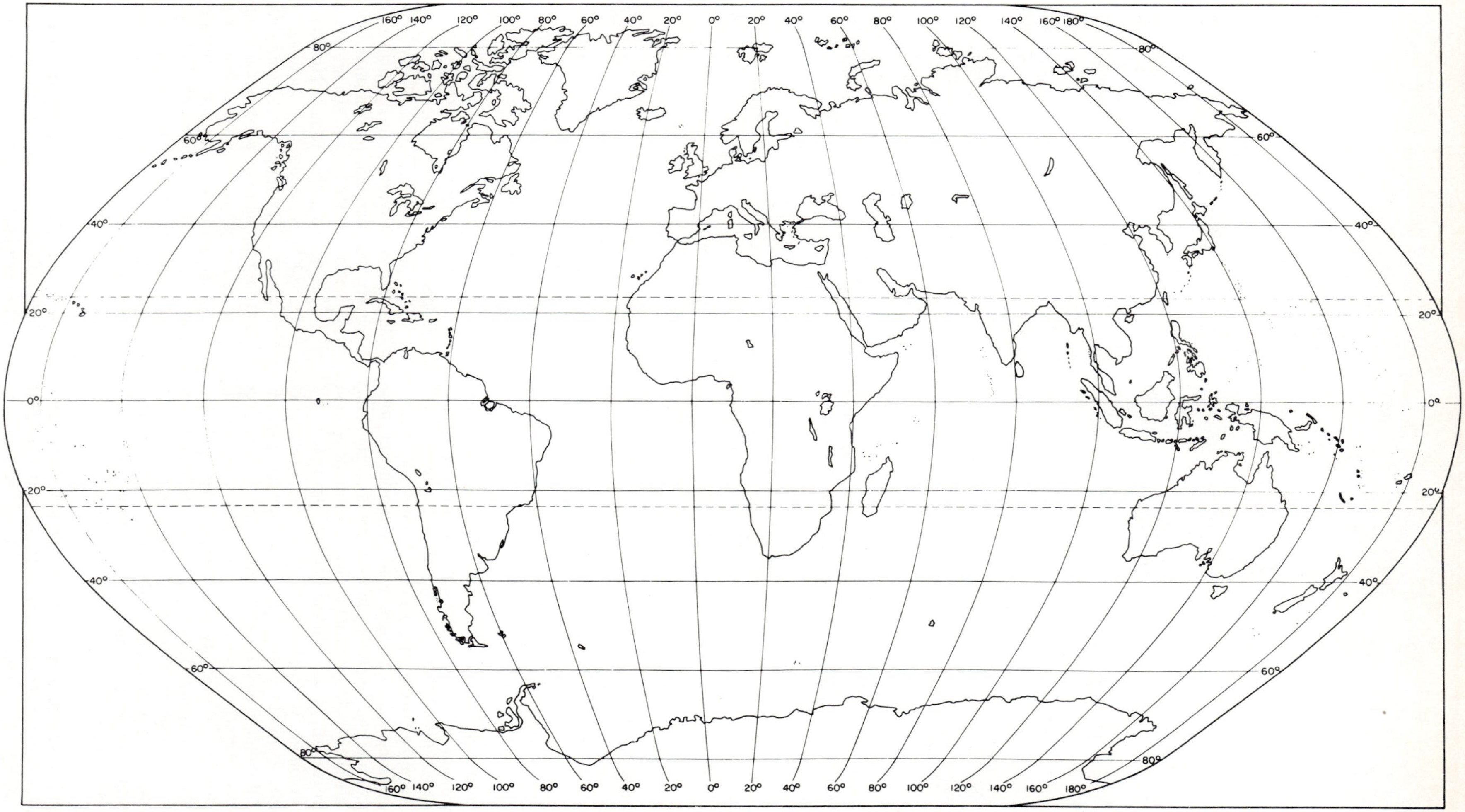

FIGURE A4-6
Goode's homolosine projection to show the three oceans to best advantage. Note that the meridians of longitude converge to points, thus not distorting the high latitudes.

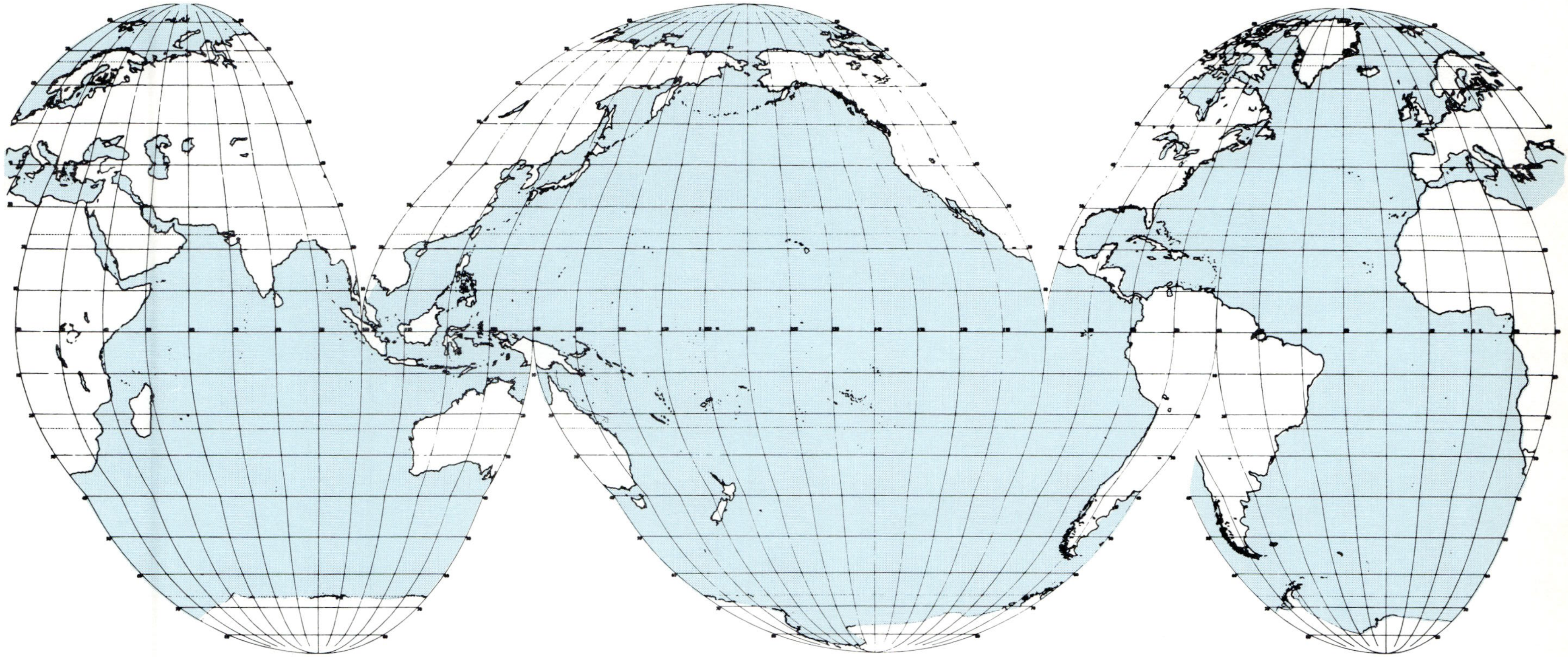

SELECTED REFERENCES

ALDRIDGE, B. G. 1968. *Mathematics for Physical Science.* Columbus, Ohio: Merrill Publishing Company. 137 pp. Elementary mathematical and data-plotting procedures.

GREENWOOD, DAVID. 1964. *Mapping.* Chicago: University of Chicago Press. 289 pp. Elementary discussion of coordinates, maps, and map projections.

Geologic Time Scale

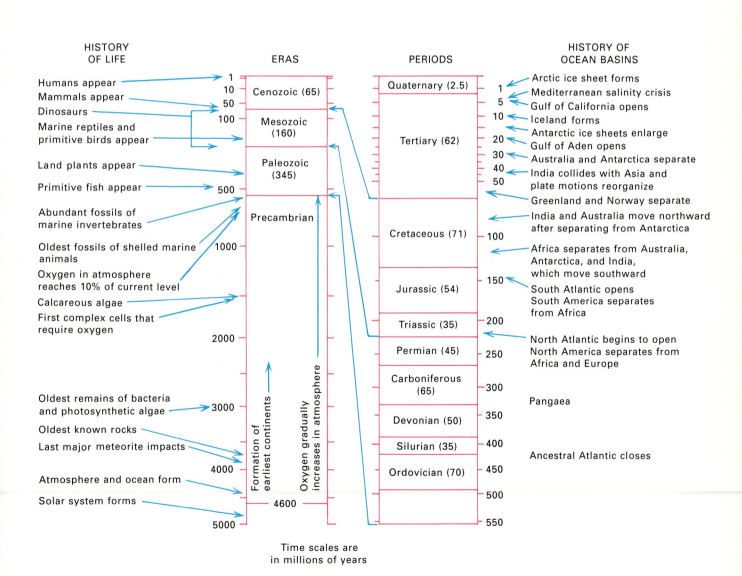

HISTORY OF LIFE	ERAS	PERIODS	HISTORY OF OCEAN BASINS

Time scales are
in millions of years

Classification of Marine Organisms*

Biologists categorize organisms using **taxonomic classification** to identify and describe similarities among marine organisms. After basic similarities in external form, internal anatomy, and biochemical characteristics are determined, groups (called **taxa,** or **taxon,** if singular) are assigned Latin names in a rigidly prescribed procedure. Finally, these groups are fitted into a system of increasingly more inclusive categories.

Taxonomic classification is used to study evolutionary relationships of organisms. It also shows the many different kinds of organisms that live in the ocean. Indeed, most of the organisms that ever lived on Earth inhabited the ocean. Life has existed in the ocean for at least 3800 million years, compared with only 450 million years on land.

The fundamental unit of taxonomy is the **species,** defined as a group of closely related individuals that can and usually do interbreed. (Many marine organisms are known only from preserved (dead) specimens.)

Some 75 million species have appeared since life began on Earth. More than 2 million are living today. In this appendix, we are primarily concerned with major groups of marine organisms, which are classified as follows:

Kingdom
  Phylum
    Subphylum
      Class
        Order
          Family
            Genus
              Species

KINGDOM MONERA—dominantly unicellular organisms, lacking nuclear membranes. Nuclear materials occur throughout the cells.

Phylum Schizophyta—smallest cells, bacteria.
Phylum Cyanophyta—blue-green algae, contain chlorophyll and other pigments.

KINGDOM PROTISTA—one-celled organisms; nuclear materials confined to nucleus by a membrane.

Phylum Chrysophyta—golden-brown algae. Includes diatoms, coccolithophores, and silicoflagellates.
Phylum Pyrrophyta—dinoflagellate algae.
Phylum Chlorophyta—green algae.
Phylum Phaeophyta—brown algae.
Phylum Rhodophyta—red algae.
Phylum Protozoa—heterotrophs.
    Class Sarcodina—ameboid, includes foraminiferans and radiolarians.
    Class Mastigophora—flagellated; includes dinoflagellates.

KINGDOM FUNGI
Phylum Mycophyta—fungi and lichens.

KINGDOM METAPHYTA—multicellular plants.
Phylum Tracheophyta—vascular plants with roots, stems, and leaves; separate liquid transport system.
    Class Angiospermae—flowering plants, with seeds.

* Adapted after H. V. Thurman and H. H. Webber. 1984. *Marine Biology*. Columbus, Ohio: Charles E. Merrill Publishing Company.

KINGDOM METAZOA—multicellular animals.

Phylum Porifera—sponges.

Class Calcarea—calcium carbonate spicules.

Class Hexactinellida—glass sponges.

Phylum Cnidaria—radially symmetrical, polyp (benthic) and medusa (planktonic) stages.

Class Hydrozoa—polyp colonies, includes Portuguese man-of-war.

Class Scyphozoa—jellyfish.

Class Anthozoa—corals and anemones.

Phylum Ctenophora—planktonic comb jellies; eight-sided radial symmetry with secondary bilateral symmetry.

Phylum Platyhelminthes—flatworms, bilateral symmetry.

Phylum Nemertea—ribbon worms, benthic and pelagic.

Phylum Nematoda—roundworms, free-living benthic; mostly meiofauna.

Phylum Rotifera—ciliated, unsegmented.

Phylum Bryozoa—moss animals, benthic, branching or encrusting.

Phylum Branchiopoda—lamp shells, benthic bivalves.

Phylum Phoronida—horsehoe worms, shallow-water benthos.

Phylum Sipuncula—peanut worms, benthic.

Phylum Echiura—spoon worms, benthic.

Phylum Pogonophora—tube-dwelling, gutless worms; absorb organic matter through skin.

Phylum Tardigrada—marine meiofauna.

Phylum Mollusca—soft bodies; posses muscular foot and mantle; usually secrete calcium carbonate shell.

Class Polyplacophora—chitons, oval, flattened body covered by eight overlapping plates.

Class Gastropoda—snails and related forms; many with spiral shell.

Class Bivalvia—clams, mussels, oysters and scallops; mostly filter feeding.

Class Aplacophora—tusk shells; benthic; feed on meiofauna.

Class Cephalopoda—octopus, squid and cuttlefish; possess no external shell (except *Nautilus*).

Phylum Annelida—segmented worms, mostly benthic.

Phylum Arthropoda—joint-legged, segmented bodies; covered by exoskeleton.

Subphylum Crustacea—calcareous exoskeletons; two pairs of antenae; includes copepods, ostracods, barnacles, shrimp, lobsters, and crabs.

Phylum Chaetognatha—arrow worms; mostly plankton.

Phylum Hemichordata—acorn worms and pterobranchs; primitive nerve chord; benthic.

Phylum Echinodermata—spiny skinned; secondary radial symmetry; water vascular system; benthic

Class Asteroidea—starfishes, flattened body with five or more rays; tube feet used for locomotion.

Class Ophiuroidea—brittle stars, basket stars; central disc with slender rays; tube feet used for feeding.

Class Echinoidea—sea urchins, sand dollars; calcium carbonate tests.

Class Holothuroidea—sea cucumbers; soft bodies with radial symmetry.

Class Crinoidea—sea lillies; cup-shaped body attached to bottom by jointed stalk or appendages.

Phylum Chordata—notochord; nerve chord and gills or gill slits.

Subphylum Vertebrata—internal skeleton; spinal column; brain.

Class Agnatha—lampreys and hagfishes; most primitive vertebrates; cartilaginous skeletons; no jaws; no scales.

Class Chondrichthyes—sharks, skates and rays; cartilaginous skeletons.

Class Osteichthyes—bony fishes; covered gill openings, swim bladder common.

Class Reptilia—snakes, turtles, lizards, and alligators.

Class Aves—birds.

Class Mammalia—warmblooded; hair; mammary glands; bear live young.

*Glossary**

Abyssal pertaining to the great depths of the ocean, generally below 3700 meters.

Acid solution in which hydrogen ion concentration exceeds hydroxyls.

Acoustic tomography technique using changes in sound velocity between acoustic transmitters and receivers in the ocean to obtain three-dimensional pictures of water-mass distributions and their movements.

Albedo the ratio of radiation reflected by a body to the amount incident upon it, commonly expressed as a percentage.

Algae marine or freshwater plants, including phytoplankton and seaweeds.

Algal ridge elevated margin of a windward reef, built by calcareous algae.

Alternation of generations mode of development, characteristic of many coelenterates, in which a sexually reproducing generation gives rise (by union of egg and sperm) to an asexually reproducing form, from which new individuals arise by budding or simple division of the ''parent'' animal.

Altimeter satellite-borne radar that measures distances between the spacecraft and the ocean surface. Ocean surface topography, calculated from the distance, indicates the current patterns; roughness of the ocean surface indicates average wave height.

Amphidromic point center of an amphidromic system; a nodal or no-tide point around which a standing-wave crest rotates once each tidal period.

Anaerobic condition in which there is no dissolved oxygen. Organisms that depend on the presence of oxygen can-

not survive. Anaerobic bacteria can live under these conditions.

Anadromous fishes that spawn in fresh water, then migrate to the ocean to mature before returning to freshwater.

Andesite volcanic rock intermediate in composition between granite and basalt, associated with partial melting of crust and mantle during subduction. Andesitic mountain ranges are often associated with subduction zones.

Anion negatively charged atom or radical.

Anoxia anoxic conditions.

Anoxic devoid of dissolved oxygen. See **Anaerobic.**

Antinode that part of a standing wave where the vertical motion is greatest and the horizontal velocities are least.

Aphotic zone that portion of the ocean where light is insufficient for photosynthesis.

Aquaculture cultivation or propagation of water-dwelling organisms.

Archipelagic plain gently sloping sea floor with a generally smooth surface, often found among islands or seamounts.

Arrow worm see **Chaetognath.**

* The glossary is compiled from B. B. Baker, W. R. Deebel, and R. D. Geisenderfer, eds. 1966. *Glossary of Oceanographic Terms, 2d ed.* U.S. Naval Oceanographic Office, Washington, D.C.; Stacy D. Hicks, 1984. *Tides and Current Glossary.* National Oceanic and Atmospheric Administration, Washington, D.C.; and Robert L. Bates and Julia A. Jackson, eds. 1980. *Glossary of Geology, 2d ed.* American Geological Institute, Falls Church, Va.

Arthropods animals with a segmented external skeleton of chitin or plates of calcium carbonate, and with jointed appendages—for example, a crab or an insect.

Asthenosphere upper zone of the earth's mantle, extending from the base of the lithosphere to about 250 kilometers beneath continents and ocean basins; relatively weak, probably partially molten.

Atmosphere gaseous outer shell of Earth.

Atoll ring-shaped organic reef that encloses a lagoon in which there is no preexisting land and that is surrounded by the open sea. Low sand islands may occur on the reef.

Autotroph an organism that manufactures its food from inorganic compounds, such as CO_2, H_2O.

Auxospore diatom cell that has shed its shell.

Azoic devoid of life.

Backshore part of a beach that is usually dry, being reached only by the highest tides; a narrow strip of relatively flat coast bordering the sea.

Baleen (whalebone) horny material growing down from the upper jaw of large plankton-feeding (baleen) whales, which forms a strainer or filtering organ consisting of numerous plates with fringed edges.

Bank large elevation of the sea floor; a submerged plateau.

Bar offshore ridge or mound that is submerged (at least at high tide).

Barrier beach bar parallel to the shore whose crest rises above high water.

Barrier island detached portion of a barrier beach between two inlets.

Barrier reef reef that is separated from a landmass by a lagoon, usually connected to the sea through passes (openings) in the reef.

Base solution in which hydroxyl concentration exceeds hydrogen ion concentration.

Basalt fine-grained igneous rock, black or greenish black, rich in iron, magnesium, and calcium.

Basin large depression of the sea floor.

Bathymetry mapping of the ocean bottom.

Baymouth bar bar extending partially or entirely across the mouth of a bay.

Beach seaward limit of the shore (limits are marked approximately by the highest and lowest water levels).

Beach cusp low mounds of beach material separated by troughs spaced at more or less regular intervals along the beach face.

Beach drift see **Littoral drift.**

Beach face highest portion of the foreshore, above the low tide terrace and exposed only to the action of wave uprush, not to the rise and fall of the tide.

Bed load see **Load.**

Benthic that portion of the marine environment inhabited by marine organisms that live permanently in or on the bottom.

Benthos bottom-dwelling marine organisms.

Berm low, nearly horizontal portion of a beach (backshore), having an abrupt fall and formed by the deposit of material by wave action. It marks the limit of ordinary high tides and waves.

Berm crest seaward limit of a berm.

Bight concavity in a coastline; a large, open bay.

Biogenous sediment sediment containing at least 30% by volume of skeletal remains of organisms.

Biogeochemical cycles paths by which elements essential to life circulate from the nonliving environment to living organisms and back again.

Bioluminescence production of light by organisms; results from chemical reactions within certain cells or organs or in some form of secretion.

Biomass amount (weight) of living matter per unit of water surface or volume.

Bird-foot delta delta formed by the outgrowth of pairs of natural levees, formed by the distributaries, making a digitate or bird-foot form.

Bivalves mollusks, generally sessile or burrowing into soft sediment, rock, wood, or other materials. Individuals possess a hinged shell and a hatchet-shaped foot, which is sometimes used in digging. Includes clams, oysters, and mussels.

Black mud (hydrogen sulfide mud) dark, fine-grained sediment. This sediment usually contains large quantities of decaying organic matter and iron sulfides and may exude hydrogen sulfide gas.

Bloom see **Plankton bloom.**

Bore see **Tidal bore.**

Bottom water water mass at the deepest part of the ocean.

Boundary currents northward- or southward-directed surface currents that flow parallel to continental margins; caused by deflection of the prevailing eastward- and westward-flowing currents by continents.

Breaker wave breaking on the shore, over a reef, etc. Breakers are classified into four kinds: **spilling breakers** break gradually over a considerable distance; **plunging breakers** tend to curl over and break with a crash; **surging breakers** peak up, but then instead of spilling or plunging, they surge up on the beach face; **collapsing breakers** break in the middle or near the bottom of the wave rather than at the top.

Breakwater structure, usually rock or concrete, protecting a shore area, harbor, anchorage, or basin from waves.

Brine water containing a higher concentration of dissolved salt than that of ordinary ocean water. Brines are produced by evaporating or freezing seawater.

Buffering any process that reduces impacts.

Calcareous algae marine plants that form a hard external covering of calcium compounds.

Calorie amount of heat required to raise the temperature of 1 gram of water by 1°C (defined on basis of water's **specific heat.**)

Canyon (submarine) see **Submarine canyon.**

Capillary wave wave in which the primary restoring force is surface tension. A water wave with wavelength less than 1.7 centimeters is considered a capillary wave.

Carbohydrate compound made of carbon, oxygen and hydrogen; starches and sugars are examples.

Carnivore animal that eats other animals.

Catadromous fishes that spawn at sea but mature in freshwater.

Cation positively charged ion.

Celsius temperature temperature based on a scale in which water freezes at 0° and boils at 100° (at standard atmospheric pressure); also called **centigrade temperature.**

Cephalopods benthic or swimming mollusks having a large head, large eyes, and a circle of arms or tentacles around the mouth; the shell is external, internal, or absent; an ink sac usually is present. They include squids and octopus.

Chaetognath small, elongate, transparent wormlike animals, pelagic in all seas from the surface to great depths.

Change of state change in the physical form of a substance, as when a liquid changes to a solid due to cooling.

Chemosynthesis carbon dioxide fixation (primary production) by certain bacteria in the absence of sunlight, using compounds such as ammonia, methane, reduced iron, hydrogen, or sulfur.

Chitin nitrogenous carbohydrate derivative forming the skeletal substance in arthropods.

Chloride atom or chlorine in solution, bearing a single negative charge.

Chlorinity measure of the chloride content, by mass, of seawater (grams per kilogram, or ‰).

Chlorophyll green pigments that occur chiefly in bodies called chloroplasts and that carry out photosynthesis.

Cilia hairlike processes of cells, which beat rhythmically and propel cells or produce currents in water.

Clay inorganic silicate material where the grains have diameters smaller than 0.004 millimeter (or 4 micrometers).

Climate meteorological conditions of a place or region, usually averaged over 30 years—in contrast to **weather,** which is the state of the atmosphere at a particular time.

Climax community end product of ecological succession in which there is a balance of production and consumption in the community and further changes in species composition take place only as a result of changes in environmental conditions or outside forces influencing the community.

Clupeoids fishes of the herring family, including sardines, anchovy, pilchard, and menhaden.

Coastal currents currents paralleling the shore, seaward of the surf zone. These currents may be caused by tides or winds or by distributions of mass in coastal waters, associated with river discharges.

Coastal ocean shallow portion of the ocean (generally overlying the continental shelf).

Coastal plain low-lying continental plain, adjacent to the ocean and extending inward to the first major change in terrain features.

Coastal plain estuary estuary in coastal plain.

Cobble a rock fragment between 64 and 256 millimeters in diameter; larger than a pebble and smaller than a boulder; usually rounded or otherwise abraded.

Coccolithophores microscopic, planktonic algae. The cells are surrounded by an envelope on which small calcareous disks or rings (coccoliths) are embedded.

Coelenterates a large, diverse group of animals possessing two cell layers and a digestive cavity with only one opening. This opening is surrounded by tentacles containing stinging cells. Some are sessile, some pelagic (medusae), and some undergo alternation of generations.

Color scanner a radiometer measures the visible and near-infrared radiation from the ocean surface. These measurements yield ocean color, from which chlorophyll concentrations and locations of turbid water masses can be determined.

Compaction decrease in volume or thickness of a sediment deposit under load through closer packing of constituent particles and accompanied by decrease in porosity, increase in density, and squeezing out of water.

Compensation depth (carbonate) depth at which carbonate produced in water column is totally dissolved; no carbonate deposition occurs below this depth.

Compensation depth (oxygen) the depth at which oxygen production by photosynthesis equals consumption by plant respiration during a 24-hour period.

Compressional wave wave that causes alternate compression and expansion of the medium through which it passes.

Conduction transfer of energy through matter by internal particle or molecular motions.

Conformal projection map projection in which the angles around any point are correctly represented.

Conservative property property whose values do not change in a particular, specified series of events or processes—for example, those properties of seawater, such as salinity, the concentrations of which are not affected by the presence or activity of organisms but are affected by diffusion and currents.

Continental block large landmass, including shallowly submerged marginal regions.

Continental climate climate characterized by cold winters and hot summers where the prevailing winds come from large land areas.

Continental crust thickened part of the crust forming continental blocks; typically about 35 kilometers thick, consisting primarily of granitic rocks.

Continental margin zone separating the land from the deep-sea bottom; generally consists of a continental shelf, slope, and rise.

Continental rise gentle slope with a generally smooth surface, rising toward the foot of the continental slope.

Continental shelf sea floor adjacent to a continent, extending from the low-water line to the change in slope, usually at about 180 meters' depth, where continental shelf and continental slope join.

Continental slope a declivity from the outer edge of the continental shelf, extending from the break in slope to the deep-sea floor.

Contour line line on a chart connecting points of equal value above or below a reference value; used to portray elevation, temperature, salinity, or other values.

Convergence area or zone where flow regimes come together or converge, usually resulting in sinking of surface waters.

Convergence zone band along which crustal area is lost. Colliding edges of crustal plates may be thickened, folded, or underthrust, or one plate may be subducted and destroyed.

Copepods minute shrimplike crustaceans; most are between about 0.5 and 10 millimeters in length.

Coral hard, calcareous skeletons of sessile, colonial coelenterate animals, or the stony solidified mass of many such skeletons; also, the entire animal, a compound polyp that produces the skeleton.

Coral reef association of bottom-living and attached calcareous, shelled marine invertebrates forming fringing reefs, barrier reefs, or atolls.

Coralline algae red algae (bushy or encrusting) that deposit calcium carbonate either on branches or as a crust on the substrate; can develop massive encrustations on coral reefs.

Core vertical, cylindrical sample of sediments; also, the central zone of the earth.

Coriolis effect apparent force acting on moving particles resulting from the Earth's rotation. It causes moving particles to be deflected to the right in the Northern Hemisphere and to the left in the Southern Hemisphere; the deflection is proportional to the speed and latitude of the moving particle. Particle speed is unchanged by the apparent deflection.

Cosmogenous sediment particles derived from outer space.

Crinoids group of echinoderms; most of them are attached by long stalks to the bottom. Species without stalks either swim or creep slowly about. Crinoids (such as sea lily, feather star) occur in shallow water as well as at great depths.

Critical depth depth above which the net effective plant production occurs in the water column; total production equals total photosynthesis.

Crust of the Earth outer shell of the solid Earth. Beneath the oceans, the outermost layer of crust is composed of sediment deposits, and basaltic rocks. The crust is approximately 5 to 7 kilometers under deep-ocean basins to 35 kilometers under continents. Its lower limit is the Mohorovičić discontinuity.

Crustaceans arthropods that breathe by means of gills or similar structures. The body is commonly covered by a hard shell or crust. The group includes barnacles, crabs, shrimps, and lobsters.

Ctenophores spherical, pear-shaped, or cylindrical animals of jellylike consistency ranging from less than 2 centimeters to about 1 meter in length. The outer surface of the body bears eight rows of comblike structures.

Current ellipse graphic representation of a rotary current in which current speed and direction at different hours of the tide cycle are represented by vectors joined at one point. A line joining the extremities of the radius vectors forms a curve roughly approximating an ellipse.

Current rose graphic representation of currents, utilizing arrows to show the direction toward which the prevailing current flows and the frequency (expressed as a percentage) of any given direction of flow.

Current tables tables that give daily predictions of times, speeds, and directions of tidal currents.

Cypris larva stage at which young tidal barnacles attach to the substrate.

Daily (diurnal) inequality difference in heights and durations of two successive high waters or of two successive low waters of each day; also, the difference in speed and direction of the two flood currents or the two ebb currents of each day.

Daily (diurnal) tide tide having only one high water and one low water each tidal day.

Debris line line near the limit of storm-wave uprush marking the landward limit of debris deposits.

Decomposers heterotrophic and chemoautotrophic organisms (chiefly bacteria and fungi) that break down nonliving matter, absorb some of the decomposition products, and release compounds usable in primary production.

Deep scattering layer stratified population of organisms in ocean waters that causes scattering of sound as recorded on an echo sounder. Such layers may be from 50 to 200 meters thick. They occur less than 200 meters below the ocean surface at night and several hundred meters below the surface during the day.

Deep-water waves water waves whose depth is greater than one-half the average wavelength.

Deep zone waters below the pycnocline.

Delta alluvial deposit formed at the mouth of a stream, tidal inlet, or river.

Demersal fishes fishes living on or near the bottom.

Density mass per unit volume of a substance, usually expressed in grams per cubic centimeter. In the centimeter-gram-second system, density is numerically equivalent to **specific gravity.**

Density current flow (caused by density differences or gravity) of one current through, under, or over another; it retains its unmixed identity from the surrounding water because of density differences.

Deposit feeding removal of edible material from sediment or detritus either by ingesting material unselectively and excreting the unusable portion or by selectively ingesting discrete particles.

Depth of no motion depth at which water is assumed to be motionless, used as a reference surface for computing geostrophic currents.

Desalination production of fresh water from seawater or brine.

Detritus loose material produced by rock disintegration. **Organic detritus** consists of decomposition or disintegration products or dead organisms, including fecal material.

Detritivore animal that eats detritus.

Diatoms microscopic phytoplankton organisms, possessing walls of overlapping halves (valves) impregnated with silica.

Diffraction bending of a wave around an obstacle.

Diffusion transfer of material (e.g., salt) or property (e.g., temperature) by eddies or molecular movement. Diffusion causes spreading or scattering of matter under the influence of a concentration gradient, with movement from the stronger to the weaker solution.

Dinoflagellates microscopic or minute organisms that possess characteristics of plants (chlorophyll and cellulose plates) and animals (ingestion of food).

Discontinuity abrupt change in a property, such as salinity or temperature, at a line or surface.

Dispersion separation of a complex wave into its component parts. Longer component parts of the wave travel faster than shorter ones.

Distributary outflowing branch of a river, usually on deltas.

Diurnal daily, especially pertaining to actions that are completed within approximately 24 hours and that recur every 24 hours.

Divergence horizontal movements in different directions from a common center or zone.

Divergence zone region along which crustal plates move apart and new lithospheric material solidifies from rising volcanic magma.

Doldrums belt of light, variable winds near the equator; an area of low atmospheric pressure.

Downwelling area of downward moving water; a convergence.

Drift bottle bottle used to study surface currents.

Drift net fishing net suspended in the water vertically so that drifting or swimming animals are trapped or entangled in the mesh.

Dune mound or ridge of sand moved by winds.

Dynamic topography configuration formed by the geopotential difference (dynamic height) between a given surface and a reference surface, usually the layer of no motion. A contour map of dynamic ''topography'' may be used to estimate geostrophic currents.

Eastern boundary current broad, shallow, slow-moving current on eastern side of ocean basin.

Ebb current tidal current directed away from shore or down a tidal stream.

Echinoderms principally benthic marine animals having calcareous plates with projecting spines forming a rigid or articulated skeleton or plates and spines embedded in the skin. They have radially symmetrical, usually five-rayed, bodies. They include starfish, sea urchins, crinoids, and sea cucumbers.

Echiuroids unsegmented, burrowing marine worms.

Echo sounding determination of water depth by measuring time intervals between emission of a sonic signal and the return of its echo from the bottom. The instrument used for this purpose is called an **echo sounder.**

Ecological efficiency ratio of the efficiency with which energy is transferred from one trophic level to the next.

Ecology study of organisms' relations to one another and to their environment.

Ecosystem ecological unit including organisms and the nonliving environment, each influencing the properties of the other and both necessary for maintenance of life.

Eddy current of air, water, or any fluid, often on the side of a main current, especially one moving in a circle—in extreme cases, a whirlpool.

Eddy viscosity turbulent transfer of momentum by eddies giving rise to an internal fluid friction.

Eelgrass seed-bearing, grasslike marine plant that grows chiefly in sand or mud-sand bottoms; most abundant in temperate waters less than 10 meters deep.

Ekman spiral representation of currents resulting from a steady wind blowing across an ocean having unlimited depth and extent and uniform viscosity. The surface layer moves 45° to the right of the wind direction in the Northern Hemisphere; water at successive depths drifts in directions more to the right until, at some depth, the water moves in a direction opposite to the wind. Speed decreases with depth throughout the spiral. The net water transport is 90° to the right of the wind in the Northern Hemisphere.

El Nino warm surface waters offshore from Peru; occurs around Christmas.

Encrusting algae see **Coralline algae, Red algae.**

Epicenter point on Earth's surface directly above an earthquake focus.

Epifauna animals that live at the water-substrate interface, attached to the bottom or moving freely over it.

Epiphytes plants that grow attached to other plants.

Equal-area projection map projection in which equal areas on the earth's surface are represented by equal areas on the map.

Equilibrium tide hypothetical semidaily tide caused by gravitational attraction of the sun and moon on a frictionless, nonrotating entirely water-covered earth.

Estuarine circulation characteristic circulation in an estuary; flow is seaward at surface, landward at depth.

Estuary a semienclosed, tidal body of saline water with free connection to the sea.

Euphausiids shrimplike, planktonic crustaceans, common in oceanic and coastal waters, especially in colder waters. They grow to 8 centimeters in length; many possess luminous organs.

Euphotic zone see **Photic zone.**

Exotic terrain fragments of continental masses, or sometimes sea floor, that have accreted to other continents.

Extratropical cyclone powerful storm that develops on a front, outside the tropics; can reach hurricane strength.

Fan gently sloping, fan-shaped feature located near the lower end of a canyon.

Fault fracture or fracture zone in rock, along which one side has moved relative to the other.

Fauna animal population of a location, region, or period.

Fecal pellets pellets of organic matter voided by marine animals, usually ovoids less than 1 mm long.

Fermentation type of respiration in which partial oxidation of organic compounds takes place in the absence of free oxygen.

Fetch ocean area where waves are generated by a wind having a constant direction and speed, also called **generating area;** also, the length of the fetch area, measured in the direction of the wind in which the seas are generated.

Filter feeding filtering or trapping edible particles from seawater; a feeding mode typical of many zooplankters and other marine organisms of limited mobility.

Fin rot disease of fishes that erodes fins.

Fjord narrow, deep, steep-walled inlet, formed either by the submergence of a glaciated mountainous coast or by entrance of the ocean into a deeply excavated glacial trough after the glacier melts.

Flagellum whiplike bit (process) of protoplasm that propels a motile cell.

Flocculation process of aggregation into small lumps, especially with regard to soils and colloids.

Floe sea ice, either as a single unbroken piece or as many individual pieces covering an area of water.

Flood current tidal current associated with the increase in the height of a tide. Flood currents generally set toward the shore or in the direction of the tide progression.

Flora plant population of a particular location.

Food chain simplification of a **food web.**

Food web interrelated food relationships in an ecosystem, including its production, consumption, and decomposition, and the energy relationships among organisms involved in the cycle.

Foraminifera benthic or planktonic protozoans possessing shells, usually of calcium carbonate.

Forced wave wave generated and maintained by a continuous force, in contrast to a **free wave,** which continues to exist after the generating force has ceased to act.

Forereef upper seaward face of a reef, extending above the lowest point of abundant living coral and coralline algae to the reef crest. This zone commonly includes a shelf, bench, or terrace that slopes to 15–30 meters, as well as the living, wave-breaking face of the reef. The terrace is an eroded surface or is veneered with organic growth. The living reef front above the terrace in some places is smooth and steep; in other places it is cut up by grooves separated by ridges, called **spur-and-groove systems,** forming comb-tooth patterns.

Forerunner low, long-period swell that commonly precedes the main swell from a distant storm.

Foreshore see **Low tide terrace.**

Foul to attach to or come to lie on the surface of submerged man-made or introduced objects, as barnacles on the hull of a ship or silt on a stationary object.

Fracture zone elongate zone of unusually irregular topography of the ocean floor characterized by seamounts, steep-sided or asymmetrical ridges, troughs, or long, steep slopes.

Free wave any wave not acted on by external forces except for the initial force that created it.

Fringing reef reef attached directly to the shore of an island or continental landmass. Its outer margin is submerged and often consists of algal limestone, coral rock, and living coral.

Front marked change in water properties.

Fully developed sea maximum height to which ocean waves can be generated by a given wind blowing over sufficient fetch, regardless of duration.

Gaia hypothesis theory that physical and chemical conditions on the Earth's surface have been controlled by the presence of life.

Gastropods mollusks that possess a distinct head, generally with eyes and tentacles and a broad, flat foot and that are usually enclosed in a spiral shell.

Geostrophic current current resulting from the balance between gravitational forces and the Coriolis effect.

Geothermal power power derived from heat energy coming from the Earth's crust.

Giant squids large cephalopods (length may be 15 meters or more) that inhabit middepths in oceanic regions but may come to the surface at night.

Gill a delicate, thin-walled structure, often an extension of the body wall, used to exchange gases with the water and sometimes for excreting wastes.

Glassworm see **Chaetognath.**

Glacial-marine sediment high-latitude, deep-ocean sediments transported from the land by glaciers or icebergs.

Glacier mass of freshwater ice, formed by recrystallization of old, compacted snow, flowing slowly from an area of accumulation to areas where snow or ice is removed.

Graben elongated crustal block, bounded by faults, that has been thrown down along the faults to form a steep-sided narrow valley.

Graded bedding type of stratification in which each stratum displays a gradation in grain size from coarse below to fine above.

Gradient rate of decrease (or increase) of one quantity with respect to another—for example, the rate of decrease of temperature with depth in the ocean.

Granite crystalline, igneous rock consisting of alkali feldspar and quartz. **Granitic** is a textural term applied to coarse- and medium-grained igneous rocks.

Gravel loose sediment with particles ranging in size from 2 to 256 millimeters.

Gravity anomaly disturbance of Earth's normal gravity field.

Gravity wave wave whose velocity of propagation is controlled primarily by gravity. Water waves longer than 1.7 centimeters are considered gravity waves.

Great circle intersection of the surface of a sphere and a plane through its center—for example, meridians of longitude and the equator are great circles on the Earth's surface.

Greenhouse effect warming of the Earth's surface caused by penetration of the atmosphere by comparatively short-wavelength solar radiation that is largely absorbed near and at the Earth's surface, whereas the relatively long-wavelength radiation emitted by the Earth is partially absorbed by water vapor, carbon dioxide, and dust in the atmosphere, thus warming the lower atmosphere.

Groin low, artificial, damlike structure of durable material placed so that it extends seaward from the land; used to slow littoral drift on beaches.

Group velocity velocity with which a wave group travels. In deep water, it is equal to one-half the individual wave velocity.

Guyot flat-topped submarine mountain or seamount.

Gyre circular or spiral form, usually applied to a very large semiclosed current system, in an open-ocean basin.

Hadley cell semiclosed system of vertical motions in the atmosphere. Warm, moist air rises in equatorial regions, flows to midlatitudes (30°N, 30°S), where it sinks and returns along the ocean surface to the equatorial zone as the trade winds.

Half-life time required for the decay of one-half the atoms of a radioactive substance.

Halocline water layer with large vertical changes in salinity.

Headland point of land extending into the sea.

Heat budget accounting for the amount of the sun's heat received on the Earth during any one year as equaling the amount lost by radiation and reflection.

Heat capacity amount of heat required to raise the temperature of a substance by a given amount.

Herbivore animal that feeds only on plants.

Heterotroph an organism which utilizes organic compounds for food.

High area of high atmospheric pressure.

High water upper limit of the surface water level reached by the rising tide; also called high tide.

Higher high water higher of the two high waters of any tidal day. The single high water occurring daily in a diurnal tide is considered a higher high water.

Higher low water higher of two low waters occurring during a tidal day.

Holoplankton organisms whose life cycle is spent in the plankton.

Hot spot area of persistent volcanic activity.

Hurricane large cyclonic storm, usually of tropical origin, containing winds of 120 kilometers per hour or higher.

Hydrogen bond relatively weak bond formed between adjacent molecules in liquid water, resulting from the mutual attractions of hydrogen and nearby atoms.

Hydrogenous sediment particles precipitated from solution in water, such as **manganese** and phosphorite **nodules.**

Hydroid polyp form of coelenterate animals that exhibit alternation of generations. It is attached, often branching, and gives rise to the pelagic, medusa form by asexual budding.

Hydrologic cycle composite cycle of interchanges of water among land, atmosphere, and ocean.

Hydrophilic having the property of attracting water.

Hydrophobic not attractive to water; unwettable.

Hydrosphere water portion of the earth, as distinguished from the solid part and the gaseous outer atmosphere. It consists of liquid water in sedimentary rocks, rivers, lakes, and oceans, as well as ice in sea ice and continental ice sheets.

Hydrothermal associated with high-temperature groundwaters—for example, alteration or precipitation of minerals and mineral ores.

Hypothesis tentative assumption made to test consequences.

Hypsographic curve representation of the elevations and depths of points on the Earth's surface above or below sea level.

Iceberg large mass of detached freshwater ice floating in the sea or stranded in shallow water.

Ice shelf thick freshwater ice formation with a fairly level surface, formed along a polar coast and in shallow bays and inlets, where it is fastened to the shore and often rests on the bottom.

Igneous rock formed by solidification of magma.

Infauna animals who live in soft sediments.

Insolation solar radiation received at Earth's surface; also, the rate at which direct solar radiation is incident upon a unit horizontal surface at any point on or above the surface of the Earth.

Instability property of a system where any disturbance grows larger instead of diminishing, so that the system never returns to the original steady state; usually refers to the vertical displacements of water parcels.

Interface surface separating two substances of different properties (such as different densities, salinities, or temperatures)—for example, the air-sea interface or the water-sediment interface.

Internal wave wave that occurs within a fluid whose density changes with depth, either abruptly at a sharp surface of discontinuity (an interface) or gradually.

Interstitial water water contained in pores between grains in rocks and sediments.

Intertidal zone (littoral zone) zone between mean high-water and mean low-water levels.

Intertropical convergence zone area toward which the Trade Winds blow.

Invertebrate animal lacking a backbone.

Ion electrically charged atom or molecule.

Ionic bond linkage between two atoms, with a separation of electric charge on the two atoms; a linkage formed by the transfer or shift of electrons from one atom to another.

Iron-manganese nodules see **Manganese nodules.**

Island-arc system group of islands, usually with a curving, archlike pattern and convex toward the open ocean, with a deep trench or trough on the convex side and usually enclosing a deep-sea basin on the concave side. Generally associated with volcanoes and subduction zones.

Isopods marine crustaceans, mostly scavengers.

Isostasy balance of portions of Earth's crust, which rise or subside until their masses are in equilibrium relationship, "floating" on the denser plastic mantle below.

Isotherm line connecting points of equal temperature.

Isothermal of the same temperature.

Isotope nuclides having the same number of protons in their nuclei and hence belonging to the same element but differing in the number of neutrons and therefore in mass number or energy content; also, a radionuclide or a preparation of an element with special isotopic composition, used principally as an isotopic tracer.

Jellyfish see **Medusa.**

Jet stream high-altitude swift air current.

Jetty a structure built to influence tidal currents, to maintain channel depths, or to protect the entrance to a harbor or river.

Lagoon shallow sound, pond, or lake, generally separated from the open ocean by a barrier beach.

Lamina sediment or sedimentary-rock layer less than 1 centimeter thick, visually separable from the material above and below.

Laminar flow flow in which fluids move smoothly in streamlines in parallel layers or sheets; a nonturbulent flow.

Land breeze wind blowing toward the sea; caused by unequal heating (and cooling) of land and water.

Langmuir circulation cellular circulation, with alternate left- and right-hand helical vortices, having axes in the direction of the wind; set up in the surface layer of a water body by winds exceeding 3.5 meters per second.

Lantern fishes (myctophids) small oceanic fishes that normally live at depths between a few hundred and a few thousand meters and characteristically have numerous small light organs on the sides of the body. Many undergo diurnal vertical migration.

Latent heat heat released or absorbed per unit mass by a system undergoing a reversible change of state at a constant temperature and pressure.

Latitude angular distance north or south of equator.

Lava fluid rock, issuing from a volcano or fissure in the earth's surface, or the same material solidified by cooling.

Layer of no motion see **Depth of no motion.**

Leeward being in or facing the direction toward which the wind is blowing; opposite to **windward.**

Levee embankment bordering one or both sides of a sea channel or delta distributary.

Lithogenous sediment sediment composed primarily of mineral grains.

Lithosphere outer, solid portion of the earth; includes the crust and part of the upper mantle.

Littoral see **Intertidal zone.**

Littoral drift sand moved parallel to the shore by wave and current action.

Load quantity of sediment transported by a current. It includes the **suspended load** of small particles that float in suspension distributed through the whole body of the current and the bottom load or **bed load** of large particles that move along the bottom by rolling and sliding.

Longitude angular distance east or west of the prime meridian.

Longshore bar see **Bar.**

Longshore current current located in the surf zone moving generally parallel to the shoreline; usually generated by waves breaking at an angle with the shoreline.

Low tide terrace zone between the ordinary high- and low-water marks; is daily traversed by the oscillating water line as the tides rise and fall. This area, together with the vertical scarp that often occurs at its upper limit, is sometimes called the **foreshore,** which ends at the highest point of normal wave uprush.

Low area of low atmospheric pressure.

Low water lowest limit of the surface-water level reached by the lowering tide; also called **low tide.**

Lower high water lower of two high tides occurring during a tidal day.

Lower low water lower of two low tides occurring during a tidal day.

Lunar tide part of the tide caused by the gravitational attraction of the moon as distinguished from that part caused by the gravitation attraction of the sun.

Lysocline depth separating well preserved carbonate shells (foraminifera, etc.) from poorly preserved forms deposited at greater depths.

Magma mobile, usually molten rock material; capable of intrusion and extrusion; forms igneous rocks when it solidifies.

Manganese nodules concretionary lumps of manganese and iron; found on the deep-ocean floor.

Mangrove tropical maritime trees that grow along shorelines.

Mantle bulk of the earth, between crust and core, from about 40 to 3500 kilometers depth. Also, the tough, protective membrane possessed by all mollusks, within which water circulates.

Map projection method of representing part or all of the surface of a sphere, such as the earth, on a plane surface.

Marginal sea semienclosed body of water adjacent to, widely open to, and connected with the ocean at the water surface but bounded at depth by submarine ridges.

Mariculture see **Aquaculture.**

Marine humus undecomposed organic matter in ocean sediment.

Marine snow particles of organic detritus and living forms. Sinking of these particles and living forms, especially in dense concentration, looks like a snowfall when viewed underwater.

Maritime climate characterized by relatively little seasonal change—warm, moist winters, cool summers; result of prevailing winds blowing from ocean to land.

Marsh area of wet land. Flat land periodically flooded by saltwater is called a **salt marsh.**

Maximum sustainable yield maximum yield of a fishery that can be sustained without depleting stocks.

Mean sea level average height of the sea surface for all stages of the tide over a 19-year period, usually determined from hourly readings of tidal height.

Mean tidal range difference in height between mean high water and mean low water, measured in feet or meters.

Meander turn or winding of a current that may become detached from the main stream; sinuous curve in a current.

Medusa (jellyfish) free-swimming coelenterates having a disk- or bell-shaped body of jellylike consistency. Many have long tentacles with stinging cells.

Meiobenthos animals in the size range 100 to 500 micrometers that live between sediment grains; includes many kinds of small invertebrates and larger protozoans.

Mélange large-scale formation containing diverse materials that were originally mixed and consolidated at great pressure, characteristic of compression or subduction zones.

Mercator projection conformal map projection.

Meridian (of longitude) great circle passing through the North and South Poles. It connects points with an equal angular separation from the Prime Meridian.

Meroplankton organisms whose early developmental stages occur in the floating state; adults are benthic.

Metabolite substance necessary to survival and growth of an organism; this includes vitamins and other materials required only in trace amounts, as well as the nutrients required for synthesis of living tissue.

Metamorphic rocks formed in the solid state, below the earth's surface, in response to changes of temperature, pressure, or chemical environment.

Microbenthos one-celled plants, animals, and bacteria living in or on the surface of bottom sediments.

Microcontinent isolated fragment of continental crust; forms oceanic plateau.

Midocean ridge great median arch or sea-bottom swell extending the length of an ocean basin and roughly paralleling the continental margins; area of oceanic crustal formation.

Midocean rise see **Oceanic rise.**

Mixed layer near-surface waters down to the pycnocline, where waters show little change in temperature or salinity with depth.

Mixed tide type of tide in which a diurnal wave produces large inequalities in heights and/or durations of successive high and/or low waters. This term applies to the tides intermediate to those predominantly semidaily and those predominantly daily.

Model system of data, inferences and relationships, presented as a description of a process or entity.

Mohorovičić discontinuity sharp discontinuity in composition between the earth's crust and mantle.

Molecular diffusion see **Diffusion.**

Monsoons seasonal winds (derived from the Arabic **mausim,** meaning season), first applied to the winds over the Arabian Sea, which blow for 6 months from the northeast and the remaining 6 months from the southwest; subsequently extended to similar seasonal winds in other parts of the world.

Mud detrital material consisting mostly of silt and clay-sized particles (less than 0.06 millimeter) but often containing varying amounts of sand and/or organic materials. It is also a general term applied to any fine-grained sediment whose particle size distribution is unknown.

Mysids elongate crustaceans usually transparent (or nearly so); benthic or deep living.

Nannoplankton plankton whose length is less than 50 micrometers.

Natural frequency characteristic frequency (number of vibrations or oscillations per unit time) of a body controlled by its physical characteristics (dimensions, density, etc.).

Nauplius limb-bearing early larval stage of many crustaceans.

Neap tide lowest range of the tide, occurring near the times of the first and last quarters of the moon.

Nekton active swimmers, pelagic animals such as most adult squids, fishes, and marine mammals.

Neritic ocean environment shallower than 200 meters.

Net primary production total amount of organic matter produced by photosynthesis minus the amount consumed by the photosynthetic organisms in their respiratory processes.

New production supported by upwelled nutrients.

Nitrogen fixation conversion of atmospheric nitrogen to oxides usable in primary food production.

Nodal line line in an oscillating area along which there is little or no rise and fall of the tide.

Nodal point no-tide point in an amphidromic region.

Node part of a standing wave where the vertical motion is least and the horizontal velocities are greatest.

Nonconservative property property whose values change in the course of a particular specified series of events or processes—for example, those properties of seawater, such as nutrient or dissolved oxygen concentrations, that are affected by biological or chemical processes.

Nonrenewable resource one that is not replenished at a rate comparable to its rate of consumption.

Nuclide species of atom characterized by the constitution of its nucleus. The nuclear constitution is specified by the number of protons, number of neutrons, and energy content—or alternatively, by the atomic number, mass number, and atomic mass.

Nudibranchs (sea slugs) gastropods in which adults have no shell.

Nutrient inorganic or organic compounds or ions necessary for the nutrition of primary producers. Nitrogen and phosphorous compounds are examples.

Obduction seafloor is welded to continent; reverse of subduction.

Ocean basin ocean floor that is more than about 2000 meters below sea level.

Oceanic crust mass of basaltic material, typically 7 kilometers thick, that lies under the ocean basins.

Oceanic (midocean) rise continuous ocean-bottom province that rises above the deep-ocean floor; area of crustal generation.

Oceanography scientific study of the ocean.

Omnivore organism that eats anything, both plant and animal.

Ooze fine-grained, deep-ocean sediment containing at least 30% (by volume) undissolved sand- or silt-sized, calcareous or siliceous skeletal remains of small marine organisms, the remainder usually being clay-sized material.

Open ocean part of the ocean that is seaward of the approximate edges of the continental shelves, usually more than 2 kilometers deep.

Ophiolitic suite assemblage of continental rocks containing deep-sea sediments, submarine lavas, and oceanic crust, apparently thrust upward during crustal plate subduction.

Oxidation loss of hydrogen or electrons; opposite of reduction.

Pack ice a rough, solid mass of broken sea-ice floes forming an obstruction to navigation.

Pangaea single large continent that split apart about 200 million years ago to form the present continents.

Paradigm a world-view.

Parasitism relationship in which the parasite harms the host from which it takes nutrition.

Partial tide one of the harmonic components comprising the tide at any point. The periods of the partial tides are derived from various combinations of the angular velocities of Earth, sun, and moon, relative to each other.

Patch reef isolated coral growths in lagoons of barriers and atolls; ranging from several kilometers across down to small coral pillars or even mushroom-shaped growths consisting of a single colony.

Pelagic oceanic environment deeper than 200 meters.

Pelagic deposits deep-ocean sediments that have accumulated by settling out of the ocean particle-by-particle.

Photic zone (euphotic zone) near-surface layer of water that receives ample sunlight for photosynthesis to exceed respiration.

Photosynthesis manufacture of carbohydrates and other compounds from carbon dioxide and water in the presence of chlorophyll by utilizing light energy and releasing oxygen.

Phytoplankton plant forms of plankton; passive drifters.

Pillow lava a characteristic form of lava extruded underwater, where the molten rock forms lumpy, rounded forms, resembling pillows.

Planetary winds see **Prevailing wind systems.**

Plankton passively drifting or weakly swimming organisms.

Plankton bloom unusually high concentration of plankton (usually phytoplankton) in an area, caused either by an explosive or gradual multiplication of organisms.

Plate tectonics theory of lithospheric plate movement caused by mantle convection.

Polder land area reclaimed from the shallow ocean bottom, usually separated from the ocean by dikes.

Polychaetes segmented marine worms; some are tube builders others are free-swimming.

Polyna persistently ice-free area surrounded by pack ice.

Polyp individual sessile coelenterate.

Prevailing wind systems (planetary winds) large, relatively constant wind systems that result from Earth's shape, inclination, revolution, and rotation, examples are the northeast and southeast trade winds, and westerlies, and the polar easterlies.

Primary production amount of organic matter synthesized by organisms from inorganic substances per unit time in a unit volume of water or in a column of water of unit area cross section and extending from the surface to the bottom; also called **gross primary production.**

Prime meridian meridian of longitude (0°), used as the reference for measurements of longitude, located on the meridian of Greenwich, England.

Production see **Primary productivity.**

Profile drawing showing a vertical section along a line.

Progressive wave wave that is manifested by progressive movements of waveforms.

Protozoa microscopic, one-celled animals.

Pseudopod an extension of protoplasm that can be projected or withdrawn by the animal for capturing food or for locomotion; characteristic of some protozoans.

Pteropods free-swimming gastropods in which the foot is modified into fins; Both shelled and nonshelled forms exist. In some shallow oceanic areas, accumulated shells of these organisms form sediment deposits called **pteropod oozes.**

Pycnocline vertical density gradient in some layer of a body of water, positive with respect to depth and appreciably greater than gradients above and below it; also, a layer in which such a gradient occurs.

Radioactivity spontaneous breakdown of an atomic nucleus, giving off energy and often particles.

Radioisotope radioactive isotope of an element.

Radiolarians single-celled planktonic protozoans possessing a skeleton of siliceous spicules and radiating threadlike pseudopodia.

Radiometer device for measuring intensity of radiation from the sea surface beneath a spacecraft or an airplane. Measurements in the infrared bands yield sea surface temperatures, and the visible bands yield color.

Radionuclide synonym for **radioactive nuclide.**

Red algae reddish, filamentous, membranous, encrusting or complexly branched plants in which the color is imparted by the predominance of a red pigment over the other pigment's present. Some are included among the coralline algae.

Red clay brown-to-red deep-sea deposit. It is the most finely divided clay material that is derived from the land and transported by ocean currents and winds, accumulating at great depths.

Red tide red or reddish brown discoloration of surface waters most frequently in coastal regions, caused by concentrations of certain microscopic organisms, particularly dinoflagellates.

Reduction gain in electrons or in hydrogen; opposite of oxidation.

Reef off-shore wave-resistant rock which is a hazard to navigation.

Reef flat (of a coral reef) flat expanse of dead reef rock that is partly or entirely dry at low tide. Shallow pools, potholes, gullies, and patches of coral debris and sand are features of the reef flat. It is divisible into inner and outer portions.

Reflection process whereby a surface or discontinuity turns back a portion of the incident radiation into the medium through which the radiation approached.

Refraction of water waves process by which the direction of a wave moving in shallow water at an angle to the contours is changed, causing the wave crest to bend toward alignment with the underwater contours; also, the bending of wave crests by currents.

Relict sediment sediment deposited on the continental shelf by processes no longer active.

Renewable resource resource replenished at a rate comparable to its rate of consumption by natural growth or by careful management of the resource.

Residence time time required for a flow of material to replace the amount of that material originally present in a given volume. Assuming a steady flow, replacement time can be calculated for any substance, such as salt or water.

Respiration oxidation-reduction process by which chemically bound energy in food is transformed into other kinds of energy on which certain processes in all living cells are dependent.

Reversing thermometer mercury-in-glass thermometer that records temperature on being inverted and thereafter retains its reading until returned to the first position.

Reversing tidal current tidal current that flows alternatively in approximately opposite directions, with a period of slack water at each reversal of direction. Reversing currents occur in rivers and straits where the flow is restricted. When the flow is toward the shore, the current is flooding; when in the opposite direction, it is ebbing.

Rift valley narrow trough formed by faulting in a divergence area.

Rill small groove, furrow, or channel made in mud or sand on a beach by tiny streams following an outflowing tide.

Ring body of water separated from surrounding waters by a strong current; formed by a meander of a boundary current.

Rip agitation of water caused by the meeting of currents or by a rapid current setting over an irregular bottom—for example, a **tide rip.**

Rip current strong current, usually of short duration, flowing seaward from the shore. It usually appears as a visible band of agitated water and is the return movement of water piled up on the shore by incoming waves and wind.

Ripple wave controlled to a significant degree by both surface tension and gravity.

Rise long, broad elevation that rises gently and generally smoothly from the ocean bottom.

River-induced upwelling upward movements of deeper water that occurs when seawater mixes with fresh water from a river—becoming less dense—and moves toward the surface.

Rotary current, tidal tidally induced current that flows continually with the direction of flow, changing through all points of the compass during the tidal period. Rotary currents are found in the ocean where the direction of flow is not restricted by any barriers.

Sabellid see **Tube worm.**

Salinity measure of the quantity of dissolved salts in seawater. Formally defined as the total amount of dissolved solids in seawater in parts per thousand (‰) by weight when all the carbonate has been converted to oxide, the bromide and iodide to chloride, and all organic matter is completely oxidized.

Salps transparent pelagic tunicates. The body is more or less cylindrical and possesses conspicuous ringlike muscle bands, which contract to propel the animal through the water.

Salt dome mass of salt, a kilometer or more in diameter, that rose through sediments deposited above it.

Salt marsh see **Marsh.**

Salt wedge estuary high-flow circulation pattern with seawater intrusion along the bottom of an estuary forming a wedge; characterized by a pronounced increase in salinity from surface to bottom.

Sand loose material that consists of grains ranging between 0.0625 and 2 millimeters in diameter.

Sargassum brown, bushy alga, substantial holdfast (rootlike structure) when attached, and a yellowish brown, greenish yellow, or orange color.

Scarp elongated, steeply sloping ocean floor.

Scattering dispersion of light when a beam strikes very small particles suspended in air or water. In light scattering there is no loss of intensity, only a redirection of light rays.

Scatterometer microwave radar to measure sea surface roughness beneath the spacecraft or aircraft. Measurements indicate the height of short surface waves caused by winds; measures surface wind velocities.

School large number of one kind of fish or other aquatic animal swimming or feeding together.

Science accumulated knowledge build up through use of scientific method.

Scour erosion of a sediment bed by waves or currents.

Scyphozoans coelenterates in which the polyp or hydroid stage is insignificant but the medusoid stage is well developed. True jellyfish belong to this group.

Sea waves generated or sustained by winds within their fetch, as opposed to **swell**; also, a subdivision of an ocean.

Sea breeze light wind blowing toward the land, caused by unequal heating (and cooling) of land and water masses.

Sea-floor spreading process by which lithosphere is generated at midocean ridges. Adjacent lithospheric plates are moved apart as new material forms.

Seamount elevation rising 900 meters or more from the ocean bottom.

Sea state (state of the sea) numerical or written description of ocean surface roughness.

Seawall rock or concrete structure built to protect a coast against wave erosion.

Sediment particulate organic and inorganic matter that accumulates in a loose, unconsolidated form. It may be chemically precipitated from solution, secreted by organisms, or transported from land by air, ice, wind, or water and deposited.

Sedimentation processes of breakup and separation of particles from the parent rock, their transportation, deposition, and consolidation into another rock.

Seiche standing wave in an enclosed or semienclosed water body that continues after the cessation of the originating force, which may have been seismic, wind, or wave induced.

Seismic reflection technique of studying buried crustal structures, by recording the travel time of acoustic energy reflected back to detectors from rock or sediment layers that have different elastic-wave velocities.

Seismic sea wave see **Tsunami.**

Semidaily (semidiurnal) tide tide having a period of approximately one-half a tidal day, with two high waters and two low waters each tidal day; most common type of tide.

Semipermeable membrane membrane through which a solvent, but not certain dissolved or colloidal substances, may pass.

Sensible heat portion of energy exchanged between ocean and atmosphere utilized in changing the temperature of the medium into which it penetrates.

Sessile permanently attached by the base of a stalk; not free to move about.

Set (current direction) direction toward which the current flows.

Shallow water waves waves in water shallower than $L/2$.

Shear wave wave that causes particles in a medium to vibrate back and forth at right angles to the direction of wave propagation.

Shoreline boundary between a body of water and the land at high tide (usually mean high water).

Significant wave height (characteristic wave height) average height of the highest one-third of waves of a given wave group.

Sill shallow portion of the ocean floor that partially restricts water flow; may be either at the mouth of an inlet, fjord, or similar structure, or at the edge of an ocean basin—for example, the Bering Sill separates the Pacific and Arctic portions of the Atlantic Ocean.

Silt particles between sands and clays in size; 4 to 62 micrometers in diameter.

Sinking (downwelling) downward movement of surface water generally caused by converging currents or as a result of a water mass becoming more dense than the surrounding water.

Siphonophores medusoid coelenterates, many of which are luminescent and some venomous. Some possess a gas-filled float; others are colonial, with polyp and medusoid individuals functioning as a single individual.

Sipunculids wormlike marine animals, unsegmented, with the mouth surrounded by tentacles. The anterior (head) end can be withdrawn into the body. They are deposit feeders.

Slack water state of a tidal current when velocities are near zero, usually when a reversing current changes direction.

Slick area of quiescent water surface, usually elongated. Slicks may form patches or weblike nets where ripple activity is greatly reduced.

Slump slippage or sliding of a mass of unconsolidated sediment down a submarine or subaqueous slope. Slumps occur frequently at the heads or along the sides of submarine canyons; triggered by any small or large earth shock, the sediment usually moves as a unit initially and may become a turbidity flow.

Solar tide partial tide caused solely by the tide-producing forces of the sun.

Solstice time when the sun is directly over the Tropic of Cancer (summer) or the Tropic of Capricorn (winter).

Sounding measurement of the water depth.

Southern Oscillation oscillation in the locations of high and low pressure areas in the Southern Hemisphere; associated with El Niños.

Specific gravity ratio of the density of a substance relative to the density of pure water at 4°C; in the centimeter-gram-second system, **density** and **specific gravity** may be used interchangeably.

Specific heat quantity of heat required to raise the temperature of 1 gram of a substance by 1°C. The common unit is calories per gram per degree Celsius.

Sperm whale see **Toothed whales.**

Spicules crystals of newly formed sea ice; also, minute, needlelike or multiradiate calcareous or siliceous bodies in many organisms.

Spit small point of land projecting into a body of water.

Spring bloom sudden proliferation of phytoplankton that occurs when the critical depth (as determined by penetration of sunlight) exceeds the depth of the mixed, stable surface layer (as determined by the pycnocline).

Spring tide tide of increased range that occurs about every 2 weeks when the moon is new or full.

Spur-and-groove structure see **Forereef.**

Stability resistance to overturning or mixing in the water column, resulting from the presence of a positive density gradient; less dense water above denser water.

Stand of the tide interval at high or low water when there is no appreciable change in the height of the tide; its duration depends on the range of the tide, being longer when the tidal range is small and shorter when the tidal range is large.

Standing crop biomass of a population present at a specified time.

Standing wave type of wave in which the surface of the water oscillates vertically between fixed points, called **nodes,** without progression. The points of maximum vertical rise and fall are called **antinodes.** At the nodes, the underlying water particles exhibit no vertical motion but maximum horizontal motion.

Steady state absence of change with time.

Still-water level level that the sea surface would assume in the absence of wind wave.

Stock an interbreeding population of animals.

Storm surge (storm wave, storm tide, tidal wave) rise or piling up of water against shore, produced by wind stresses and atmospheric low pressures in a storm.

Stratosphere part of the Earth's atmosphere between the troposphere and the upper layer (ionosphere).

Subduction zone inclined plane descending away from a trench, separating a sinking oceanic plate from an overriding plate; usually associated with a **trench** and active volcanoes.

Sublimation transition of the solid phase of certain substances into a gas—and vice versa—without passing through the liquid phase.

Submarine canyon submarine valley.

Subsurface current current usually flowing below the pycnocline, generally at slower speeds and frequently in a different direction from the currents near the surface.

Subtropical high one of the semipermanent highs of the subtropical high-pressure belt.

Succession, ecological process of community change whereby communities replace one another.

Surf collective term for breakers; also, the wave activity in the area between the shoreline and the outermost limit of breakers.

Surf zone area between the outermost breaker and the limit of wave uprush.

Surface-active agent substance, usually in solution, that can markedly change the surface or interfacial properties of the liquid, even when present in minute amounts.

Surface tension (surface energy, capillary forces, interfacial tension) phenomenon peculiar to liquid surfaces caused by a strong attraction (toward the interior) acting on molecules at or near the surface in such a way as to reduce the surface area.

Surface zone (mixed zone) water above the pycnocline where waves and convection mix the water; results in uniform temperatures and salinities within the mixed zone.

Surge horizontal oscillation of water with comparatively short period accompanying a seiche (see also **Storm surge**).

Suspension feeding feeding by removing food particles from water.

Swallow float tubular buoy, usually made of aluminum, that can be adjusted to remain at a selected density level to drift with the motion of that water mass. The float is tracked by listening devices.

Swash rush of water up onto the beach following the breaking of a wave.

Swell waves that have traveled out of their generating area.

Swim bladder gas-filled sac lying in the body cavity between the vertebral column and the alimentary tract of certain fishes. It serves a hydrostatic function in most fishes that possess it; in some, it participates in sound production.

Symbiosis relationship between two species in which one or both members benefit and neither is harmed.

Synthetic aperture radar radar technique that provides high-resolution images of ocean surface features, such as swell, internal waves, current boundaries, and sea-ice distributions and movements.

Tablemount see **Guyot.**

Technology application of science and engineering.

Tectonic estuary estuary occupying a basin formed by mountain building.

Theory tentative conclusion, supported by some evidence.

Thermocline marked vertical temperature change in a body of water; also a layer in which such a temperature change occurs.

Thermohaline circulation circulation induced by differences on water density which is controlled primarily by temperature and salinity.

Tidal bore very rapid rise of the tide in which the advancing water forms an abrupt front; occurs in certain shallow estuaries having a large tidal range.

Tidal bulge (tidal crest) long-period wave associated with the tide-producing forces of the moon and sun; identified with the rising and falling of the tide. The trough located between the two tidal bulges present at any given time on the earth is known as the **tidal trough.**

Tidal constituent an element in a mathematical expression for the tide-generating forces, tides, or tidal currents. Each constituent represents a periodic change or variation in the relative positions of the Earth, moon, and sun.

Tidal current alternating horizontal movement of water associated with the rise and fall of the tide.

Tidal day interval between two successive upper transits of the moon over a location. A **mean tidal day,** sometimes called a **lunar day,** is 24 hours, 50 minutes.

Tidal flats marshy or muddy areas which are covered and uncovered by the rise and fall of the tide; also called **tidal marshes.** Usually covered by plants.

Tidal period elapsed time between successive high or low waters.

Tidal range difference in height between consecutive high and low waters.

Tidal trough see **Tidal bulge.**

Tide periodic rise and fall of the ocean and atmosphere, caused by the gravitational attraction of moon and sun acting on the Earth.

Tide curve presentation of the rise and fall of tide; time (in hours or days) is plotted against height of the tide.

Tide-producing forces slight local difference between the gravitational attraction of two astronomical bodies and the centrifugal force that holds them apart. Gravitational attraction predominates at the surface point nearest to the other body while centrifugal "repulsion" predominates at the surface point farthest from the other body.

Tide pool depression—usually water filled—in the intertidal zone, alternately submerged and exposed by the rise and fall of the tide or wave action.

Tide rip see **Rip.**

Tide tables tables that predict the times and heights of tidal phenomena at specified locations.

Tide wave long-period gravity wave that has its origin in the tide-producing force; manifests itself in the rising and falling of the tide.

Tombolo area of unconsolidated material deposited by wave or current action that connects a rock, island, and the like to the main shore or other body of land.

Toothed whales dolphins, porpoises, killer whales, and sperm whales.

Trade winds wind system, in most of the tropics, which blows from the subtropical highs toward the equatorial lows. Trade winds are from the northeast in the Northern Hemisphere, from the southeast in the Southern Hemisphere.

Transform fault fault along which lithospheric plates move past each other.

Trawl bag or funnel-shaped net for catching bottom fish by dragging along the bottom; also, a large net for catching zooplankton and fishes by towing in intermediate depths.

Trench long, narrow, deep depressions of the ocean floor.

Triple junction intersection of 3 plates; may involve any combination of **trenches, mid-ocean ridges** and **transform faults**

Trochophore free-swimming pelagic stage of some segmented worms and mollusks.

Trophic level successive stage of nourishment as represented by links of the food chain. Primary producers constitute the first trophic level, herbivores the second, and carnivores the third and higher trophic levels.

Tropics equatorial region between Tropic of Cancer (north) and Tropic of Capricorn (south); climate found in the belt close to the equator. Daily variations in temperature exceed seasonal variations; generally high rainfall.

Tropopause upper limit of the troposphere.

Troposphere portion of the atmosphere next to the Earth's surface where temperature generally rapidly decreases with altitude; clouds form, and convection is active.

Tsunami (seismic sea wave) long-period sea waves produced by submarine earthquakes, volcanic explosions, or slumps.

Tube worm polychaetes, chiefly sabellids and related groups, that build calcareous or leathery tubes.

Tunicates globular or cylindrical, often saclike, animals. Some are sessile; others are planktonic.

Turbidite turbidity-current deposit characterized by vertically and horizontally graded bedding.

Turbidity reduced water clarity resulting from suspended or dissolved matter.

Turbidity current gravity current resulting from a density increase caused by suspended materials.

Turbulence irregular motions of air or water; marked departure from a smooth flow.

Turbulent flow flow characterized by random velocity fluctuations.

Upwelling process by which water rises from a lower to a higher depth, usually caused by divergence.

van der Waals forces weak attractive forces between molecules that arise from interactions between the atomic nuclei of one molecule and the electrons of another molecule.

Veliger planktonic larval stage of many gastropods.

Viscosity internal resistance-to-flow, property of fluids that enables them to support certain stresses and thus resist deformation for a finite time.

Volcanic island island formed by the top of a volcano or solidified volcanic material.

Water budget accounts for interchanges of water among the land, atmosphere, and ocean.

Water mass water body usually identified by its temperature and salinity or by some other tracer.

Water parcel water mass with a certain temperature and salinity, separated from surrounding waters.

Wave disturbance that moves through or over the ocean surface.

Wave age state of development of a wind-generated sea surface wave, conveniently expressed by the ratio of wave speed to wind speed. Wind speed is usually measured at about 8 meters above still-water level.

Wave energy capacity of a wave to do work. In a deep-water wave, about half the energy is kinetic, associated with water movement, and about half is potential energy, associated with the elevation of water above the still-water level in the crest or its depression below still-water level in the trough.

Wave group series of waves in which the wave direction, wavelength, and wave height vary only slightly.

Wave height vertical distance between crest and preceding trough in a wave.

Wavelength horizontal distance between successive wave crests measured perpendicular to the crests.

Wave period time required for two successive wave crests to pass a fixed point.

Wave spectrum distribution of wave energy (square of wave height) with wave frequency (1/period). The square of the wave height is related to the potential energy of the sea surface, so the spectrum can also be called the **energy spectrum.**

Wave steepness ratio of wave height to wavelength.

Wave train series of waves from the same direction.

Wave velocity speed at which individual waveforms advance; also, a vector quantity that specifies the speed and direction with which a wave travels through a medium.

Weathering destruction or partial destruction of rock by thermal, chemical, and mechanical processes.

Western boundary current strong, narrow, deep currents on western side of ocean basins.

Wetland see **Marsh.**

Whitecaps white froth on crests of waves in a wind.

Wind-driven circulation surface-current system driven by winds.

Wind mixing mechanical stirring of water due to motions, induced by the surface wind.

Wind waves waves formed and growing in height under the influence of wind; any wave generated by winds.

Windrows rows of floating debris, aligned in the wind direction, formed on the surface of a lake or ocean by Langmuir cells.

Windward being in or facing the direction from which the wind is blowing; opposite to **leeward.**

Year class organisms of a particular species spawned during a single year or breeding season.

Zonation organization of a habitat into bands of distinctive plant and animal associations where conditions for survival are optimal.

Zone layer that encompasses some feature, structure, depth, or property.

Zooplankton animal forms of plankton.

Index

Note: Page numbers appearing in boldface indicate figures.